Reactive Oxygen Species and Male Fertility

Reactive Oxygen Species and Male Fertility

Special Issue Editor
Cristian O'Flaherty

MDPI • Basel • Beijing • Wuhan • Barcelona • Belgrade

Special Issue Editor
Cristian O'Flaherty
Department of Surgery
(Urology Division),
Faculty of Medicine, McGill University
Canada

Editorial Office
MDPI
St. Alban-Anlage 66
4052 Basel, Switzerland

This is a reprint of articles from the Special Issue published online in the open access journal *Antioxidants* (ISSN 2076-3921) from 2019 to 2020 (available at: https://www.mdpi.com/journal/antioxidants/special_issues/Oxygen_Male_Fertility).

For citation purposes, cite each article independently as indicated on the article page online and as indicated below:

LastName, A.A.; LastName, B.B.; LastName, C.C. Article Title. *Journal Name* **Year**, *Article Number*, Page Range.

ISBN 978-3-03936-024-6 (Pbk)
ISBN 978-3-03936-025-3 (PDF)

Cover image courtesy of Eleonora Scarlata.

© 2020 by the authors. Articles in this book are Open Access and distributed under the Creative Commons Attribution (CC BY) license, which allows users to download, copy and build upon published articles, as long as the author and publisher are properly credited, which ensures maximum dissemination and a wider impact of our publications.
The book as a whole is distributed by MDPI under the terms and conditions of the Creative Commons license CC BY-NC-ND.

Contents

About the Special Issue Editor vii

Cristian O'Flaherty
Reactive Oxygen Species and Male Fertility
Reprinted from: *Antioxidants* **2020**, *9*, 287, doi:10.3390/antiox9040287 1

Cristian O'Flaherty, Annie Boisvert, Gurpreet Manku and Martine Culty
Protective Role of Peroxiredoxins against Reactive Oxygen Species in Neonatal Rat Testicular Gonocytes
Reprinted from: *Antioxidants* **2020**, *9*, 32, doi:10.3390/antiox9010032 6

Anaïs Noblanc, Alicia Klaassen and Bernard Robaire
The Exacerbation of Aging and Oxidative Stress in the Epididymis of *Sod1* Null Mice
Reprinted from: *Antioxidants* **2020**, *9*, 151, doi:10.3390/antiox9020151 20

Pei You Wu, Eleonora Scarlata and Cristian O'Flaherty
Long-Term Adverse Effects of Oxidative Stress on Rat Epididymis and Spermatozoa
Reprinted from: *Antioxidants* **2020**, *9*, 170, doi:10.3390/antiox9020170 32

Karolina Nowicka-Bauer and Brett Nixon
Molecular Changes Induced by Oxidative Stress that Impair Human Sperm Motility
Reprinted from: *Antioxidants* **2020**, *9*, 134, doi:10.3390/antiox9020134 48

Lauren E. Hamilton, Michal Zigo, Jiude Mao, Wei Xu, Peter Sutovsky, Cristian O'Flaherty and Richard Oko
GSTO2 Isoforms Participate in the Oxidative Regulation of the Plasmalemma in Eutherian Spermatozoa during Capacitation
Reprinted from: *Antioxidants* **2019**, *8*, 601, doi:10.3390/antiox8120601 70

Ana Izabel Silva Balbin Villaverde, Jacob Netherton and Mark A. Baker
From Past to Present: The Link Between Reactive Oxygen Species in Sperm and Male Infertility
Reprinted from: *Antioxidants* **2019**, *8*, 616, doi:10.3390/antiox8120616 88

Robert J. Aitken and Joel R. Drevet
The Importance of Oxidative Stress in Determining the Functionality of Mammalian Spermatozoa: A Two-Edged Sword
Reprinted from: *Antioxidants* **2020**, *9*, 111, doi:10.3390/antiox9020111 107

Joël R. Drevet and Robert John Aitken
Oxidation of Sperm Nucleus in Mammals: A Physiological Necessity to Some Extent with Adverse Impacts on Oocyte and Offspring
Reprinted from: *Antioxidants* **2020**, *9*, 95, doi:10.3390/antiox9020095 126

Fernando J. Peña, Cristian O'Flaherty, José M. Ortiz Rodríguez, Francisco E. Martín Cano, Gemma L. Gaitskell-Phillips, María C. Gil and Cristina Ortega Ferrusola
Redox Regulation and Oxidative Stress: The Particular Case of the Stallion Spermatozoa
Reprinted from: *Antioxidants* **2019**, *8*, 567, doi:10.3390/antiox8110567 140

David Martin-Hidalgo, Maria Julia Bragado, Ana R. Batista, Pedro F. Oliveira and Marco G. Alves
Antioxidants and Male Fertility: From Molecular Studies to Clinical Evidence
Reprinted from: *Antioxidants* **2019**, *8*, 89, doi:10.3390/antiox8040089 163

About the Special Issue Editor

Cristian O'Flaherty received a DVM and Ph.D. degrees from the University of Buenos Aires, Argentina. He completed his postdoctoral training in sperm physiology and toxicology at McGill University. Dr. O'Flaherty is an associate professor in the Department of Surgery (Division of Urology) and an associate member of the Department of Pharmacology and Therapeutics, at the Faculty of Medicine, McGill University. He is also a medical scientist at the Research Institute of the McGill University Health Centre. Dr. O'Flaherty is the Co-Director of the McGill Centre for Research in Reproduction and Development. He is serving as associate editor of Andrology and a member of the Editorial Board of Biology of Reproduction. His research program focuses on male reproduction, particularly on the molecular mechanisms involved in the production and function of healthy spermatozoa, sperm activation and the role of reactive oxygen species in sperm physiology and toxicology. His current work includes the use of mouse models and clinical studies in infertile men to elucidate the role of peroxiredoxins in the regulation of sperm function using genomic, proteomic and biochemical approaches. Moreover, Dr. O'Flaherty's laboratory is working on the molecular mechanism that drives the formation and function of the acrosome and developing novel diagnostic and therapeutic tools for infertile men. His research program has been and is currently funded by different agencies, including the Canadian Institutes of Health Research, FRQS and the Fonds de Recherche Nature et technologies.

Editorial
Reactive Oxygen Species and Male Fertility

Cristian O'Flaherty [1,2,3,*]

1. Department of Surgery (Urology Division), McGill University, Montréal, QC H4A 3J1, Canada
2. Department of Pharmacology and Therapeutics, McGill University, Montréal, QC H3G 1Y6, Canada
3. The Research Institute, McGill University Health Centre, Montréal, QC H4A 3J1, Canada

Received: 17 March 2020; Accepted: 25 March 2020; Published: 29 March 2020

Human infertility affects ~15% of couples worldwide, and it is now recognized that in half of these cases, the causes of infertility can be traced to men [1,2]. The spermatozoon is a terminal cell with the unique goal of delivering the paternal genome into the oocyte. This essential task for any species survival can be threatened by environmental pollutants, chemicals, drugs, smoke, toxins, radiation, diseases, and lifestyles. Oxidative stress is a common feature of the mechanism of action of these factors and conditions that negatively impact male fertility [3–7]. Indeed, the reactive oxygen species (ROS)-mediated damage to spermatozoa is a significant contributing factor to infertility in 30-80% of infertile men [8–12].

Spermatozoa are terminally differentiated cells produced in the testes during the hormone-regulated process of spermatogenesis. Two somatic cell types are critical to this process: Sertoli cells protect and support the germ cell development, whereas interstitial Leydig cells produce necessary intratesticular steroids (e.g., testosterone) [13,14]. After their release from the testis, spermatozoa complete their maturation during the transit through the epididymis. There, they acquire the potential for motility and fertility through extensive morphological and biochemical modifications [15]. After ejaculation, spermatozoa must undergo the complex and timely process of capacitation that involves ionic, metabolic, and membrane changes, including the production of ROS at low concentrations [16,17]. Capacitation allows spermatozoa to bind to the zona pellucida that surrounds the oocyte and induce the acrosome reaction [18], an exocytotic event by which proteolytic enzymes (e.g., acrosin and hyaluronidase) are released. Thus, the spermatozoon penetrates the zona pellucida and reaches and fuses with the oocyte. Failure to undergo sperm capacitation and/or acrosome reaction is associated with infertility [19,20].

A key feature of sperm capacitation is the production of ROS at very low and controlled levels by the spermatozoon. This essential phenomenon for the acquisition of the fertility ability was first reported in humans [21], bovine [22], and equine [23] spermatozoa, and then confirmed by others (see [24] for more information). ROS play the role of second messengers and act in most of the known signal transduction pathways involved in this complex phenomenon [25–27]. Sperm capacitation is a redox-regulated process. Peroxiredoxins (PRDXs) play a crucial role in maintaining low levels of intracellular ROS to allow the achievement of fertilizing ability by the spermatozoon [28,29].

Oxidative stress can promote detrimental changes during spermatogenesis, epididymal maturation, and sperm capacitation that can lead to infertility [30–35]. Lipid peroxidation of the sperm plasma membrane is one of the first described oxidative damage associated with low sperm quality and infertility [36,37]. 4-hydroxynonenal, a subproduct of lipid peroxidation, forms adducts with proteins and DNA, impairing sperm mitochondrial function and promoting mutations in the sperm genome [38]. PRDX6, a unique antioxidant with peroxidase and calcium-independent phospholipase A_2, is a key element in the antioxidant defence to protect sperm membranes and DNA from oxidative damage [39,40]. Oxidative damage to the paternal genome has been reported in humans and animals and associated with fertility failure [41–44]. Redox-dependent protein modifications are related to low sperm quality and infertility [45–47]. The reproductive system is equipped with antioxidant enzymes to avoid the adverse effects of high levels of ROS during the production and maturation of spermatozoa. Different studies have addressed the consequences of the absence of antioxidant enzymes on male reproduction (for details see [48]), and from knockout mouse models, we learn the critical need of superoxide dismutase, glutathione peroxidases, thioredoxins, and PRDXs to produce a healthy and fertile spermatozoa in young adult and aging males [32,41,49–52].

The decline of fertility and the increase of abnormalities in the semen of men that is observed over more than two decades is worrisome. However, we still do not have sufficient information to understand why this is happening [53–55]. The exposure environmental toxicants could partly explain this phenomenon, but there is still a significant amount of uncertainty regarding the possible causes of the decrease in sperm quality over the years [53,55]. It is an established trend that men are delaying fatherhood due to professional reasons, and scientific data from studies in animals and humans revealed that sperm quality worsens as men age. There is increasing evidence that children fathered by 50-year-old-or older men are prone to manifest a variety of disorders that can be linked to significant mutations of the paternal genome [56].

This Special Issue on ROS and male fertility is composed of four original contributions that provide new data on the role of SOD [57] and PRDXs [58] as essential antioxidants in the epididymis, and PRDXs as novel players in the protection of gonocytes against oxidative stress [59] and the participation of glutathione-S transferase omega 2 in the regulation of sperm function during capacitation [60]. The Special Issue is completed with six review articles that provide an update on the molecular changes induced by oxidative stress that impairs human sperm motility [61], the importance of ROS in determining the functionality of spermatozoa and situations in which oxidative stress occurs and impacts on male fertility [24], the oxidation of the sperm nucleus and the impact of oxidative stress on the paternal genome [62], the importance of using appropriate tools to study the role of ROS in sperm and infertility [63], and an exhaustive evaluation of ROS production and its relevance in male fertility and antioxidant therapy [64]. The characterization of the redox regulation and effects of oxidative stress in equine spermatozoa is also included in this Special Issue [65].

Many groundbreaking studies have contributed to our understanding of the field of ROS in male fertility. It is undeniable that these reactive molecules play an essential role in both physiological and pathological mechanisms of the male reproductive system. Antioxidant therapy is a common clinical strategy to improve sperm quality and function in infertile men [66,67]. However, there are controversial results regarding the efficacy of these treatments. Some controlled trials suggested that such treatments are beneficial to achieve live births [66], whereas others did not show any benefit [68]. This discrepancy is likely based on the lack of tools in clinics to establish whether oxidative stress is responsible for the infertility of a given patient. Thus, more fundamental and clinical research is needed to comprehend how ROS modulate male fertility to design better diagnostic tools and therapeutic strategies to fight against male infertility.

Funding: Original research by the author of this editorial was supported by the Canadian Institutes of Health Research (CIHR), operating grant # MOP133661 to CO.

Conflicts of Interest: The author declares no conflict of interest.

References

1. Bushnik, T.; Cook, J.L.; Yuzpe, A.; Tough, S.; Collins, J. Estimating the prevalence of infertility in Canada. *Hum. Reprod.* **2012**, *7*, 738–746. [CrossRef] [PubMed]
2. World Health Organization. Towards more objectivity in diagnosis and management of male fertility. *Int. J. Androl.* **1997**, *7*, 1–53.
3. Anderson, J.B.; Williamson, R.C. Testicular torsion in Bristol: A 25-year review. *Br.J. Surg.* **1988**, *75*, 988–992. [CrossRef] [PubMed]
4. Hasegawa, M.; Wilson, G.; Russell, L.D.; Meistrich, M.L. Radiation-induced cell death in the mouse testis: Relationship to apoptosis. *Radiat. Res.* **1997**, *147*, 457–467. [CrossRef] [PubMed]

5. Brennemann, W.; Stoffel-Wagner, B.; Helmers, A.; Mezger, J.; Jager, N.; Klingmuller, D. Gonadal function of patients treated with cisplatin based chemotherapy for germ cell cancer. *J. Urol.* **1997**, *158 Pt 1*, 844–850. [CrossRef]
6. Smith, R.; Kaune, H.; Parodi, D.; Madariaga, M.; Rios, R.; Morales, I.; Castro, A. Increased sperm DNA damage in patients with varicocele: Relationship with seminal oxidative stress. *Hum. Reprod.* **2006**, *21*, 986–993. [CrossRef]
7. Turner, T.T. The study of varicocele through the use of animal models. *Hum. Reprod. Update* **2001**, *7*, 78–84. [CrossRef]
8. Agarwal, A.; Gupta, S.; Sikka, S. The role of free radicals and antioxidants in reproduction. *Curr. Opin. Obstet. Gynecol.* **2006**, *18*, 325–332. [CrossRef]
9. Aitken, R.J.; Baker, M.A. Oxidative stress, sperm survival and fertility control. *Mol. Cell. Endocrinol.* **2006**, *250*, 66–69. [CrossRef]
10. Gagnon, C.; Iwasaki, A.; De Lamirande, E.; Kovalski, N. Reactive oxygen species and human spermatozoa. *Ann. N. Y. Acad. Sci.* **1991**, *637*, 436–444. [CrossRef]
11. De Lamirande, E.; Gagnon, C. Impact of reactive oxygen species on spermatozoa: A balancing act between beneficial and detrimental effects. *Hum. Reprod.* **1995**, *10* (Suppl. 1), 15–21. [CrossRef] [PubMed]
12. Tremellen, K. Oxidative stress and male infertility: A clinical perspective. *Hum. Reprod. Update* **2008**, *14*, 243–258. [CrossRef] [PubMed]
13. Clermont, Y. Kinetics of spermatogenesis in mammals: Seminiferous epithelium cycle and spermatogonial renewal. *Physiol. Rev.* **1972**, *52*, 198–236. [CrossRef] [PubMed]
14. Stocco, D.M.; McPhaul, M.J. Physiology of Testicular Steroidogenesis. In *Knobil and Neill's Physiology of Reproduction*, 3rd ed.; Jimmy, D.N., Tony, M.P., Donald, W.P., John, R.G.C., David, M.d.K., JoAnne, S.R., Paul, M.W., Eds.; Academic Press: St Louis, MO, USA, 2006; pp. 977–1016.
15. Robaire, B.; Hinton, B.T. *The Epididymis, in Knobil and Neill's Physiology of Reproduction*, 4th ed.; Plant, T.M., Zeleznik, A.J., Eds.; Academic Press: San Diego, CA, USA, 2015; pp. 691–771.
16. De Lamirande, E.; O'Flaherty, C. Sperm activation: Role of reactive oxygen species and kinases. *Biochim. Biophys. Acta* **2008**, *1784*, 106–115. [CrossRef]
17. De Lamirande, E.; O'Flaherty, C. *Sperm Capacitation as An Oxidative event, in Studies on Men's Health and Fertility, Oxidative Stress in Applied Basic Research and Clinical Practice*; Aitken, J., Alvarez, J., Agawarl, A., Eds.; Springer Science: Berlin/Heidelber, Germany, 2012; pp. 57–94.
18. Yanagimachi, R. *Mammalian fertilization, in The Physiology of Reproduction*; Knobil, E., Neill, D., Eds.; Raven Press: New York, NY, USA, 1994; pp. 189–318.
19. Aitken, R.J. Sperm function tests and fertility. *Int. J. Androl.* **2006**, *29*, 69–75. [CrossRef]
20. Buffone, M.G.; Calamera, J.C.; Verstraeten, S.V.; Doncel, G.F. Capacitation-associated protein tyrosine phosphorylation and membrane fluidity changes are impaired in the spermatozoa of asthenozoospermic patients. *Reproduction* **2005**, *129*, 697–705. [CrossRef]
21. De Lamirande, E.; Gagnon, C. Capacitation-associated production of superoxide anion by human spermatozoa. *Free Radic. Biol. Med.* **1995**, *18*, 487–495. [CrossRef]
22. O'Flaherty, C.; Beorlegui, N.; Beconi, M.T. Participation of superoxide anion in the capacitation of cryopreserved bovine sperm. *Int. J. Androl.* **2003**, *26*, 109–114. [CrossRef]
23. Burnaugh, L.; Sabeur, K.; Ball, B.A. Generation of superoxide anion by equine spermatozoa as detected by dihydroethidium. *Theriogenology* **2007**, *67*, 580–589. [CrossRef]
24. Aitken, R.J.; Drevet, J.R. The Importance of Oxidative Stress in Determining the Functionality of Mammalian Spermatozoa: A Two-Edged Sword. *Antioxidants* **2020**, *9*, 111. [CrossRef]
25. Aitken, R.J.; Harkiss, D.; Knox, W.; Paterson, M.; Irvine, D.S. A novel signal transduction cascade in capacitating human spermatozoa characterised by a redox-regulated, cAMP-mediated induction of tyrosine phosphorylation. *J. Cell Sci.* **1998**, *111 Pt 5*, 645–656. [PubMed]
26. O'Flaherty, C.; de Lamirande, E.; Gagnon, C. Reactive oxygen species modulate independent protein phosphorylation pathways during human sperm capacitation. *Free Radic. Biol. Med.* **2006**, *40*, 1045–1055. [CrossRef] [PubMed]
27. O'Flaherty, C.; de Lamirande, E.; Gagnon, C. Positive role of reactive oxygen species in mammalian sperm capacitation: Triggering and modulation of phosphorylation events. *Free Radic. Biol. Med.* **2006**, *41*, 528–540. [CrossRef] [PubMed]

28. Lee, D.; Moawad, A.; Morielli, T.; Fernandez, M.; O'Flaherty, C. Peroxiredoxins prevent oxidative stress during human sperm capacitation. *Mol. Hum. Reprod.* **2017**, *23*, 106–115. [CrossRef] [PubMed]
29. O'Flaherty, C. Redox regulation of mammalian sperm capacitation. *Asian J. Androl.* **2015**, *17*, 583–590. [CrossRef] [PubMed]
30. Morielli, T.; O'Flaherty, C. Oxidative stress impairs function and increases redox protein modifications in human spermatozoa. *Reproduction* **2015**, *149*, 113–123. [CrossRef] [PubMed]
31. Aitken, R.J.; Harkiss, D.; Buckingham, D. Relationship between iron-catalysed lipid peroxidation potential and human sperm function. *J. Reporod. Fertil.* **1993**, *98*, 257–265. [CrossRef]
32. Moawad, A.R.; Fernandez, M.C.; Scarlata, E.; Dodia, C.; Feinstein, S.I.; Fisher, A.B.; O'Flaherty, C. Deficiency of peroxiredoxin 6 or inhibition of its phospholipase A2 activity impair the in vitro sperm fertilizing competence in mice. *Sci. Rep.* **2017**, *7*, 12994. [CrossRef]
33. Aitken, R.J.; Gordon, E.; Harkiss, D.; Twigg, J.P.; Milne, P.; Jennings, Z.; Irvine, D.S. Relative impact of oxidative stress on the functional competence and genomic integrity of human spermatozoa. *Biol. Reprod.* **1998**, *59*, 1037–1046. [CrossRef]
34. Noblanc, A.; Kocer, A.; Drevet, J.R. Post-testicular protection of male gametes from oxidative damage. The role of the epididymis. *Med. Sci.* **2012**, *28*, 519–525.
35. Iwasaki, A.; Gagnon, C. Formation of reactive oxygen species in spermatozoa of infertile patients. *Fertil. Steril.* **1992**, *57*, 409–416. [CrossRef]
36. Jones, R.; Mann, T.; Sherins, R. Peroxidative breakdown of phospholipids in human spermatozoa, spermicidal properties of fatty acid peroxides, and protective action of seminal plasma. *Fertil. Steril.* **1979**, *31*, 531–537. [CrossRef]
37. Alvarez, J.G.; Touchstone, J.C.; Blasco, L.; Storey, B.T. Spontaneous lipid peroxidation and production of hydrogen peroxide and superoxide in human spermatozoa. Superoxide dismutase as major enzyme protectant against oxygen toxicity. *J. Androl.* **1987**, *8*, 338–348. [PubMed]
38. Aitken, R.J.; Whiting, S.; De Iuliis, G.N.; McClymont, S.; Mitchell, L.A.; Baker, M.A. Electrophilic Aldehydes Generated by Sperm Metabolism Activate Mitochondrial Reactive Oxygen Species Generation and Apoptosis by Targeting Succinate Dehydrogenase. *J. Biol. Chem.* **2012**, *287*, 33048–33060. [CrossRef]
39. O'Flaherty, C. Peroxiredoxin 6: The Protector of Male Fertility. *Antioxidants* **2018**, *7*, 173. [CrossRef]
40. Fernandez, M.C.; O'Flaherty, C. Peroxiredoxin 6 activates maintenance of viability and DNA integrity in human spermatozoa. *Hum. Reprod.* **2018**, *33*, 1394–1407. [CrossRef]
41. Ozkosem, B.; Feinstein, S.I.; Fisher, A.B.; O'Flaherty, C. Absence of Peroxiredoxin 6 Amplifies the Effect of Oxidant Stress on Mobility and SCSA/CMA3 Defined Chromatin Quality and Impairs Fertilizing Ability of Mouse Spermatozoa. *Biol. Reprod.* **2016**, *94*, 1–10. [CrossRef]
42. Aitken, R.J.; De Iuliis, G.N.; Finnie, J.M.; Hedges, A.; McLachlan, R.I. Analysis of the relationships between oxidative stress, DNA damage and sperm vitality in a patient population: Development of diagnostic criteria. *Hum. Reprod.* **2010**, *25*, 2415–2426. [CrossRef]
43. Noblanc, A.; Damon-Soubeyrand, C.; Karrich, B.; Henry-Berger, J.; Cadet, R.; Saez, F.; Guiton, R.; Janny, L.; Pons-Rejraji, H.; Alvarez, J.G.; et al. DNA oxidative damage in mammalian spermatozoa: Where and why is the male nucleus affected? *Free Radic. Biol. Med.* **2013**, *65*, 719–723. [CrossRef]
44. Cambi, M.; Tamburrino, L.; Marchiani, S.; Olivito, B.; Azzari, C.; Forti, G.; Baldi, E.; Muratori, M. Development of a specific method to evaluate 8-hydroxy,2-deoxyguanosine in sperm nuclei: Relationship with semen quality in a cohort of 94 subjects. *Reproduction* **2013**, *145*, 227–235. [CrossRef]
45. O'Flaherty, C.; Matsushita-Fournier, D. Reactive oxygen species and protein modifications in spermatozoa. *Biol. Reprod.* **2017**, *97*, 577–585. [CrossRef] [PubMed]
46. Salvolini, E.; Buldreghini, E.; Lucarini, G.; Vignini, A.; Di Primio, R.; Balercia, G. Nitric oxide synthase and tyrosine nitration in idiopathic asthenozoospermia: An immunohistochemical study. *Fertil. Steril.* **2012**, *97*, 554–560. [CrossRef] [PubMed]
47. Morielli, T.; O'Flaherty, C. Oxidative stress promotes protein tyrosine nitration and S-glutathionylation impairing motility and capacitation in human spermatozoa. *Free Radic. Biol. Med.* **2012**, *53* (Suppl. 2), S137. [CrossRef]
48. Scarlata, E.; O'Flaherty, C. Antioxidant Enzymes and Male Fertility: Lessons from Knockout Models. *Antioxid. Redox Signal.* **2020**, *32*, 569–580. [CrossRef] [PubMed]

49. Selvaratnam, J.S.; Robaire, B. Effects of Aging and Oxidative Stress on Spermatozoa of Superoxide-Dismutase 1- and Catalase-Null Mice1. *Biol. Reprod.* **2016**, *95*, 60–61. [CrossRef] [PubMed]
50. Smith, T.B.; Baker, M.A.; Connaughton, H.S.; Habenicht, U.; Aitken, R.J. Functional deletion of Txndc2 and Txndc3 increases the susceptibility of spermatozoa to age-related oxidative stress. *Free Radic. Biol. Med.* **2013**, *65*, 872–881. [CrossRef]
51. Chabory, E.; Damon, C.; Lenoir, A.; Kauselmann, G.; Kern, H.; Zevnik, B.; Garrel, C.; Saez, F.; Cadet, R.; Henry-Berger, J.; et al. Epididymis seleno-independent glutathione peroxidase 5 maintains sperm DNA integrity in mice. *J. Clin. Investig.* **2009**, *119*, 2074–2085. [CrossRef]
52. Schneider, M.; Forster, H.; Boersma, A.; Seiler, A.; Wehnes, H.; Sinowatz, F.; Neumuller, C.; Deutsch, M.J.; Walch, A.; Hrabe de Angelis, M.; et al. Mitochondrial glutathione peroxidase 4 disruption causes male infertility. *FASEB J.* **2009**, *23*, 3233–3242. [CrossRef]
53. Smarr, M.M.; Sapra, K.J.; Gemmill, A.; Kahn, L.G.; Wise, L.A.; Lynch, C.D.; Factor-Litvak, P.; Mumford, S.L.; Skakkebaek, N.E.; Slama, R.; et al. Is human fecundity changing? A discussion of research and data gaps precluding us from having an answer. *Hum. Reprod.* **2017**, *32*, 499–504. [CrossRef]
54. Levine, H.; Jørgensen, N.; Martino-Andrade, A.; Mendiola, J.; Weksler-Derri, D.; Mindlis, I.; Pinotti, R.; Swan, S.H. Temporal trends in sperm count: A systematic review and meta-regression analysis. *Hum. Reprod. Update* **2017**, *23*, 646–659. [CrossRef]
55. Skakkebaek, N.E.; Meyts, E.R.-D.; Louis, G.M.B.; Toppari, J.; Andersson, A.-M.; Eisenberg, M.L.; Jensen, T.K.; Jørgensen, N.; Swan, S.H.; Sapra, K.J.; et al. Male Reproductive Disorders and Fertility Trends: Influences of Environment and Genetic Susceptibility. *Physiol. Rev.* **2016**, *96*, 55–97. [CrossRef] [PubMed]
56. Paul, C.; Robaire, B. Ageing of the male germ line. *Nat. Rev. Urol.* **2013**, *10*, 227–234. [CrossRef] [PubMed]
57. Noblanc, A.; Klaassen, A.; Robaire, B. The Exacerbation of Aging and Oxidative Stress in the Epididymis of Sod1 Null Mice. *Antioxidants* **2020**, *9*, 151. [CrossRef] [PubMed]
58. Wu, P.Y.; Scarlata, E.; O'Flaherty, C. Long-Term Adverse Effects of Oxidative Stress on Rat Epididymis and Spermatozoa. *Antioxidants* **2020**, *9*, 170. [CrossRef]
59. O'Flaherty, C.; Boisvert, A.; Manku, G.; Culty, M. Protective Role of Peroxiredoxins against Reactive Oxygen Species in Neonatal Rat Testicular Gonocytes. *Antioxidants* **2019**, *9*, 32. [CrossRef]
60. Hamilton, L.E.; Zigo, M.; Mao, J.; Xu, W.; Sutovsky, P.; O'Flaherty, C.; Oko, R. GSTO2 Isoforms Participate in the Oxidative Regulation of the Plasmalemma in Eutherian Spermatozoa during Capacitation. *Antioxidants* **2019**, *8*, 601. [CrossRef]
61. Nowicka-Bauer, K.; Nixon, B. Molecular Changes Induced by Oxidative Stress that Impair Human Sperm Motility. *Antioxidants* **2020**, *9*, 134. [CrossRef]
62. Drevet, J.R.; Aitken, R.J. Oxidation of Sperm Nucleus in Mammals: A Physiological Necessity to Some Extent with Adverse Impacts on Oocyte and Offspring. *Antioxidants* **2020**, *9*, 95. [CrossRef]
63. Villaverde, A.I.S.B.; Netherton, J.; Baker, M.A. From Past to Present: The Link Between Reactive Oxygen Species in Sperm and Male Infertility. *Antioxidants* **2019**, *8*, 616. [CrossRef]
64. Martin-Hidalgo, D.; Bragado, M.J.; Batista, A.R.; Oliveira, P.F.; Alves, M.G. Antioxidants and Male Fertility: From Molecular Studies to Clinical Evidence. *Antioxidants* **2019**, *8*, 89. [CrossRef] [PubMed]
65. Peña, F.J.; O'Flaherty, C.; Ortiz Rodríguez, J.M.; Martín Cano, F.E.; Gaitskell-Phillips, G.L.; Gil, M.C.; Ortega Ferrusola, C. Redox Regulation and Oxidative Stress: The Particular Case of the Stallion Spermatozoa. *Antioxidants* **2019**, *8*, 567. [CrossRef] [PubMed]
66. Showell, M.G.; Mackenzie-Proctor, R.; Brown, J.; Yazdani, A.; Stankiewicz, M.T.; Hart, R.J. Antioxidants for male subfertility. *Cochrane Database Syst. Rev.* **2014**, *12*, CD007411. [CrossRef] [PubMed]
67. Windsor, B.; Popovich, I.; Jordan, V.; Showell, M.; Shea, B.; Farquhar, C. Methodological quality of systematic reviews in subfertility: A comparison of Cochrane and non-Cochrane systematic reviews in assisted reproductive technologies. *Hum. Reprod.* **2012**, *27*, 3460–3466. [CrossRef] [PubMed]
68. Zini, A.; Al-Hathal, N. therapy in male infertility: Fact or fiction? *Asian J. Androl.* **2011**, *13*, 374–381. [CrossRef] [PubMed]

© 2020 by the author. Licensee MDPI, Basel, Switzerland. This article is an open access article distributed under the terms and conditions of the Creative Commons Attribution (CC BY) license (http://creativecommons.org/licenses/by/4.0/).

Article

Protective Role of Peroxiredoxins against Reactive Oxygen Species in Neonatal Rat Testicular Gonocytes

Cristian O'Flaherty [1,2], Annie Boisvert [1], Gurpreet Manku [1,3] and Martine Culty [1,3,4,*]

1. The Research Institute of the McGill University Health Centre, Montreal, QC H4A 3J1, Canada; cristian.oflaherty@mcgill.ca (C.O.); annieboisvert@hotmail.com (A.B.); gurpreet.manku@mail.mcgill.ca (G.M.)
2. Department of Surgery (Urology Division), McGill University, Montreal, QC H4A 3J1, Canada
3. Department of Medicine, McGill University, Montreal, QC H4A 3J1, Canada
4. Department of Pharmacology and Pharmaceutical Sciences, School of Pharmacy, University of Southern California School of Pharmacy, Los Angeles, CA 90089, USA
* Correspondence: culty@usc.edu; Tel.: +1-323-865-1677

Received: 2 December 2019; Accepted: 25 December 2019; Published: 30 December 2019

Abstract: Peroxiredoxins (PRDXs) are antioxidant enzymes that protect cells from oxidative stress and play a role in reactive oxygen species (ROS)-mediated signaling. We reported that PRDXs are critical for human fertility by maintaining sperm viability and regulating ROS levels during capacitation. Moreover, studies on $Prdx6^{-/-}$ mice revealed the essential role of PRDX6 in the viability, motility, and fertility competence of spermatozoa. Although PRDXs are abundant in the testis and spermatozoa, their potential role at different phases of spermatogenesis and in perinatal germ cells is unknown. Here, we examined the expression and role of PRDXs in isolated rat neonatal gonocytes, the precursors of spermatogonia, including spermatogonial stem cells. Gene array, qPCR analyses showed that PRDX1, 2, 3, 5, and 6 transcripts are among the most abundant antioxidant genes in postnatal day (PND) 3 gonocytes, while immunofluorescence confirmed the expression of PRDX1, 2, and 6 proteins. The role of PRDXs in gonocyte viability was examined using PRDX inhibitors, revealing that the 2-Cys PRDXs and PRDX6 peroxidases activities are critical for gonocytes viability in basal condition, likely preventing an excessive accumulation of endogenous ROS in the cells. In contrast to its crucial role in spermatozoa, PRDX6 independent phospholipase A_2 (iPLA$_2$) activity was not critical in gonocytes in basal conditions. However, under conditions of H_2O_2-induced oxidative stress, all these enzymatic activities were critical to maintain gonocyte viability. The inhibition of PRDXs promoted a two-fold increase in lipid peroxidation and prevented gonocyte differentiation. These results suggest that ROS are produced in neonatal gonocytes, where they are maintained by PRDXs at levels that are non-toxic and permissive for cell differentiation. These findings show that PRDXs play a major role in the antioxidant machinery of gonocytes, to maintain cell viability and allow for differentiation.

Keywords: testis; gonocytes; peroxiredoxins; oxidative stress; ROS; differentiation

---

1. Introduction

Peroxiredoxins are found in all living organisms, from bacteria, plants, yeasts to animals, where they act as scavengers of hydrogen peroxide (H_2O_2), lipid peroxides and peroxynitrite. Mammalian Peroxiredoxins (PRDXs) are important, not only as antioxidant enzymes preventing reactive oxygen species (ROS)-induced cell damage, but also as physiological regulators and sensors in a variety of cell and tissue types [1]. Indeed, the activation of several phosphatases, kinases, and tumor suppressor proteins have been shown to require a certain level of H_2O_2 acting as a second messenger in the vicinity of the enzymes, which is achieved by the transient and localized inhibition of PRDXs [2,3]. PRDXs are classified depending on the cysteine residues (Cys) in their active site, that will react with peroxides. They comprise the 2-Cys PRDX1 to 4, the atypical 2-Cys PRDX5, and the 1-Cys PRDX6, which has the particularity of being bifunctional, with both peroxidase and calcium-independent phospholipase A_2 (iPLA_2) activities [4]. The 2-Cys PRDX1 to 4 are homodimers in which the thiol of a cysteine residue of one PRDX subunit gets oxidized, then further reacts with the thiol group of the catalytic cysteine of the other subunit, forming a disulfide bond between the two subunits. By contrast, in the atypical PRDX5, 2 Cys of the same chain react upon oxidation to form an intrasubunit disulfide bond. Inactive PRDXs are then reactivated by a reduction of the disulfide bonds by thioredoxin (TRX), itself further reactivated by TRX reductase (TRD), using NADPH as a reducing equivalent. In the case of PRDX6, since the enzyme has only one catalytic Cys, the oxidized thiol will be reduced by the glutathione-GSH-transferase P1 (GSTP1) system [4–6].

Studies in PRDXs knockout mice have support the understanding of the diverse roles of these enzymes by highlighting the different defects in mice deficient for a specific PRDX [7]. In particular, spermatozoa have been shown to express the six PRDX isoforms [4], which act as ROS scavengers and are required to maintain viability as well as fertilizing competence [3,6,8–10]. While ROS are needed for sperm capacitation, due to their regulatory role in the phosphorylation of key proteins, their levels must be tightly controlled to prevent damaging oxidative stress, mainly by PRDX1 and 6 in rat [8–10]. In mice, PRDX6 deficiency or inhibition of its PLA_2 activity were found to impair in vitro sperm fertilizing competence [11]. Low levels of PRDX6 were observed in infertile men, positioning PRDX6 as the first line of defense against oxidative stress in human spermatozoa [12]. Although spermatogenesis occurs in the testes of PRDX6 KO mice, these animals are subfertile, with defective and underperforming spermatozoa, suggesting potential alterations of some of the processes leading to sperm formation. While the importance of PRDXs on sperm integrity and function is clear, little is known on the role of PRDXs in germ cells from primordial germ cells to spermatids.

The goal of this study was to examine the expression and role of PRDXs in neonatal gonocytes (also called pre-/pro-spermatogonia), the direct precursors of spermatogonial stem cells and first wave spermatogonia [13,14]. Gonocytes differentiate from primordial germ cells in the fetal gonad primordium, and undergo distinct phases of development, including successive phases of proliferation and quiescence in the fetus, resuming mitosis at postnatal day (PND) 3 in the rat, and simultaneously migrating toward the basement membrane of the seminiferous tubules where they differentiate to spermatogonia around PND6 [15]. We have previously shown that rat neonatal gonocyte differentiation is regulated by all trans-retinoic acid (RA) [16,17]. Extensive cell remodeling takes place during the proliferation, relocation and differentiation of neonatal gonocytes, in part regulated by the ubiquitin proteasome system [18]. We recently reported that neonatal gonocytes express high levels of cyclooxygenase 2 (COX2) and produce prostaglandins [19]. While COX2 and prostaglandins were reported to regulate ROS production in Sertoli cells [20], in other cell types such as the kidney mesanglial cell, ROS were shown to regulate COX2 expression and prostaglandin synthesis [21]. However, nothing is known on ROS formation and the antioxidant machinery in neonatal gonocytes. The present study demonstrates that PRDXs are essential for maintaining ROS homeostasis and cell viability in neonatal gonocytes, and that the iPLA_2 activity of PRDX6 in these cells is not as critical as it is in spermatozoa, suggesting differential role for these antioxidant enzymes at different phases of germ cell development.

2. Materials and Methods

2.1. Chemicals

Conoidin A, an inhibitor of 2-cystein PRDX1-5 peroxidase activities was purchased from Cayman Chemical (Ann Arbor, MI, USA). MJ33 (1-Hexadecyl-3-(trifluoroethyl)-sn-glycero-2-phosphomethanol lithium), competitive inhibitor of the phospholipase A_2 activity of PRDX6 was from Sigma-Aldrich (Milwaukee, WI, USA). Ezatiostat, a glutathione analog inhibitor of the Glutathione S-transferase P1 (GSTP1), required for the re-activation of the peroxidase activity of PRDX6, but not other PRDXs, was purchased from Sigma-Aldrich (Milwaukee, WI, USA). All-trans-retinoic acid, H_2O_2 and common reagents were from Sigma-Aldrich (Milwaukee, WI, USA).

2.2. Animals

PND2 newborn male Sprague Dawley rats were purchased from Charles Rivers Laboratories (Saint-Constant, Quebec, Canada). The pups were handled and euthanized according to the protocols approved by the McGill University Health Centre Animal Care Committee and the Canadian Council on Animal Care. USC Institutional Animal Care and Use Committee; Martine Culty protocol #20792-AM001 (Physiology and toxicology of male reproductive system).

2.3. Gonocyte Isolation and in Vitro Culture and Treatments

Neonatal gonocytes were isolated by performing sequential enzymatic tissue dissociation together with mechanical dissociation of the pooled testes from 40 PND3 pups per experiment. This was followed by a step of differential overnight adhesion at 37 °C in medium containing 5% fetal bovine serum (FBS), and cell separation of the non-adherent cells on a 2–4% bovine serum albumin (BSA) gradient in serum-free medium on the next morning [22,23]. Enriched gonocyte preparations at 70–80% purity were obtained by pooling fractions containing high proportions of gonocytes, according to size and appearance, while a gonocyte purity above 95% was used for gene array analysis [18]. Freshly isolated gonocytes were cultured at 20 to 30,000 cells per well in 500 µL of RPMI 1640 containing 2.5% FBS, antibiotics, alone or with the PRDX inhibitors conoidin A, MJ33 and ezatiostat, and/or H_2O_2, at different concentrations, for 2 to 18 h, in 3.5% CO_2, at 37 °C. Cell differentiation was examined by treating the gonocytes with 10^{-6} M retinoic acid (RA), in the absence or presence of the PRDX inhibitors.

2.4. Cell Viability

Cell viability was assessed using a Trypan blue exclusion assay, by counting live and trypan blue-positive gonocytes as previously described [19], and viability was expressed as the mean ± SEM of the percentage of live cells against the total number of gonocytes in 3 independent experiments, each performed with triplicates.

2.5. RNA Extraction and Real-Time Quantitative PCR (Q-PCR) Analysis

Total RNA was extracted from cell pellets using the PicoPure RNA isolaton kit (Arcturus, Mountain View, CA, USA) and digested with DNase I (Qiagen, Santa Clarita, CA, USA), followed by cDNA synthesis with a single-strand cDNA transcriptor synthesis kit (Roche Diagnostics, Indianapolis, IN, USA), as previously described [18,19]. Quantitative real-time PCR (qPCR) was performed using SYBRgreen PCR Master Mix kit (Bio-Rad, Hercules, CA, USA) on a LightCycler 480 (LC480, Roche Diagnostics) [18]. The forward and reverse primers used are provided in Table 1. The comparative Ct method was used to calculate the relative expression of the differentiation marker Stra8, and PRDX 1 to 6, using 18S rRNA as housekeeping gene for data normalization. Changes in Stra8 gene expression are expressed as percent of the control values; and given as means ± SEM from 3 or 4 experiments, each using triplicates. PRDX data are expressed in relative gene expression and shown as the means ± SEM from 3 experiments.

2.6. Gene Array Analysis

Briefly, the RNA samples isolated from three independent PND3 gonocytes and PND8 spermatogonia cell preparations, each made from multiple rat testes (each sample corresponding to the RNA of 60–90 pups for gonocytes, and 10 pup rats for spermatogonia) were analyzed using Illumina RatRef-12 Expression BeadChips, as previously described [18]. The relative levels of antioxidant genes were expressed in arbitrary units and presented as the means ± SEM of all data.

As a comparison, the RNA from an enriched germ cell population prepared from the pooled testes of 4 PND60 adult rats was also analyzed in the gene arrays, as described before [24].

Table 1. Primer sets for qPCR analysis of genes in rat gonocytes. The underlined bases correspond to bases added to the gene sequence by the primer design program to generate better-balanced primers.

Gene	Accession Number	Forward and Reverse Primers
18S rRNA	X01117.1	Cgggtgctcttagctgagtgtcccg ctcgggcctgctttgaacac
Stra8	XM_575429.2	tgcttttgatgtggcgagct gcgctgatgttagacagacgct
Peroxiredoxin 1 (PRDX1)	NM_057114.1	acctgtagctcgactctgctg aacagccgtggctttgaa
PRDX2	NM_017169.1	gactctcagttcacccacctg tattcagtgggcccaagc
PRDX3	NM_022540.1	agaagaacctgcttgacagaca caggggtgtggaatgaaga
PRDX4	NM_053512.2	tgagacactgcgtttggttc tgtttcactaccaggtttccag
PRDX5	NM_053610.1	agtgccgcggtgactatg caaaacacctttcttgtccttga
PRDX6	NM_053576.2	ttgattgctctttcaatagactctg ctgcaccattgtaagcattga

2.7. Lipid Peroxidation Measurement by Bodipy Labelling

Lipid peroxidation was measured using the fluorescent lipid peroxidation sensor BODIPY 581/591 C_{11}, a reporter fatty acid labelled with bodipy (4,4-difluoro-3a,4adiaza-s-indacene) fluorophore, which can enter the cells and is used as a surrogate for endogenous lipids (Bodipy; Life Technologies (Burlington, Ontario, ON, Canada)). The peroxidation of the reporter fatty acid leads to a shift in the fluorescence of BODIPY from red to green in the cells. Thus, cells presenting green fluorescence corresponds to cells in which lipid peroxidation occurred, which can be quantified by assessing their proportion in each sample. Following 1.5 h of gonocyte treatments with either medium, PRDXs inhibitors and/or H_2O_2, the Bodipy reagent was added to the wells at 20 µM final concentration for an additional 30 min at 37 °C. The reactions were stopped by collecting and centrifuging the cells at 425× g for 10 min at 4 °C. The pellets were washed with PBS, the cells fixed with paraformaldehyde (3.5% final) for 7 min, washed and further collected by cytospin centrifugation on microscopic slides. For each slide, 5 pictures were taken using FITC (oxidized Bodipy reporter) and Texas Red (non-oxidized reporter) fluorescence, on a Leica fluorescent microscope. Lipid peroxidation was measured as the percent of Bodipy-positive green fluorescent cells over the total gonocyte numbers. Data are shown as the means ± SEM of samples from 3 different experiments.

2.8. Immunohistochemistry (IHC)/Immunocytochemistry (ICC)

For IHC, PND3 testis samples were fixed with 4% paraformaldehyde solution (PFA), paraffin embedded, and 4 to 6 µm thick sections were made. IHC and ICC analyses were performed as previously described [18,19,23,25]. Briefly, the slides were dewaxed, rehydrated, treated for antigen retrieval, and then with a blocking reagent in PBS. The slides were then incubated overnight at 4 °C in Anti PRDX6 primary antibody diluted (1:100) in PBS containing 0.1% Triton X-100 and serum. This was followed by washes and 1-h incubation with a biotinylated secondary antibody (dilution 1:100) in 0.1% Triton X-100 and 10% BSA, at room temperature, then 15 min of treatment with Streptavidin-horse radish peroxidase and 15 min with an AEC Chromogen solution. Mayer's hematoxylin was used for counterstaining, and Clear-Mount for coating. The slides were examined using bright-field microscopy. As negative controls, some slides were treated with non-specific immunoglobulin G instead of specific antibody.

For ICC, the protein expression of PRDX1, 2, and 6 were examined in low purity gonocyte fractions pooled, washed with PBS, and fixed with 3.5% paraformaldehyde right after the BSA gradient. The fixed cells were collected by cytospin centrifugation, the slides dried and treated with acetone:methanol (60:40), followed by the antigen retrieval solution. The ICC reactions were similar to those described above for IHC, except for the use of fluorescent secondary antibodies. DAPI was used as a nuclear signal. Rabbit and mouse IgG were used as negative controls and gave no fluorescent signal (data not shown). Pictures of fluorescent signals and bright field were taken, using the same time of exposure for the specific antibodies and non-specific IgGs. Representative samples are shown.

2.9. Statistical Analysis

Statistical analysis was performed using the unpaired Student's t test or one-way ANOVA with post-hoc Tukey multiple comparison analysis, using Prism version 5.04 (GraphPad Software, San Diego, CA, USA). Changes with p values ≤ 0.05 were considered significant.

3. Results

3.1. PRDXs Are among the Highest Expressed Antioxidant Genes in Neonatal Gonocytes and in Spermatogonia

Gene array analysis of antioxidant genes in highly purified rat PND3 gonocytes and PND8 spermatogonia showed that, at both ages, Prdxs were among the most abundant antioxidant genes (Figure 1A, Table S1). Comparing the signal intensities of all genes in the arrays showed that thioredoxin 1 (*Txn1*), *Prdx1* and superoxide dismutase (*Sod*) 1 were among the 1% most abundant genes in these cells, comprising relative signal intensities around and above 2000. Next, *Prdx2*, *Prdx5*, *Prdx6*, *Sod2*, *Gstp1*, and *Gpx4* were among the most abundant genes, with relative signal intensities from 600 to 1200 (Table S1). Other highly expressed genes (signal intensities between 300 and 600) included *Gsto1*, *Prdx3*, *Txnl1*, *Mgst1*, *Txnrd1*, *Gstp2*, *Gpx1*, and *Nrf2*. As a comparison, these genes had much higher signal intensities than 65 percent of the genes in the cells, which presented intensities below or around 20, levels found for Sod3 and Catalase. The ranking of expression for PRDXs were *Prdx1* > *Prdx2* > *Prdx5* > *Prdx6* > *Prdx3*, in both germ cell types. Measurement of the relative expression of the PRDXs by qPCR in PND3 gonocytes indicated a similar ranking of expression levels, where *Prdx1* and *Prdx2* were the most abundant, followed by *Prdx6*, *Prdx5* and *Prdx3*, and finally *Prdx4* present at very low levels (Figure 1B). The similarities in expression profiles of antioxidant genes between gonocytes and spermatogonia suggested a conserved antioxidant machinery between the two phases of germ cell development.

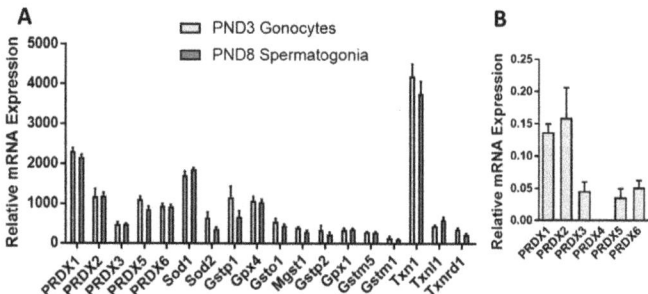

Figure 1. Relative expression of antioxidant genes in postnatal day 3 (PND3) gonocytes and PND8 spermatogonia. (**A**) gene array analysis of PND3 gonocytes and PND8 spermatogonia antioxidant genes. (**B**) qPCR analysis of PRDXs expression in PND3 gonocytes. Results represent the means ± SEM from 3 or 4 experiments, each using multiples animals.

This was different from the levels found in enriched PND60 adult rat germ cells, in which the relative gene expression of *Gstm5* (6235) was the highest, followed by *Gpx4* (5507), *Hagh* (1141), *Sod1* (1136), *mGst1* (446), *Prdx1* (407), *Prdx6* (345), *Gstm1* (341), *Txn1* (335), *Prdx2* (223), *Prdx5* (204), *Gsto1* (188), *Prdx3* (144), *Gstt2* (140), *Txnl1* (132), *Txn2* (81), *Sod3* (73), *Gsta2* (64), *Gstp1* (65), and *Sod2* (44), the remaining genes being at very low expression levels.

The immunological analysis of the protein expression of PRDX1, 2, and 6 in mixed suspensions of PND3 gonocytes, Sertoli and myoid cells confirmed that the three PRDX proteins were expressed in gonocytes, Sertoli, and myoid cells, but at lower levels in gonocytes than in most of the somatic cells (Figure 2A). PRDX1 showed a slightly stronger signal than PRDX2, both being higher than PRDX6 in gonocytes, in agreement with the transcript levels. PRDX6 protein was also visible by IHC analysis in the cytoplasm of gonocytes in PND3 testis sections, but it was lower than the levels observed in Sertoli cells and some interstitial cells (Figure 2B).

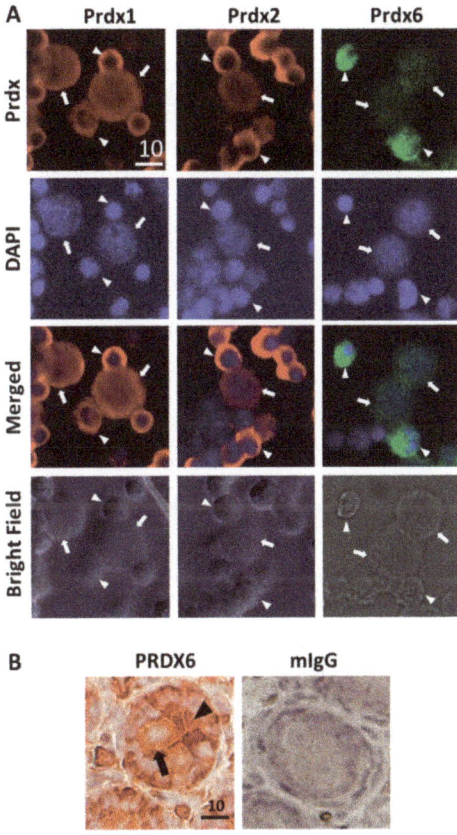

Figure 2. Protein expression of PRDX1, 2, and 6 in PND3 gonocytes. (**A**) Immunocytochemistry (ICC) analysis of PRDXs expression in cells collected right after the cell isolation procedure on cytospin slides. Low purity pooled fractions were used. White arrow: gonocytes. White arrowhead: somatic cells (Sertoli and peritubular myoid cells). (**B**) Immunohistochemistry (IHC) analysis of PND3 testis section. Representative pictures are shown. Black arrow: gonocytes. Black arrowhead: somatic cells. Scales are in µm.

3.2. PRDXs Are Required for PND3 Gonocyte Survival in Basal Conditions

In view of the abundance of PRDXs in gonocytes, and knowing the importance of these peroxidases for spermatozoa viability and function, we examined whether inhibiting PRDXs would have an impact on the survival of PND3 gonocytes. For that, we performed dose-response and time-course studies in basal conditions where only endogenous ROS may be formed, using either the inhibitor of 2-Cys PRDXs, Conoidin A; the inhibitor of GSTP1 (that promotes PRDX6 glutathione-dependent re-activation), Ezatiostat; or the inhibitor of the phospholipase A_2 activity of PRDX6, MJ33. Conoidin A induced a rapid dose-dependent decrease in gonocyte viability with concentrations of 5–10 µM, inducing 50% cell death after one hour of treatment (Figure 3A). There were no surviving cells with conoidin A concentrations of 1–10 µM after 18 h of treatments. By contrast, after 18 h of treatment with MJ33, there was no or minimally detrimental effects on cell viability with concentration from 1 to 20 µM (Figure 3B). However, at 50 µM, MJ33 did induce cell death after 18 h. Eztiostat dose responses were performed for two hours of treatment, because preliminary studies showed that longer time-periods induced the loss of a large amount of cells. Eztiostat inhibited cell survival in a dose-dependent manner (Figure 3C), with a concentration of 50 µM inducing nearly 50% of cell death. These data suggest that the peroxidase activity of 2-cys PRDXs and PRDX6 play a critical role in gonocyte protection against excessive endogenous ROS, and that inhibiting these enzymes even for a short time is sufficient to induce significant gonocyte cell death. Thus, PRDXs are required to maintain ROS homeostasis in neonatal gonocytes. By contrast, inhibiting the PLA2 activity of PRDX6 with MJ33 did not exert rapid deleterious effects on cell viability.

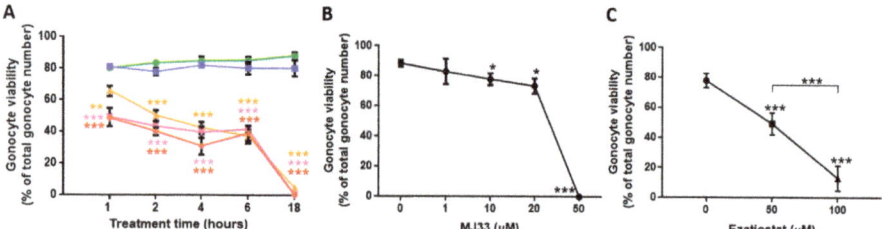

Figure 3. Effects of PRDX inhibitors on PND3 gonocyte viability. Cell viability was determined by trypan blue exclusion method. (**A**) Dose response and time course of conoidin A effects. Control: green; Conoidin A at 0.1 µM (blue); 1 µM (orange); 5 µM (pink); 10 µM (red). (**B**) MJ33 dose response at 18 h treatments. (**C**) Eztiostat dose response at two hours of treatment. Results represent the means ± SEM from at least three experiments. Statistical significance: * $p \leq 0.05$; ** $p \leq 0.01$; *** $p \leq 0.001$.

PRDXs are required to prevent oxidative stress-induced cell death in PND3 gonocytes. Oxidative stress can be generated by external factors, as a result of exposure of the animals and their reproductive system to oxidative agents, or through the production of ROS by other testicular cell types in the vicinity of the gonocytes. Thus, we tested the ability of PRDXs to protect PND3 gonocytes from exogenous ROS, using H_2O_2 as an oxidative agent. The treatment of gonocytes with H_2O_2 at 100 and 200 µM for two hours induced a dose-dependent decrease in cell viability, with a 40% and 50% reduction in viability, respectively (Figure 4).

These damaging effects were significantly worsened by treating the cells with PRDXs inhibitors, with MJ33 having the mildest and ezatiostat the worst effects (Figure 4). These results indicate that the oxidative stress generated by H_2O_2 exerts an adverse effect on gonocyte viability, and that 2 Cys PRDX and PRDX6 peroxidase activities are required to protect gonocytes from oxidative stress. This further implies that other antioxidant proteins expressed in gonocytes, even those present at high levels such as Txn1, Sod1, and Sod2, are not able to rescue gonocytes from exposure to exogenous oxidants. Moreover, the fact that ezatiostat and MJ33 exacerbated the adverse effect of H_2O_2 on gonocyte viability suggests that both the peroxidase and phospholipase A_2 activity of PRDX6 are important for gonocyte survival.

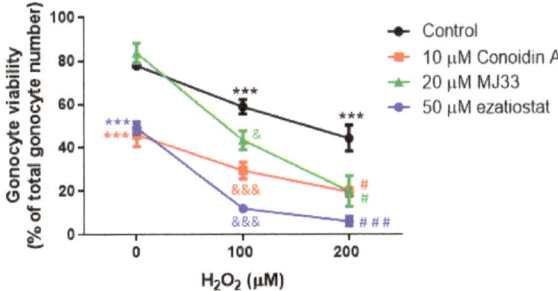

Figure 4. Effects of H_2O_2 and PRDX inhibitors on PND3 gonocyte viability. H_2O_2 and PRDX inhibitors were added to the cells at the indicated concentrations, alone or together for 2 h. Cell viability was determined by trypan blue exclusion method. Results represent the means ± SEM from at least three experiments. Statistical significance: *** $p \leq 0.001$ treatment vs control; & $p \leq 0.05$, &&& $p \leq 0.001$ treatment vs 100 µM H_2O_2; # $p \leq 0.05$, ### $p \leq 0.001$ treatment vs 200 µM H_2O_2.

3.3. H_2O_2 Induces Lipid Peroxidation, and PRDXs Prevent Endogenous ROS-Induced Lipid Peroxidation in PND3 Gonocytes

Lipid peroxidation has been known for decades to be one of the sources of spermatozoa damage in infertile men [26]. Thus, we examined whether lipid peroxidation could also play a role in the deleterious effects of H_2O_2 and PRDX inhibitors on gonocyte viability, using a fluorescent fatty acid reporter system that allowed quantifying the proportion of cells where lipid peroxidation had occurred. While lipid peroxidation was detectable in 40% of the control gonocytes, a condition corresponding to 80% cell viability, treatment with H_2O_2 (a condition decreasing cell viability by 50%) doubled the numbers of gonocytes presenting lipid peroxidation to nearly 80% of positive cells in gonocytes (Figure 5). To examine the effects of PRDX inhibitors, we used concentrations of conoidin A and ezatiostat decreasing viability by ~50% (10 and 50 µM respectively), and a concentration of MJ33 that did not significantly alter cell viability compared to controls (20 µM). Treatments with conoidin A alone had the same effect as H_2O_2 on lipid peroxidation, doubling the percent of gonocytes presenting lipid peroxidation, while MJ33 and ezatiostat induced significant but lower increases in lipid peroxidation, reaching 70% of positive cells, a 1.55 increase over the levels in control cells (Figure 5).

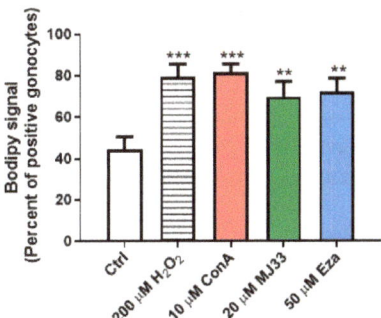

Figure 5. Effects of H_2O_2 and PRDX inhibitors on lipid peroxide formation in PND3 gonocytes. H_2O_2 and PRDX inhibitors were added for two hours on gonocytes. Lipid peroxidation was measured using the Bodipy assay. Results are the means ± SEM from at least three experiments. Statistical significance: ** $p \leq 0.01$; *** $p \leq 0.001$.

Combining PRDX inhibitors with H_2O_2 did not further increase lipid peroxidation, which already affected most of the cells with individual treatments (data not shown). These results show that an oxidative stressor such as H_2O_2 induces similar levels of lipid peroxidation as endogenous ROS accumulating in the absence of PRDX activities in PND3 gonocytes. This suggests that lipid peroxides are produced physiologically in gonocytes, and that PRDXs are critical for preventing excessive levels that could affect gonocyte survival. The data also imply that PRDXs protect gonocytes from excessive endogenous ROS-induced lipid peroxidation. Moreover, the fact that 40% of control gonocytes showed lipid peroxidation without affecting their viability, and that lipid peroxidation was not proportional to viability in cells treated with PRDX inhibitors, suggest that gonocytes can tolerate a certain level of lipid peroxidation. The finding that MJ33 treatment increased lipid peroxidation suggests that the PRDX6 iPLA$_2$ activity also plays a role in repairing oxidized membranes [27].

3.4. PRDX Inhibition Blocks RA-Induced Differentiation in PND3 Gonocytes

Considering the physiological roles played by ROS in many tissues and cell types, including in spermatozoa, and our finding that inhibiting PRDXs increases endogenous ROS formation in PND3 gonocytes, we examined whether blocking PRDXs could affect RA-induced gonocytes differentiation. First, we determined cell viability in cells treated for two hours with RA alone or together with two different concentrations of PRDX inhibitors, in order to select concentrations of inhibitors that would not overly decrease cell survival. RA at 1 µM did not have an effect on cell survival. Similarly, concentrations of 20 µM MJ33 or 10 µM ezatiostat did not affect cell viability (Figure 6). While 0.5 µM conoidin A decreased cell viability by 9% when used alone, and 17% when added with RA in comparison to control and RA treatments, these was still relatively minor effects.

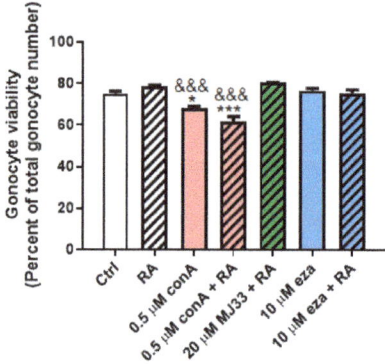

Figure 6. Effects of trans-retinoic acid (RA) and PRDX inhibitors on PND3 gonocyte viability. RA (1 µM) and PRDX inhibitors at the indicated concentrations were added alone or together for two hours on gonocytes. Cell viability was determined by the trypan blue exclusion method. Results represent the means ± SEM from at least 3 experiments. Statistical significance: * $p \leq 0.05$, *** $p \leq 0.001$, treatment against control; &&& $p \leq 0.001$, treatment against RA alone.

Next, we measured the mRNA expression of Stra8, previously found to be a good marker for assessing neonatal gonocyte differentiation [15–17]. While two hours of treatment with RA significantly increased Stra8 expression by over three-fold, there was no effect of conoidin A, MJ33 or ezatiostat added alone to the cells (Figure 7). However, both conoidin A and ezatiostat significantly repressed RA-induced Stra8 increases, indicating inhibitory effects of PRDX inhibitors on gonocytes differentiation. However, MJ33 did not alter RA effect on Stra8 levels. These data suggest that the oxidative stress resulting from PRDXs inhibition is detrimental to gonocytes differentiation, whereas PRDX6 iPLA activity does not appear to be required for gonocyte differentiation.

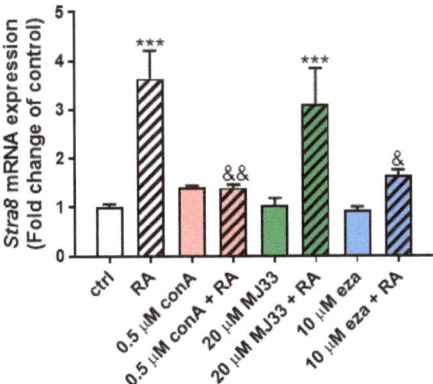

Figure 7. Effects of trans-retinoic acid (RA) and PRDX inhibitors on PND3 gonocyte differentiation. RA (1 µM) and PRDX inhibitors at the indicated concentrations were added alone or together for two hours on gonocytes. *Stra8* mRNA expression was measured by qPCR analysis. Results represent the means ± SEM from three experiments. Statistical significance: *** $p \leq 0.001$, treatment against control; & $p \leq 0.05$, && $p \leq 0.01$, treatment against RA alone.

4. Discussion

The goal of the present study was to determine whether peroxiredoxins play a role in the survival of neonatal male germ cells, as it is the case in spermatozoa. We hypothesized that neonatal germ cells and spermatozoa would likely require different antioxidant machineries, in view of their considerable molecular, morphological and functional differences, as well as their distinct cellular environments. Indeed, the most abundant antioxidants genes in PND3 gonocytes were *Txn1*, *Prdx1*, *Sod1*, *Prdx2*, *Prdx6*, and *Prdx5*, whereas a mixed population of adult germ cells expressed *Sod1* at a similar level as these genes, followed by *mGst1*, *Prdx1*, *Prdx6*, *Txn1*, *Prdx2* and *Prdx5* at lower levels. These data indicate that neonatal gonocytes express very high levels of antioxidant genes relative to adult germ cells, although one cannot exclude that some of these genes might be highly expressed in discreet subsets of adult spermatogonia, spermatocytes, or spermatids. The requirement for high levels of antioxidant genes in neonatal gonocytes might reflect the production of ROS during their multiple functions, including DNA methylation, cell proliferation, migration, and differentiation, all processes requiring energy and the generation of ROS [13,14].

Indeed, our finding that the direct inhibition of 2-Cys PRDX with conoidin A and the specific inhibition of GSTP1, enzyme necessary for the re-activation of PRDX6 peroxidase activity, with ezatiostat [28–31] induce rapid and extensive cell death in PND3 gonocytes implies that high levels of ROS are formed in these cells in physiological conditions. This further suggests that PRDXs play an essential role in maintaining ROS at levels required for physiological functions, but not high enough to induce cell damage. Our results showing that lipid peroxidation is greatly increased by these inhibitors and is associated with increased cell death suggest that lipid peroxides are in part responsible for the adverse effects on gonocytes. In this context, high levels of antioxidant genes, in particular PRDXs, protect neonatal gonocytes from damaging levels of ROS. Moreover, the data clearly show that other antioxidant genes are not capable of rescuing gonocytes from the deleterious effects of ROS accumulation in the absence of PRDXs. In this regard, PRDXs are clearly essential to the maintenance on ROS homeostasis and gonocytes survival, as they are in spermatozoa [4,6].

Recently, it was reported that as human PRXD6, targeted to the yeast mitochondrial matrix, elicited glutathione disulfide (GSSG) formation upon treatment of cells with H_2O_2 [32]. Because yeast lack GSTP1, it was suggested that other enzymes may re-activate the peroxidase activity of PRDX6. Although we observed that glutaredoxin 1 (GRX1) and other GSTs were present in gonocytes, the level of their expression was lower than that of GSTP1 (Figure 1; Table S1). This finding and the decrease in gonocyte survival due to the inhibition of GSTP1 by ezatiostat, indicate that the re-activation of PRDX6 peroxidase activity is probably accomplished by GSTP1 and GSH in gonocytes, as described in other mammalian cells [33].

The finding, that conoidin A and ezatiostat both greatly aggravated H_2O_2 adverse effects, suggests that 2-Cys PRDXs, likely PRDX1 and 2 which were the most abundant at the protein level, as well as PRDX6 are the peroxidases that actively remove ROS such as H_2O_2 from neonatal gonocytes exposed to oxidative stress. Taken together with the fact that Sertoli cells produce H_2O_2 [20], these data suggest that the exposure of neonatal gonocytes to ROS produced by Sertoli cells might be part of a physiological process, as it happens in various tissues and cell types [8,34–36]. However, as shown by the effects of PRDX inhibitors, ROS levels need to be tightly regulated in order to maintain ROS at non-toxic levels in gonocytes.

The fact that the PRDX inhibitors conoidin A and ezatiostat both decrease RA-induced gonocyte differentiation suggests that PRDXs are essential to control ROS levels, preventing cell death and allowing some of the cells to undergo gonocyte differentiation. Moreover, the finding that PRDX inhibition promoted high levels of ROS, which prevented gonocyte differentiation without impairing viability, suggests that one way by which the Sertoli cells could control the timing of gonocyte differentiation would be by producing ROS at levels that would prevent or allow gonocyte differentiation. Further studies are necessary to confirm this possibility. Considering recent studies showing that gonocytes are heterogeneous, similarly to spermatogonia [14,37,38], it would be interesting to study whether neonatal testes contain sub-sets of gonocytes with different PRDX activities, corresponding to different functional states.

One of the main differences we found with our studies on PRDXs in spermatozoa is related to the role of the PRDX6 iPLA$_2$ activity. While this activity is critical for the survival and fertilizing ability of spermatozoa by repairing oxidized membranes [10–12], our results with MJ33 suggest that the PLA$_2$ activity of PRDX6 plays a role in gonocyte survival mainly in the presence of exogenous oxidative stress. Yet, the increased cell death observed after treating gonocytes for 18 h with 50 µM MJ33 suppressing PRDX6 iPLA$_2$ activity implies that impairing the repair of oxidized membranes in gonocytes can jeopardize PND3 gonocyte survival over time. We recently reported that exogenous addition of arachidonic acid or lysophosphatidic acid (LPA) prevented cell death in spermatozoa treated with MJ33 [39]. The addition of H_2O_2 to cells where PRDX6 iPLA$_2$ is inactive and possibly arachidonic acid levels were depleted is sufficient to induce a large increase in cell death in only two hours. Thus, excessive levels of ROS combined with an increased lipid peroxidation and depletion of arachidonic acid and/or LPA appear to aggravate the fate of the cells.

Interestingly, we recently reported that neonatal gonocytes express COX2 and other enzymes of the prostaglandin pathway, and produce prostaglandins (PG) E2 and PGF2a [19]. Moreover, blocking PGE2 and PGF2a synthesis with ibuprofen for 24 h correlated with a partial decrease in RA-induced differentiation in PND3 gonocytes [19]. However, inhibiting the PRDX6 iPLA$_2$ activity for 2 h does not seem to affect gonocyte differentiation. Since arachidonic acid is the precursor of prostaglandins, the present data suggest that a short two-hour treatment with MJ33 may not deplete arachidonic acid sufficiently to compromise cell differentiation, or that this process involves other PLA$_2$ enzymes expressed in gonocytes [19].

5. Conclusions

In conclusion, the present study identified PRDX1, 2, 6, and 5 as major antioxidant genes and proteins in neonatal gonocytes, essential for the survival of the cells, especially under conditions of oxidative stress. Moreover, gonocytes were more sensitive to Conoidin A effects than spermatozoa, whereas they were less sensitive to the inhibition of the PLA$_2$ activity of PRDX6 than spermatozoa, suggesting a difference in the antioxidant machinery of gonocyte and spermatozoa. Further studies will be needed to examine the possibility that ROS, endogenous or produced by Sertoli cells, at non-toxic levels, might play a role in the regulation of neonatal gonocyte differentiation.

Supplementary Materials: The following are available online at http://www.mdpi.com/2076-3921/9/1/32/s1, Table S1: Relative gene expression of antioxidant genes expressed in rat PND3 gonocytes and PND8 spermatogonia. The results represent the mean ± SEM of 3 independent RNA samples for each age, each made from cells isolated in multiple animals. There was no significant difference between the two types of germ cells for any of the genes. A few genes presenting a signal intensity below 20 were not included.

Author Contributions: A.B. performed the experiments presented in Figures 2–7, while G.M. performed the gene array experiments presented in Figure 1. M.C. and C.O. designed the experiments, analyzed the results, and M.C. prepared the manuscript. A.B., C.O. and M.C. revised the final version of the manuscript. All authors have read and agreed to the published version of the manuscript.

Funding: This work was supported in part by funds from a grant from the Canadian Institutes of Health Research (CIHR) (Operating grant # MOP-133456 to MC and # MOP-133661 to CO), a grant from the Natural Sciences and Engineering Research Council of Canada (NSERC) (Discovery Grant # 386038-2013) and funds from the USC School of Pharmacy to MC.

Acknowledgments: We thank Maria Celia Fernandez for her assistance in setting up the bodipy method with A.B.

Conflicts of Interest: The authors declare no conflict of interest.

References

1. Rhee, S.G.; Kil, I.S. Multiple Functions and Regulation of Mammalian Peroxiredoxins. *Annu. Rev. Biochem.* **2017**, *86*, 749–775. [CrossRef] [PubMed]
2. Rhee, S.G. Overview on Peroxiredoxin. *Mol. Cells* **2016**, *39*, 1–5. [PubMed]
3. O'Flaherty, C. Redox regulation of mammalian sperm capacitation. *Asian J. Androl.* **2015**, *17*, 583–590. [CrossRef] [PubMed]
4. O'Flaherty, C. Peroxiredoxins: Hidden players in the antioxidant defence of human spermatozoa. *Basic Clin. Androl.* **2014**, *24*, 4. [CrossRef]
5. Karlenius, T.C.; Tonissen, K.F. Thioredoxin and Cancer: A Role for Thioredoxin in all States of Tumor Oxygenation. *Cancers (Basel)* **2010**, *2*, 209–232. [CrossRef]
6. O'Flaherty, C. Peroxiredoxin 6: The Protector of Male Fertility. *Antioxidants* **2018**, *7*, 173. [CrossRef]
7. Ozkosem, B.; Feinstein, S.I.; Fisher, A.B.; O'Flaherty, C. Absence of Peroxiredoxin 6 Amplifies the Effect of Oxidant Stress on Mobility and SCSA/CMA3 Defined Chromatin Quality and Impairs Fertilizing Ability of Mouse Spermatozoa. *Biol. Reprod.* **2016**, *94*, 68. [CrossRef]
8. O'Flaherty, C.; de Lamirande, E.; Gagnon, C. Positive role of reactive oxygen species in mammalian sperm capacitation: Triggering and modulation of phosphorylation events. *Free Radic. Biol. Med.* **2006**, *41*, 528–540. [CrossRef]
9. Liu, Y.; O'Flaherty, C. In vivo oxidative stress alters thiol redox status of peroxiredoxin 1 and 6 and impairs rat sperm quality. *Asian J. Androl.* **2017**, *19*, 73–79.
10. Lee, D.; Moawad, A.R.; Morielli, T.; Fernandez, M.C.; O'Flaherty, C. Peroxiredoxins prevent oxidative stress during human sperm capacitation. *Mol. Hum. Reprod.* **2017**, *23*, 106–115. [CrossRef]
11. Moawad, A.R.; Fernandez, M.C.; Scarlata, E.; Dodia, C.; Feinstein, S.I.; Fisher, A.B.; O'Flaherty, C. Deficiency of peroxiredoxin 6 or inhibition of its phospholipase A$_2$ activity impair the in vitro sperm fertilizing competence in mice. *Sci. Rep.* **2017**, *7*, 12994. [CrossRef]
12. Gong, S.; San Gabriel, M.C.; Zini, A.; Chan, P.; O'Flaherty, C. Low amounts and high thiol oxidation of peroxiredoxins in spermatozoa from infertile men. *J. Androl.* **2012**, *33*, 1342–1351. [CrossRef]
13. Culty, M. Gonocytes, the forgotten cells of the germ cell lineage. *Birth Defects Res. Part C* **2009**, *87*, 1–26. [CrossRef] [PubMed]
14. Culty, M. Gonocytes, from the Fifties to the Present: Is There a Reason to Change the Name? *Biol. Reprod.* **2013**, *89*, 46. [CrossRef] [PubMed]
15. Manku, G.; Culty, M. Mammalian gonocyte and spermatogonia differentiation: Recent advances and remaining challenges. *Reproduction* **2015**, *149*, R139–R157. [CrossRef] [PubMed]

16. Wang, Y.; Culty, M. Identification and distribution of a novel platelet-derived growth factor receptor beta variant. Effect of retinoic acid and involvement in cell differentiation. *Endocrinology* **2007**, *148*, 2233–2250. [CrossRef]
17. Manku, G.; Wang, Y.; Merkbaoui, V.; Boisvert, A.; Ye, X.; Blonder, J.; Culty, M. Role of Retinoic Acid and Platelet-Derived Growth Factor Receptor crosstalk in the regulation of neonatal gonocyte and embryonal carcinoma cell differentiation. *Endocrinology* **2015**, *156*, 346–359. [CrossRef]
18. Manku, G.; Wing, S.S.; Culty, M. Expression of the Ubiquitin Proteasome System in Neonatal Rat Gonocytes and Spermatogonia: Role in Gonocyte Differentiation. *Biol. Reprod.* **2012**, *87*, 44. [CrossRef]
19. Manku, G.; Papadopoulos, P.; Boisvert, A.; Culty, M. Cyclooxygenase 2 (COX2) Expression and prostaglandin synthesis in Neonatal Rat Testicular Germ Cells: Effects of Acetaminophen and Ibuprofen. *Andrology* **2019**, in press. [CrossRef]
20. Rossi, S.P.; Windschüttl, S.; Matzkin, M.E.; Rey-Ares, V.; Terradas, C.; Ponzio, R.; Puigdomenech, E.; Levalle, O.; Calandra, R.S.; Mayerhofer, A.; et al. Reactive oxygen species (ROS) production triggered by prostaglandin D2 (PGD2) regulates lactate dehydrogenase (LDH) expression/activity in TM4 Sertoli cells. *Mol. Cellular Endoc.* **2016**, *434*, 154–165. [CrossRef]
21. Kiritoshi, S.; Nishikawa, T.; Sonoda, K.; Kukidome, D.; Senokuchi, T.; Matsuo, T.; Matsumura, T.; Tokunaga, H.; Brownlee, M.; Araki, E. Reactive oxygen species from mitochondria induce cyclooxygenase-2 gene expression in human mesangial cells: Potential role in diabetic nephropathy. *Diabetes* **2003**, *52*, 2570–2577. [CrossRef]
22. Thuillier, R.; Mazer, M.; Manku, G.; Boisvert, A.; Wang, Y.; Culty, M. Interdependence of PDGF and estrogen signaling pathways in inducing neonatal rat testicular gonocytes proliferation. *Biol. Reprod.* **2010**, *82*, 825–836. [CrossRef]
23. Manku, G.; Mazer, M.; Culty, M. Neonatal testicular gonocytes isolation and processing for immunocytochemical analysis. *Methods Mol. Biol.* **2012**, *825*, 17–29. [PubMed]
24. Culty, M.; Liu, Y.; Manku, G.; Chan, W.-Y. Papadopoulos. Expression of steroidogenesis-related genes in murine male germ cells. *Steroids* **2015**, *103*, 105–114. [CrossRef] [PubMed]
25. Boisvert, A.; Jones, S.; Issop, L.; Erythropel, H.C.; Papadopoulos, V.; Culty, M. In vitro Functional screening as a means to identify new plasticizers devoid of reproductive toxicity. *Environ. Res.* **2016**, *150*, 496–512. [CrossRef]
26. Aitken, R.J. Free radicals, lipid peroxidation and sperm function. *Reprod. Fertil. Dev.* **1995**, *7*, 659–668. [CrossRef]
27. Fernandez, M.C.; O'Flaherty, C. Peroxiredoxin 6 activates maintenance of viability and DNA integrity in human spermatozoa. *Hum. Reprod.* **2018**, *33*, 1394–1407. [CrossRef]
28. Allocati, N.; Masulli, M.; Di Ilio, C.; Federici, L. Glutathione transferases: Substrates, inihibitors and pro-drugs in cancer and neurodegenerative diseases. *Oncogenesis* **2018**, *7*, 8. [CrossRef]
29. Crawford, L.A.; Weerapana, E. A tyrosine-reactive irreversible inhibitor for glutathione S-transferase Pi (GSTP1). *Mol. Biosyst.* **2016**, *12*, 1768–1771. [CrossRef]
30. Liu, X.; An, B.H.; Kim, M.J.; Park, J.H.; Kang, Y.S.; Chang, M. Human glutathione S-transferase P1 1 functions as an estrogen receptor alpha signaling modulator. *Biochem. Biophys. Res. Commun.* **2014**, *452*, 840–844. [CrossRef]
31. Mahadevan, D.; Sutton, G.R. Ezatiostat hydrochloride for the treatment of myelodysplastic syndromes. *Expert Opin. Investig. Drugs* **2015**, *24*, 725–733. [CrossRef] [PubMed]
32. Calabrese, G.; Peker, E.; Amponsah, P.S.; Hoehne, M.N.; Riemer, T.; Mai, M.; Bienert, G.P.; Deponte, M.; Morgan, B.; Riemer, J. Hyperoxidation of mitochondrial peroxiredoxin limits H_2O_2-induced cell death in yeast. *EMBO J.* **2019**, *38*, e101552. [CrossRef] [PubMed]
33. Zhou, S.; Lien, Y.C.; Shuvaeva, T.; DeBolt, K.; Feinstein, S.I.; Fisher, A.B. Functional interaction of glutathione S-transferase pi and peroxiredoxin 6 in intact cells. *Int. J. Biochem. Cell Biol.* **2013**, *45*, 401–407. [CrossRef] [PubMed]
34. Wilson, C.; Munoz-Palma, E.; González-Billault, C. From birth to death: A role for reactive oxygen species in neuronal development. *Semin. Cell Dev. Biol.* **2018**, *80*, 43–49. [CrossRef] [PubMed]
35. Weidinger, A.; Kozlov, A.V. Biological activities of reactive oxygen and nitrogen species: Oxidative stress versus signal transduction. *Biomolecules* **2015**, *5*, 472–484. [CrossRef]
36. Ray, P.D.; Huang, B.W.; Tsuji, Y. Reactive oxygen species (ROS) homeostasis and redox regulation in cellular signaling. *Cell. Signal.* **2012**, *24*, 981–990. [CrossRef]

37. Niedenberger, B.A.; Busada, J.T.; Geyer, C.B. Marker expression reveals heterogeneity of spermatogonia in the neonatal mouse testis. *Reproduction* **2015**, *149*, 329–338. [CrossRef]
38. Hermann, B.P.; Mutoji, K.N.; Velte, E.K.; Ko, D.; Oatley, J.M.; Geyer, C.B.; McCarrey, J.R. Transcriptional and Translational Heterogeneity among Neonatal Mouse Spermatogonia. *Biol. Reprod.* **2015**, *92*, 54. [CrossRef]
39. Fernandez, M.C.; Yu, A.; Moawad, A.R.; O'Flaherty, C. Peroxiredoxin 6 regulates the phosphoinositide 3-kinase/AKT pathway to maintain human sperm viability. *Mol. Hum. Reprod.* **2019**, *25*, 787–796. [CrossRef]

© 2019 by the authors. Licensee MDPI, Basel, Switzerland. This article is an open access article distributed under the terms and conditions of the Creative Commons Attribution (CC BY) license (http://creativecommons.org/licenses/by/4.0/).

Article

The Exacerbation of Aging and Oxidative Stress in the Epididymis of *Sod1* Null Mice

Anaïs Noblanc [1], Alicia Klaassen [1] and Bernard Robaire [1,2,*

[1] Department of Pharmacology and Therapeutics, McGill University, Montreal, QC H3G 1Y6, Canada; anaisnoblanc@hotmail.fr (A.N.); alicia.klaassen@mail.mcgill.ca (A.K.)
[2] Department of Obstetrics and Gynecology, McGill University, Montreal, QC H4A 3J1, Canada
* Correspondence: bernard.robaire@mcgill.ca

Received: 23 December 2019; Accepted: 10 February 2020; Published: 11 February 2020

Abstract: There is growing evidence that the quality of spermatozoa decreases with age and that children of older fathers have a higher incidence of birth defects and genetic mutations. The free radical theory of aging proposes that changes with aging are due to the accumulation of damage induced by exposure to excess reactive oxygen species. We showed previously that absence of the superoxide dismutase 1 (*Sod1*) antioxidant gene results in impaired mechanisms of repairing DNA damage in the testis in young $Sod1^{-/-}$ mice. In this study, we examined the effects of aging and the $Sod^{-/-}$ mutation on mice epididymal histology and the expression of markers of oxidative damage. We found that both oxidative nucleic acid damage (via 8-hydroxyguanosine) and lipid peroxidation (via 4-hydroxynonenal) increased with age and in $Sod1^{-/-}$ mice. These findings indicate that lack of SOD1 results in an exacerbation of the oxidative damage accumulation-related aging phenotype.

Keywords: aging; epididymis; spermatozoa; oxidative stress; reactive oxygen species; superoxide dismutase; 4-hydroxynonenal; 8-hydroxyguanosine

1. Introduction

Reactive oxygen species (ROS) generate chain reactions that can affect numerous biomolecules, including lipids, RNA, DNA, proteins [1–4]. The consequences of these biochemical reactions are part of the normal cellular function but, in excess, they can be deleterious and severely damage major cellular processes. We will focus on the impact of oxidative stress during aging in the male reproductive system and, specifically, in the epididymis. Most studies on the effects of aging on the epididymis were done using the Brown Norway rat model, a reliable model of aging [5–8]. The major characteristics of aging in the epididymis are a thickening of the basement membrane along the epididymis, a decreased integrity and functionality of the blood-epididymis barrier [9], a segment-dependent variation of each cell type distribution; in particular a decrease in the proportion of principal, clear, and basal cells and an increase in the number of halo cells in the entire epididymis was found [10]. Halo cells are immune cells and their increased number in the epididymal epithelium may reflect a lack of balance of the immune steady-state of the epididymis, in accordance with the proposed general hallmarks of aging [11].

Aging is a dynamic and continuous process with time-dependent physiological and molecular changes. Some of the major hallmarks of aging include changes in telomere length, genetic instability and an accumulation of epigenetic modifications, mitochondrial dysfunction, a decrease in the stem cell reserve, and chronic inflammation [11–13]. In 1956, a "Free Radical Theory of Aging" was proposed that states that aging is the consequence of an accumulation of oxidative damage on biomolecules [14,15].

In most species, aging is associated with an accumulation of oxidized biomolecules (proteins, nucleic acids, and lipids), resulting in the dysregulation of metabolism, with an increase in the leak of superoxide anions by the mitochondria and a decrease in the ability of cells to scavenge ROS and to repair oxidative damage. ROS play a major role in the aging process via their action on various signaling pathways, including insulin/insulin like growth factor 1 (IGF1), mammalian target of rapamycin kinase (mTOR) and AMP kinase (AMPK) [16–21]. The null mutation of *Igf1* in mice induces a shrinking of the epididymis from the corpus epididymidis to the vas deferens due to a major decrease in the sperm cell content of the tubule [22]. *Igf1* gene expression is induced in the epididymis of orchidectomized adult rats and is under an androgenic control [23].

ROS play essential roles in male reproduction. One of these is the production of disulfide bridges between cysteine thiol groups; this biochemical reaction is mandatory for the post-translational maturation of numerous proteins and various enzymatic reactions. During spermiogenesis, sperm DNA is compacted first through replacement of histones by transition proteins and then by cysteine-rich protamines [24,25]. Therefore, numerous disulfide bonds are generated when condensing sperm nucleus undergo 3D re-organization during spermatid elongation and spermatozoal maturation in the epididymis. Mice knocked out for the mitochondrial isoform of *Gpv4* (mGPX4) are infertile, due to non-motile spermatozoa with abnormal mid-piece and broken flagella [26,27]. Another form of this enzyme that is considered to play a major role in this process is the sperm nuclear isoform of GPX4 (snGPX4) [28–31]. The $snGpx4^{-/-}$ mice are viable and fertile, but their sperm chromatin shows a delay in chromatin condensation during maturation in the epididymis that is linked to a decrease in disulfide bridging in the caput epididymal sperm. Finally, the acrosome matrix is structured and held by numerous disulfide bridges and the inhibition of the disulfide isomerase enzymatic activity by sperm pre-treatment decreases the fertilization level during sperm-egg fusion in vitro [32].

ROS are also an essential component of the redox signaling pathways during sperm capacitation in the female tract. This process involves a complex group of sperm modifications to prepare the spermatozoa to undergo flagellar hyperactivity and the acrosome reaction, both necessary to the journey through the female tract and for sperm-egg recognition and fusion. Numerous reviews have provided excellent description of this process [33,34] and the role of ROS in this phenomenon [35–37].

However, despite the obvious need of ROS for sperm production, maturation, and function, spermatozoa are highly sensitive to oxidative damage. ROS can modify purine and pyrimidine bases, break nucleic acid backbone, and create covalent bonds intra-/inter-strands, and between nucleic acids and proteins [38]. The main target of ROS is guanine due to its lower redox potential, with the major by-product of the DNA oxidation being 7,8-dihydro-8-oxo-2′-deoxyguanosine (8-OHdG); this oxidative DNA damage can be lethal for cells. While spermatogonia and spermatocytes have extensive DNA repair machinery, condensing spermatids and spermatozoa lose their DNA repair capacity [39–41].

Lipids are also targets of ROS, especially sterols and polyunsaturated fatty acids; these lipids affect the fluidity of the sperm cytoplasmic membrane. Lipid peroxidation can result in the generation of numerous toxic side products (alkanes, aldehydes, and acids), including 4-hydroxy-2-nonenal (4-HNE) which is highly cytotoxic, a molecule that is commonly used as a marker of lipid peroxidation. Lipid peroxidation decreases the fluidity of cell membranes, a major concern for sperm motility and ability to undergo capacitation and sperm-egg fusion [42,43].

Redox balance must be tightly maintained to limit the negative effects of ROS on reproductive tissues and spermatozoa, particularly during epididymal maturation when sperm are devoid of the ability to react to their environment. Consequently, the epididymal epithelium must play a protective role. A wide array of antioxidant enzymes is present in the epididymis, both in the epithelium and the luminal fluid; these include superoxide dismutases (SODs), catalase, glutathione transferases, glutathione peroxidases, and peroxiredoxins [44].

SODs are a family of metalloproteins composed of three different members, located in different cellular compartments and associated with different metal ions. SOD1, or Cu/ZnSOD is located in the intermembrane space of mitochondria, in the nucleus and in the cytoplasm [45]. Mutations of the *Sod1* gene have been associated with human pathologies such as amyotrophic lateral sclerosis [46]. Male mice lacking the ability to synthesize SOD1 are fertile but in vitro studies demonstrate that several characteristics of spermatozoa from $Sod1^{-/-}$ mice are affected, including higher lipid peroxidation, decreased motility, impaired capacitation ability, and a low fertilizing ability in in vitro fertilization [47].

$Sod1^{-/-}$ mice with a C57Bl/6 genetic background provide an interesting rodent model to test the free radical theory of aging and the mechanisms involved in the aging process. The phenotype of C57Bl/6 mice during aging is well characterized, as this animal model is often used for aging studies [48,49]. Despite a relatively normal phenotype and health during development and young adulthood, $Sod1^{-/-}$ mice have a reduced lifespan of 20.8 ± 0.7 (mean ± S.D.) months compared to the lifespan of wild-type mice of 29.8 ± 2.1 months [50]. The absence of SOD1 leads to extensive oxidative damage in various tissues including the intestine, liver, ovary, and testis [46–53] as early as 3 months of age and even in fetal fibroblasts [54]. In previous studies of the effects of the lack of SOD1 on the male reproductive system, we reported an impact on the male germline with a decreased ability to scavenge ROS, an increased number of oxidized spermatozoa, and a reduced DNA repair machinery [53]. In the present study, we examine the effects of aging on the epididymis of wild-type and $Sod1^{-/-}$ mice.

2. Materials and Methods

2.1. Animal Model

B6.129S7-Sod1^{tm1Leb}/DnJ mice were purchased from the Jackson Laboratory (#003881, Bar Harbor, ME, USA). They were backcrossed with C57Bl/6NJ mice (#005304, Bar Harbor, ME, USA) and bred in-house at McGill University in the McIntyre Comparative Medicine and Animal Resources Centre. Mice were kept on a 12L:12D cycle; food and water were provided *ad libitum*, and the temperature was maintained at 22 °C. Young (3 to 4-month-old) and aged (18-month-old) wild-type (WT) and *Sod1* null ($Sod1^{-/-}$) mutants mice were selected for this study. The number of mice in each group was from 3 to 6. Mice were euthanized by CO_2 asphyxiation followed by cervical dislocation. Both epididymides were harvested. All animal studies were conducted in accordance with the principles and procedures outlined in the Guide to the Care and Use of Experimental Animals prepared by the Canadian Council on Animal Care (McGill Animal Care Committee protocol #4687).

2.2. Tissue Fixation

Epididymides were fixed using Modified Davidson's fluid (12% formaldehyde, 15% ethanol, 5% glacial acetic acid) for 8 h at 4 °C [55]. Tissues were then washed in 70% ethanol, dehydrated, embedded in paraffin and sectioned at 4 µm at the Goodman Cancer Research Center Histology Facility (McGill University, Montreal, QC, Canada).

2.3. Histology

Sections were stained with toluidine blue. Tissues were deparaffinized (Histoclear; Diamed Inc., Mississauga, ON Canada), rehydrated and stained with 0.1% (w/v) Toluidine Blue O (#198161, Sigma) in 0.25% acid alcohol (HCl) solution. The slides were mounted with Permount™ mounting medium (Fisher Scientific, Montreal, QC, Canada). Sections were observed under bright field microscopy using a Leica LB2 microscope at 400× magnification.

2.4. Immunofluorescence

Tissue sections were deparaffinized and rehydrated. Heat-induced antigen retrieval in 10mM sodium citrate, pH 6.0, 0.05% Tween 20 was done prior to immunostaining for 4-hydroxynonenal (4-HNE) protein adducts and non-receptor tyrosine kinase SRC (SRC), but not for 8-hydroxyguanosine (8-OHG). The slides were washed in Tris Buffered Saline 1×, 0.02% Tween 20 (TBS-T) and were incubated in TBS-T, 5% normal goat serum (NGS) at room temperature for 30 min. Tissue sections were incubated overnight at 4 °C with the primary antibodies diluted in TBS-T, 1% NGS (Table 1). They were washed twice in TBS-T and incubated at room temperature for 1 hour with the secondary antibodies diluted in TBS-T, 1% NGS, 1 µg/mL 4′,6-diamidine-2′-phenylindole dihydrochloride (DAPI). After one wash in TBS-T and two washes in TBS 1X, slides were mounted with PermaFluor™ mounting medium (ThermoFisher Scientific, Montreal, QC, Canada). For the negative controls, the primary antibodies were omitted.

Table 1. List of antibodies used for immunofluorescence experiments.

Target	Conjugation	Clonality	Dilution	Company	Catalog
Primary Antibodies					
8-OHG (DNA/RNA)	None	Mouse Monoclonal (15A3)	1/1000	Novus Biologicals	NB110-96878
4-HNE	None	Rabbit polyclonal	1/500	Abcam	ab46545
SRC, C-term	None	Mouse monoclonal	1/250	Santa Cruz	sc8056
Secondary Antibodies					
Mouse whole IgG	Alexa Fluor 488	Goat polyclonal	1/500	Invitrogen	a11029
Rabbit whole IgG	Alexa Fluor 546	Goat polyclonal	1/1000	Invitrogen	a11010
Mouse whole IgG	Alexa Fluor 633	Goat polyclonal	1/2000	Invitrogen	a21046

8-hydroxyguanosine (8-OHG), 4-hydroxynonenal (4-HNE), proto-oncogene tyrosine-protein kinase (SRC).

2.5. Immunofluorescence Imaging and Quantification

Imaging of each full section was done with the Opera Phenix™ high-content screening system and Harmony software (Perkin Elmer, Montreal, QC, Canada) with a field overlap of 5%, at a 40× magnification. The images were further analyzed using Columbus™ system (Perkin Elmer, Montreal, QC, Canada) to quantify the signal of the immunofluorescent staining. Analyses were applied which recognized DAPI as a nucleus marker and SRC as a cytoplasmic marker to segment the cells for quantification of fluorescence intensity in each cellular compartment. The 4-HNE and 8-OHG intensities were used to determine oxidative damage in membranes and DNA, respectively. Analyses were done separately in the caput, corpus and cauda epididymides, as well as in the epithelium and in the interstitial cells (endothelial, smooth muscle, and conjunctive cells).

2.6. Statistical Analyses

All graphical data are represented as the means ± SEM. Prior to analysis, data were determined to be normally distributed. All data were then analyzed using a 3-way analysis of variance ANOVA with genotype, epididymal segment and age as variables, followed by Tukey's multiple comparisons test. All statistical analyses were done with the software Prism8 v8.0.1 (GraphPad Software, La Jolla, CA, USA). Differences among samples were considered to be significant when the $p < 0.05$.

3. Results

3.1. Histology

Representative sections of the initial segment, caput, corpus and cauda regions of the epididymis of young and old wild-type and $Sod^{-/-}$ mice are shown in Figure 1. No major differences were observed between young $Sod1^{-/-}$ and WT mice (3 months) with respect to either the epididymal tubule histology (Figure 1) or the interstitial tissue histology (smooth muscle cells, conjunctive tissue, vascular system) (data not shown). In old (18 months) $Sod1^{-/-}$ and WT mice distinctive features included an increase in the tubule diameter all along the epididymis, a decrease in the height of epididymal cells, as well as an accumulation of spermatozoa in the initial segments, where they are usually absent. There was also an increase in luminal round cells (circles). Further, we observed the presence of large vacuoles in some epithelial cells in the corpus and cauda epididymides (arrows). The thickness of the myoid layer surrounding the epididymal tubule was increased only in the cauda epididymidis. Among these features, only two were enhanced by the absence of the SOD1 protein in this tissue. The accumulation of cellular vacuoles in the cauda epididymidis was increased in old $Sod1^{-/-}$ mice compared to old WT mice. Moreover, the thickening of the myoid layer was greater in old $Sod1^{-/-}$ mice than in old WT mice, but this is also visible in cauda epididymidis of young $Sod1^{-/-}$ mice. In one old $Sod1^{-/-}$ mouse, the accumulation of an amorphous matter in the epididymal lumen of the caput, corpus, and cauda epididymides was associated with round cells only in the cauda epididymides [53]. Thus, overall our observations indicate that aging affects the histology of the epididymis and this phenotype is enhanced by the absence of the SOD1 protein.

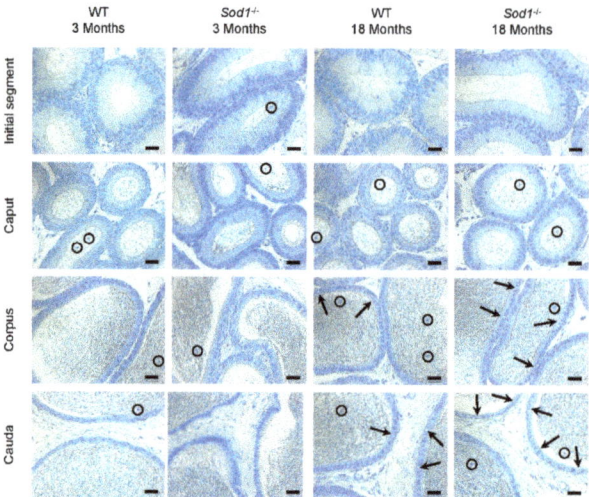

Figure 1. The histology of the epididymis is affected by aging in wild-type mice and even more so in $Sod1^{-/-}$ mice. Representative pictures of epididymal sections (initial segment, caput, corpus, and cauda epididymides) stained with toluidine blue from 3-month-old and 18-month-old wild-type and $Sod1^{-/-}$ mice. An increase in luminal round cells (circles) and in the number and size of clear vesicles (arrows) were observed in the epididymal epithelium of the corpus and cauda epididymides (n = 3). Scale bar: 40 μm.

3.2. DNA Oxidation

DNA oxidation was assessed by quantitative immunostaining of the oxidized guanine in the nucleus (8-OHdG). Immunofluorescent staining was detected in both epithelial and interstitial cells of the different regions of the epididymis (initial segment, caput, corpus, and cauda) (Figure 2). In 18-month-old mice, staining was more intense in the cauda epididymidis than in the rest of the tissue; further, this increase was more pronounced in the distal cauda epididymidis than in the proximal cauda epididymidis. Quantification of nuclear 8-OHdG immunofluorescent staining (Figure 2 bar graph) indicated that the major increase in DNA oxidation was observed in the cauda epididymidis of 18-month-old $Sod1^{-/-}$ mice. Statistical analysis by 3-way ANOVA (age, genotype, region, Table 2) revealed that the main source of variation in the DNA oxidation levels was the $Sod1^{-/-}$ genotype; in addition, there were interactions between epididymal regions and age and between age and genotype.

Table 2. Statistical analysis of the staining quantification. The 3-way ANOVA was followed by Tukey's multiple comparisons test performed using GraphPad Prism version 8.0.1 for Windows, GraphPad Software, San Diego, CA, USA.

Staining	8-Hydroxyguanine			4-Hydroxynonenal		
Source of Variation	% of Total Variation	p Value	Significance	% of Total Variation	p Value	Significance
Region	10.23	0.0122	*	5.78	0.3546	ns
Age	1.07	0.1604	ns	2.18	0.0880	ns
Genotype	14.00	<0.0001	****	4.47	0.0153	*
Region × Age	12.50	0.0030	**	0.31	0.9997	ns
Region × Genotype	5.38	0.1988	ns	1.84	0.9251	ns
Age × Genotype	5.37	0.0020	**	8.69	0.0008	***
Region × Age × Genotype	4.63	0.2897	ns	1.06	0.9833	ns

* $p < 0.05$, ** $p < 0.01$, *** $p < 0.001$, **** $p < 0.0001$.

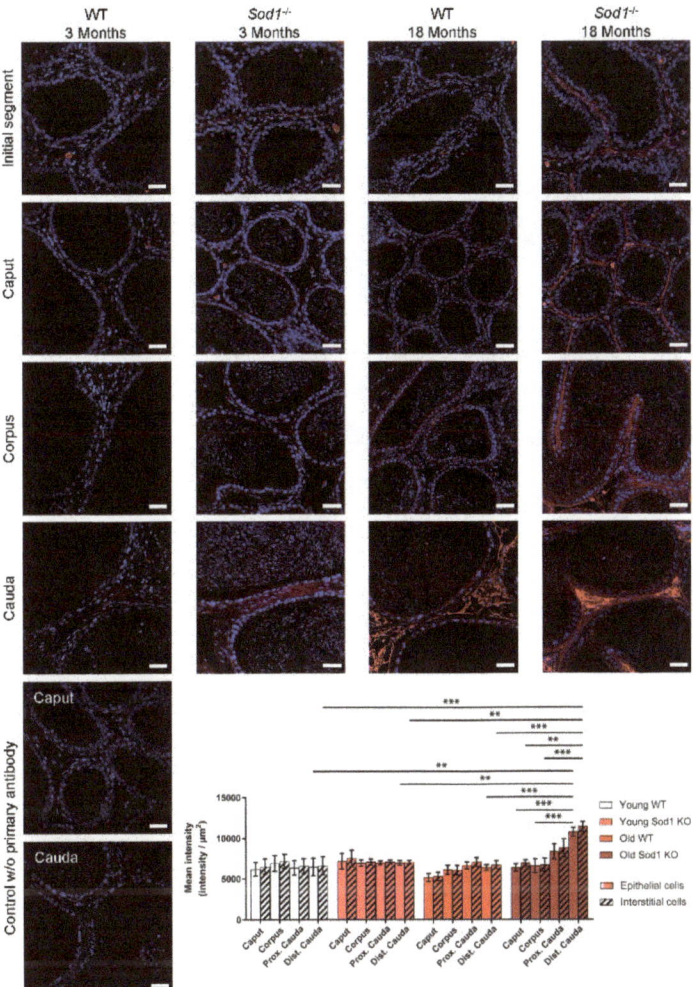

Figure 2. Aging induces an accumulation the nucleic acid damage in the epididymis (initial segment, caput, corpus, and cauda epididymides) which is enhanced in $Sod1^{-/-}$ mice. Representative pictures of sections of the epididymis from 3-month-old and 18-month-old wild-type and $Sod1^{-/-}$ mice after immunostaining of oxidized nucleic acids (8-OHG, red); the nuclei are counterstained with DAPI (blue). Immunostaining negative controls (no primary antibody) are displayed for the caput and cauda epididymides. Scale bar: 40 μm. 8-OHG staining was quantified separately in the epididymal epithelium (clear histograms) and in the interstitial tissue (dashed histograms). ** $p < 0.01$, *** $p < 0.001$ (3-way ANOVA, n = 4–5).

3.3. Lipid Peroxidation

Lipid peroxidation was evaluated by immunostaining of 4-hydroxynonenal (4-HNE) protein adducts, a by-product of lipid peroxidation. Longitudinal sections of the epididymides of 18-month-old WT and $Sod1^{-/-}$ mice revealed major differences in 4-HNE staining intensity and its regional distribution (Figure 3). Staining was absent in the initial segment and the caput epididymidis and was found almost exclusively in the corpus and the cauda epididymides. Further, in 18-month-old $Sod1^{-/-}$ mice, it appeared to be higher in the distal than in the proximal cauda epididymidis. The 4-HNE staining was detected mainly on cytoplasmic membranes, particularly on the apical membranes; in highly stained cells, it was also visible on intracellular and nuclear membranes. Quantitative analyses were done for staining on all cell surfaces. In the caput epididymidis, staining was indistinguishable to that seen in the control sections without primary antibody. The 3-way ANOVA analysis on quantitative data clearly indicated that there was a significant difference between the WT and $Sod1^{-/-}$ genotypes and that this effect was amplified by the age of the mice (Table 2).

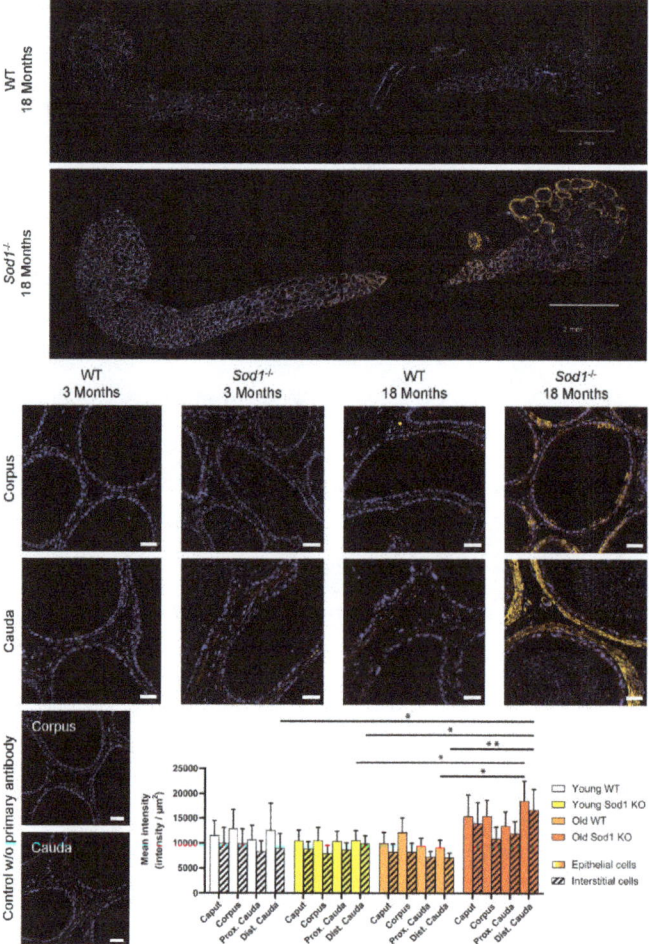

Figure 3. Lipid peroxidation as assessed by 4-HNE immunostaining is increased in the corpus and cauda epididymides with aging and this phenotype is enhanced in the absence of SOD1 expression. Representative pictures of whole epididymis sections of 18-month-old wild-type and $Sod1^{-/-}$ mice after the immunostaining of the 4-HNE (yellow) and the counterstaining of the nucleus (DAPI, blue). Scale bar: 2 mm. Observations have been carried out in epididymis sections of 3-month-old and 18-month-old wildtype and $Sod1^{-/-}$ mice. Staining was strongest in the corpus and cauda epididymides on the apical membrane of epithelial cells (arrow) and on intracellular membranes (arrow head). The negative controls (no primary antibody) for immunostaining are displayed for the corpus and cauda epididymides. Scale bar: 40 μm. The 4-HNE staining has been quantified in the epididymal epithelium (clear histograms) and in the interstitial tissue (dashed histograms) separately. * $p < 0.05$; ** $p < 0.01$ (3-way ANOVA, n = 4–5).

4. Discussion

Various theories have been proposed to explain the complex physiological and molecular modifications that occur during aging [13]. Among the potential causes and/or consequences of aging are an increase in oxidative stress and the accumulation of oxidative damage on many cellular components. Thus, to study the effects and mechanisms of aging in the male reproductive system, we have investigated various animal models, particularly rats and mice, with or without a genetically induced oxidative stress [53,56]. In this study, we analyzed the consequences of aging and oxidative stress in the epididymis, the site where spermatozoa become mature, using $Sod^{-/-}$ mice. $Sod1^{-/-}$ mice have a decreased ability to protect against ROS attack, leading to an accumulation of oxidative damage even in young animals, and have a reduced lifespan [46–49].

We analyzed the histology of the epididymis of 3- and 18-month-old wild-type and $Sod1^{-/-}$ mice. We identified some features that are indicative of a decline in epididymis structure and function with aging. The most obvious phenotypic change observed in old WT and $Sod1^{-/-}$ mice is the appearance of vacuoles in the epididymal epithelial cells, selectively in the corpus and the cauda epididymides. This feature was described previously in the corpus and proximal cauda epididymides of old Brown Norway rats [10]. The vacuoles found in principal cells were identified as giant lysosomes and lipid droplets, suggesting that the digestion/recycling system and/or the intracellular trafficking are dysregulated during aging. Further analyses in these old Brown Norway rats demonstrated a variation of the expression of glutathione S-transferases, enzymes that play an important role in detoxification of electrophiles. Thus, it would be interesting to study the effects of aging on the expression of genes/proteins in the epididymis of the aging $Sod1^{-/-}$ mice to determine if the same process occurs in mice.

Interestingly, we observed an increase in the number of round cells in the epididymal lumen of old mice compared to young mice; the nature of these cells is not clear. These cells were identified as round spermatids in some animal models of spermatogenic arrest but previous analyses of the testis in $Sod1^{-/-}$ mice did not reveal this phenotype [49]. Chronic and systemic inflammation is a hallmark of aging [11,12]. The increase in the number of halo cells (resident immune cells of the epididymis) along the epithelium during aging [57], in conjunction with the demonstration of dendritic cells [58], indicates a close interaction between the immune system and the epididymis. Therefore, it is interesting to speculate that these intraluminal round cells could be infiltrating immune cells, passed through the less impermeable blood-epididymis barrier of old animals [9]. An analysis of the blood-epididymis barrier and of the potential immune markers on the intraluminal round cells would be necessary to test this hypothesis,

The increase in the tubule diameter along the epididymis in old (18 month) $Sod1^{-/-}$, accompanied by a decrease in epithelial cell height, is consistent with previous observations in C57Bl/6NJ mice [53]. Surprisingly, opposite histological features of aging have been observed in old Brown Norway rats [6]. In the hamster, the diameter of the lumen also decreases whereas epithelial height does not change between young and old animals [59].

Spermatozoa were present consistently in the lumen of the initial segment of the epididymis of young and old $Sod1^{-/-}$ mice compared to wild-type animals. This may be due to a decrease in the ability of the epididymis to transport spermatozoa. This transport is dependent on peristaltic contractions of the smooth muscle layer surrounding the tubule, pressure from the testicular fluid and new spermatozoa, and the movement of stereocilia at the apical pole of epithelial cells [60]. Another indication that the ability of the epididymis to transport spermatozoa is affected by aging is the increase in the smooth muscle layer in the distal cauda epididymidis where spermatozoa are stored between two ejaculations.

A progressive increase in the nucleic acid oxidation in the lumen of the tubule, from the caput epididymidis to the distal cauda epididymidis, was observed in 18-month-old mice, but not in 3-month-old wild-type mice. This phenomenon could be due to the increased quantity of spermatozoa contained in the lumen as long as they travel down the epididymis and to the increased generation of ROS by spermatozoa during aging [61].

Finally, we assessed 4-HNE immunofluorescent staining as a marker of lipid peroxidation. Lipid peroxidation appeared to be absent in the initial segment and the caput epididymidis and was only visible in the corpus and the cauda epididymidis. The staining was low in 3-month-old mice and increased with aging. As for the DNA oxidation, it increased along the tubule. The apical membranes of epithelial cells appeared to be more sensitive to lipid peroxidation. The reasons for the apical susceptibility to oxidative damage is unknown. It could be due to the direct contact with the luminal fluid and spermatozoa, a potential source of ROS, or to the high activity of this membrane (exocytosis, endocytosis, sensing of the luminal contents) which generates numerous reactions, another possible source of ROS.

The absence of SOD1 in the mouse epididymis did not affect overall oxidation as assessed by markers of DNA and lipid oxidation in 3-month-old mice. Other members of the SOD family may compensate for this loss and the repair machinery of the various cell components seems able to deal with the low level of oxidized biomolecules. However, oxidized nucleic acids and peroxidized lipids increased strikingly in the epididymal tissue of 18-month-old $Sod1^{-/-}$ mice, even compared to the old wild-type mice. Histological analyses revealed an increase in the quantity and size of vacuoles in the epididymal epithelial cells of the corpus and the cauda epididymides and an increase in the thickening of the myoid cell layer in the distal cauda epididymidis in old $Sod1^{-/-}$ mice compared to old wild-type mice. Thus, there is clearly a worsening of the aging phenotype in the mouse epididymis in the absence of SOD1. These findings suggest that $Sod1^{-/-}$ mice constitute a valuable model for better understanding aging in the epididymis.

5. Conclusions

Aging is a progressive biological process that is characterized by an accumulation of various physiological and molecular changes, including an increase in oxidative damage to different biomolecules, leading to cell and tissue dysfunctions. In this study, we evaluated the impact on the epididymis of removing SOD1 using $Sod1^{-/-}$ mice. We demonstrated that, using this rodent model, the aging phenotype is exacerbated by increased oxidative damage.

Author Contributions: Conceptualization, B.R. A.N. and A.K.; Methodology, A.N. and A.K.; Data Curation, A.N.; Writing—Original Draft Preparation, A.N.; Writing—Review and Editing, B.R. and A.N.; Supervision, Project Administration, and Funding Acquisition, B.R. All authors have read and agreed to the published version of the manuscript.

Funding: These studies were funded by a grant from the Canadian Institutes of Health Research (CIHR-TE1-138298).

Acknowledgments: We thank the Nicolas Audet, from the Imaging and Molecular Biology Platform of the Department of Pharmacology and Therapeutics in McGill University (Montreal, QC, Canada), for his advice about high-throughput screening and image analysis.

Conflicts of Interest: None of the authors has any conflict of interest.

References

1. Collin, F. Chemical Basis of Reactive Oxygen Species Reactivity and Involvement in Neurodegenerative Diseases. *Int. J. Mol. Sci.* **2019**, *20*, 2407. [CrossRef] [PubMed]
2. Ylä-Herttuala, S. Oxidized LDL and Atherogenesis. *Ann. New York Acad. Sci.* **1999**, *874*, 134–137. [CrossRef] [PubMed]
3. Stadtman, E.R.; Levine, R.L. Protein Oxidation. *Ann. New York Acad. Sci.* **2000**, *899*, 191–208. [CrossRef]
4. Marnett, L.J. Oxyradicals and DNA Damage. *Carcinogenesis* **2000**, *21*, 361–370. [CrossRef] [PubMed]
5. Wright, W.W.; Fiore, C.; Zirkin, B.R. The Effect of Aging on the Seminiferous Epithelium of the Brown Norway Rat. *J. Androl.* **1993**, *14*, 110–117.
6. Wang, C.; Hikim, A.S.; Ferrini, M.; Bonavera, J.J.; Vernet, L.; Leung, A.; Lue, Y.-H.; Gonzalez-Cavadid, N.F.; Swerdloff, R.S. Male Reproductive Ageing: Using the Brown Norway Rat as a Model for Man. *Novartis Found. Symp.* **2002**, *242*, 82–97.

7. Robaire, B. Aging of the Epididymis. In *The Epididymis: From Molecules to Clinical Practice*; Robaire, B., Hinton, B.T., Eds.; Springer: Boston, MA, USA, 2002; pp. 285–296.
8. Beattie, M.; Adekola, L.; Papadopoulos, V.; Chen, H.; Zirkin, B. Leydig Cell Aging and Hypogonadism. *Exp. Gerontol.* **2015**, *68*, 87–91. [CrossRef]
9. Levy, S.; Robaire, B. Segment-Specific Changes with Age in the Expression of Junctional Proteins and the Permeability of the Blood-Epididymis Barrier in Rats. *Boil. Reprod.* **1999**, *60*, 1392–1401. [CrossRef]
10. Serre, V. Segment-Specific Morphological Changes in Aging Brown Norway Rat Epididymis. *Boil. Reprod.* **1998**, *58*, 497–513. [CrossRef]
11. López-Otín, C.; Blasco, M.A.; Partridge, L.; Serrano, M.; Kroemer, G. The Hallmarks of Aging. *Cell* **2013**, *153*, 1194–1217. [CrossRef]
12. McHugh, D.; Gil, J. Senescence and Aging: Causes, Consequences, and Therapeutic Avenues. *J. Cell Biol.* **2018**, *217*, 65–77. [CrossRef] [PubMed]
13. Lipsky, M.S.; King, M. Biological Theories of Aging. *Dis. Mon.* **2015**, *61*, 460–466. [CrossRef] [PubMed]
14. Harman, D. Aging: A Theory Based on Free Radical and Radiation Chemistry. *J. Gerontol.* **1956**, *11*, 298–300. [CrossRef] [PubMed]
15. Terman, A.; Brunk, U.T. Oxidative Stress, Accumulation of Biological 'Garbage', and Aging. *Antioxid. Redox Signal.* **2006**, *8*, 197–204. [CrossRef]
16. Sarbassov, D.D.; Sabatini, D.M. Redox Regulation of the Nutrient-Sensitive Raptor-mTOR Pathway and Complex. *J. Boil. Chem.* **2005**, *280*, 39505–39509. [CrossRef]
17. Yoshida, S.; Hong, S.; Suzuki, T.; Nada, S.; Mannan, A.M.; Wang, J.; Okada, M.; Guan, K.-L.; Inoki, K. Redox Regulates Mammalian Target of Rapamycin Complex 1 (mTORC1) Activity by Modulating the TSC1/TSC2-Rheb GTPase Pathway. *J. Boil. Chem.* **2011**, *286*, 32651–32660. [CrossRef]
18. Ramírez-Rangel, I.; Bracho-Valdés, I.; Vázquez-Macías, A.; Carretero-Ortega, J.; Reyes-Cruz, G.; Vazquez-Prado, J. Regulation of mTORC1 Complex Assembly and Signaling by GRp58/ERp57. *Mol. Cell. Boil.* **2011**, *31*, 1657–1671. [CrossRef]
19. Wang, Q.; Liang, B.; Shirwany, N.A.; Zou, M.-H. 2-Deoxy-D-Glucose Treatment of Endothelial Cells Induces Autophagy by Reactive Oxygen Species-Mediated Activation of the AMP-Activated Protein Kinase. *PLOS ONE* **2011**, *6*, e17234. [CrossRef]
20. Chandrasekaran, A.; Idelchik, M.D.P.S.; Melendez, J.A. Redox Control of Senescence and Age-Related Disease. *Redox Biol.* **2017**, *11*, 91–102. [CrossRef]
21. Weichhart, T. MTOR as Regulator of Lifespan, Aging, and Cellular Senescence: A Mini-Review. *Gerontology* **2018**, *64*, 127–134. [CrossRef]
22. Baker, J.; Hardy, M.P.; Zhou, J.; Bondy, C.; Lupu, F.; Bellvé, A.R.; Efstratiadis, A. Effects of an Igf1 Gene Null Mutation on Mouse Reproduction. *Mol. Endocrinol.* **1996**, *10*, 903–918. [PubMed]
23. Hamzeh, M.; Robaire, B. Identification of Early Response Genes and Pathway Activated by Androgens in the Initial Segment and Caput Regions of the Regressed Rat Epididymis. *Endocrinology* **2010**, *151*, 4504–4514. [CrossRef] [PubMed]
24. Balhorn, R. The Protamine Family of Sperm Nuclear Proteins. *Genome Boil.* **2007**, *8*, 227. [CrossRef] [PubMed]
25. Ward, W.S. Organization of Sperm DNA by the Nuclear Matrix. *Am. J. Clin. Exp. Urol.* **2018**, *6*, 87–92. [PubMed]
26. Maiorino, M.; Roveri, A.; Benazzi, L.; Bosello, V.; Mauri, P.; Toppo, S.; Tosatto, S.C.E.; Ursini, F. Functional Interaction of Phospholipid Hydroperoxide Glutathione Peroxidase with Sperm Mitochondrion-associated Cysteine-rich Protein Discloses the Adjacent Cysteine Motif as a New Substrate of the Selenoperoxidase. *J. Boil. Chem.* **2005**, *280*, 38395–38402. [CrossRef] [PubMed]
27. Ursini, F. Dual Function of the Selenoprotein PHGPx During Sperm Maturation. *Science* **1999**, *285*, 1393–1396. [CrossRef]
28. Conrad, M.; Moreno, S.G.; Sinowatz, F.; Ursini, F.; Kölle, S.; Roveri, A.; Brielmeier, M.; Wurst, W.; Maiorino, M.; Bornkamm, G.W. The Nuclear Form of Phospholipid Hydroperoxide Glutathione Peroxidase Is a Protein Thiol Peroxidase Contributing to Sperm Chromatin Stability. *Mol. Cell. Boil.* **2005**, *25*, 7637–7644. [CrossRef]
29. Noblanc, A.; Peltier, M.; Damon-Soubeyrand, C.; Kerchkove, N.; Chabory, E.; Vernet, P.; Saez, F.; Cadet, R.; Janny, L. and Pons-Rejraji, H. et al. Epididymis Response Partly Compensates for Spermatozoa Oxidative Defects in snGPx4 and GPx5 Double Mutant Mice. *PLoS ONE* **2012**, *7*, e38565. [CrossRef]

30. Pfeifer, H. Identification of a Specific Sperm Nuclei Selenoenzyme Necessary for Protamine Thiol Cross-Linking during Sperm Maturation. *FASEB J.* **2001**, *15*, 1236–1238. [CrossRef]
31. Puglisi, R.; Maccari, I.; Pipolo, S.; Conrad, M.; Mangia, F.; Boitani, C. The Nuclear Form of Glutathione Peroxidase 4 Is Associated with Sperm Nuclear Matrix and Is Required for Proper Paternal Chromatin Decondensation at Fertilization. *J. Cell. Physiol.* **2012**, *227*, 1420–1427. [CrossRef]
32. Ellerman, D.A.; Myles, D.G.; Primàkoff, P. A Role for Sperm Surface Protein Disulfide Isomerase Activity in Gamete Fusion: Evidence for the Participation of ERp57. *Dev. Cell* **2006**, *10*, 831–837. [CrossRef] [PubMed]
33. Gangwar, D.; Atreja, S. Signalling Events and Associated Pathways Related to the Mammalian Sperm Capacitation. *Reprod. Domest. Anim.* **2015**, *50*, 705–711. [CrossRef] [PubMed]
34. Molina, L.C.P.; Luque, G.M.; Balestrini, P.A.; Marín-Briggiler, C.I.; Romarowski, A.; Buffone, M.G. Molecular Basis of Human Sperm Capacitation. *Front. Cell Dev. Boil.* **2018**, *6*, 72. [CrossRef] [PubMed]
35. De Lamirande, E.; O'Flaherty, C. Sperm Activation: Role of Reactive Oxygen Species and Kinases. *Biochim. Biophys. Acta (BBA) Proteins Proteom.* **2008**, *1784*, 106–115. [CrossRef] [PubMed]
36. De Lamirande, E.; Zini, A.; Gabriel, M.C. Human Sperm Chromatin Undergoes Physiological Remodeling During in Vitro Capacitation and Acrosome Reaction. *J. Androl.* **2012**, *33*, 1025–1035. [CrossRef] [PubMed]
37. O'Flaherty, C.; De Lamirande, E.; Gagnon, C. Positive Role of Reactive Oxygen Species in Mammalian Sperm Capacitation: Triggering and Modulation of Phosphorylation Events. *Free. Radic. Boil. Med.* **2006**, *41*, 528–540. [CrossRef]
38. Jena, N.R. DNA Damage by Reactive Species: Mechanisms, Mutation and Repair. *J. Biosci.* **2012**, *37*, 503–517. [CrossRef]
39. Iguchi, N.; Tobias, J.W.; Hecht, N.B. Expression Profiling Reveals Meiotic Male Germ Cell mRNAs that Are Translationally Up- and Down-Regulated. *Proc. Natl. Acad. Sci.* **2006**, *103*, 7712–7717. [CrossRef]
40. Hao, S.-L.; Ni, F.-D.; Yang, W.-X. The Dynamics and Regulation of Chromatin Remodeling during Spermiogenesis. *Gene* **2019**, *706*, 201–210. [CrossRef]
41. Cavé, T.; Desmarais, R.; Lacombe-Burgoyne, C.; Boissonneault, G. Genetic Instability and Chromatin Remodeling in Spermatids. *Genes* **2019**, *10*, 40. [CrossRef]
42. Aitken, R.J.; Clarkson, J.S.; Fishel, S. Generation of Reactive Oxygen Species, Lipid Peroxidation, and Human Sperm Function. *Boil. Reprod.* **1989**, *41*, 183–197. [CrossRef] [PubMed]
43. Gomez, E.; Irvine, D.S.; Aitken, R.J. Evaluation of a Spectrophotometric Assay for the Measurement of Malondialdehyde and 4-hydroxyalkenals in Human Spermatozoa: Relationships with Semen Quality and Sperm Function. *Int. J. Androl.* **1998**, *21*, 81–94. [CrossRef] [PubMed]
44. O'Flaherty, C. Orchestrating the Antioxidant Defenses in the Epididymis. *Andrology* **2019**, *7*, 662–668. [CrossRef] [PubMed]
45. Mccord, J.M.; Fridovich, I. The Utility of Superoxide Dismutase in Studying Free Radical Reactions. II. The Mechanism of the Mediation of Cytochrome c Reduction by a Variety of Electron Carriers. *J. Boil. Chem.* **1970**, *245*, 1374–1377.
46. Rosen, D.R.; Siddique, T.; Patterson, D.; Figlewicz, D.A.; Sapp, P.; Hentati, A.; Donaldson, D.; Goto, J.; O'Regan, J.P.; Deng, H.X. Mutations in Cu/Zn Superoxide Dismutase Gene Are Associated with Familial Amyotrophic Lateral Sclerosis. *Nature* **1993**, *362*, 59–62. [CrossRef]
47. Tsunoda, S.; Kawano, N.; Miyado, K.; Kimura, N.; Fujii, J. Impaired Fertilizing Ability of Superoxide Dismutase 1-Deficient Mouse Sperm During in Vitro Fertilization1. *Boil. Reprod.* **2012**, *87*, 121. [CrossRef]
48. Brayton, C.F.; Treuting, P.M.; Ward, J.M. Pathobiology of Aging Mice and GEM: Background Strains and Experimental Design. *Vet. Pathol.* **2012**, *49*, 85–105. [CrossRef]
49. Pettan-Brewer, C.; Treuting, P.M. Practical Pathology of Aging Mice. *Pathobiol. Aging Age Relat. Dis.* **2011**, *1*, 7202. [CrossRef]
50. Elchuri, S.; Oberley, T.D.; Qi, W.; Eisenstein, R.S.; Jackson Roberts, L.; Van Remmen, H.; Epstein, C.J.; Huang, T.-T. CuZnSOD Deficiency Leads to Persistent and Widespread Oxidative Damage and Hepatocarcinogenesis Later in Life. *Oncogene* **2005**, *24*, 367–380. [CrossRef]
51. Kruidenier, L. Attenuated Mild Colonic Inflammation and Improved Survival from Severe DSS-Colitis of Transgenic Cu/Zn-SOD Mice. *Free. Radic. Boil. Med.* **2003**, *34*, 753–765. [CrossRef]
52. Matzuk, M.M.; Dionne, L.; Guo, Q.; Kumar, T.R.; Lebovitz, R.M. Ovarian Function in Superoxide Dismutase 1 and 2 Knockout Mice. *Endocrinology* **1998**, *139*, 4008–4011. [CrossRef] [PubMed]

53. Selvaratnam, J.S.; Robaire, B. Effects of Aging and Oxidative Stress on Spermatozoa of Superoxide-Dismutase 1- and Catalase-Null Mice1. *Boil. Reprod.* **2016**, *95*, 60. [CrossRef] [PubMed]
54. Huang, T.-T.; Yasunami, M.; Carlson, E.J.; Gillespie, A.M.; Reaume, A.G.; Hoffman, E.K.; Chan, P.H.; Scott, R.W.; Epstein, C.J. Superoxide-Mediated Cytotoxicity in Superoxide Dismutase-Deficient Fetal Fibroblasts. *Arch. Biochem. Biophys.* **1997**, *344*, 424–432. [CrossRef] [PubMed]
55. Latendresse, J.R.; Warbrittion, A.R.; Jonassen, H.; Creasy, D.M. Fixation of Testes and Eyes Using a Modified Davidson's fluid: Comparison with Bouin's Fluid and Conventional Davidson's Fluid. *Toxicol. Pathol.* **2002**, *30*, 524–533. [CrossRef]
56. Selvaratnam, J.; Paul, C.; Robaire, B. Male Rat Germ Cells Display Age-Dependent and Cell-Specific Susceptibility in Response to Oxidative Stress Challenges1. *Boil. Reprod.* **2015**, *93*, 72. [CrossRef]
57. Serre, V. Distribution of Immune Cells in the Epididymis of the Aging Brown Norway Rat Is Segment-Specific and Related to the Luminal Content. *Boil. Reprod.* **1999**, *61*, 705–714. [CrossRef]
58. Da Silva, N.; Barton, C.R. Macrophages and Dendritic Cells in the Post-Testicular Environment. *Cell Tissue Res.* **2016**, *363*, 97–104. [CrossRef]
59. Calvo, A.; Pastor, L.M.; Roca, J.; Martínez, E.; Vázquez, J.M. Age-Related Changes in the Hamster Epididymis. *Anat. Rec.* **1999**, *256*, 335–346. [CrossRef]
60. Robaire, B.; Hinton, B. The Epididymis. In *Knobil and Neill's Physiology of Reproduction*, 4th ed.; Plant, T., Zeleznik, A., Eds.; Elsevier: Amsterdam, The Netherlands, 2014; pp. 691–771.
61. Weir, C.P.; Robaire, B. Spermatozoa Have Decreased Antioxidant Enzymatic Capacity and Increased Reactive Oxygen Species Production during Aging in the Brown Norway rat. *J. Androl.* **2007**, *28*, 229–240. [CrossRef]

© 2020 by the authors. Licensee MDPI, Basel, Switzerland. This article is an open access article distributed under the terms and conditions of the Creative Commons Attribution (CC BY) license (http://creativecommons.org/licenses/by/4.0/).

Article

Long-Term Adverse Effects of Oxidative Stress on Rat Epididymis and Spermatozoa

Pei You Wu [1,2], Eleonora Scarlata [1,3] and Cristian O'Flaherty [1,2,3,*]

[1] Department of Surgery (Urology Division), McGill University, Montréal, QC H4A 3J1, Canada; pei.y.wu@mail.mcgill.ca (P.Y.W.); eleonora.scarlata@mail.mcgill.ca (E.S.)
[2] Department of Pharmacology and Therapeutics, McGill University, Montréal, QC H3G 1Y6, Canada
[3] The Research Institute, McGill University Health Centre, Montréal, QC H4A 3J1, Canada
* Correspondence: cristian.oflaherty@mcgill.ca; Tel.: +1-514-934-1934 (ext. 35410)

Received: 31 January 2020; Accepted: 17 February 2020; Published: 19 February 2020

Abstract: Oxidative stress is a common culprit of several conditions associated with male fertility. High levels of reactive oxygen species (ROS) promote impairment of sperm quality mainly by decreasing motility and increasing the levels of DNA oxidation. Oxidative stress is a common feature of environmental pollutants, chemotherapy and other chemicals, smoke, toxins, radiation, and diseases that can have negative effects on fertility. Peroxiredoxins (PRDXs) are antioxidant enzymes associated with the protection of mammalian spermatozoa against oxidative stress and the regulation of sperm viability and capacitation. In the present study, we aimed to determine the long-term effects of oxidative stress in the testis, epididymis and spermatozoa using the rat model. Adult male rats were treated with tert-butyl hydroperoxide (t-BHP) or saline (control group), and reproductive organs and spermatozoa were collected at 3, 6, and 9 weeks after the end of treatment. We determined sperm DNA oxidation and motility, and levels of lipid peroxidation and protein expression of antioxidant enzymes in epididymis and testis. We observed that cauda epididymal spermatozoa displayed low motility and high DNA oxidation levels at all times. Lipid peroxidation was higher in caput and cauda epididymis of treated rats at 3 and 6 weeks but was similar to control levels at 9 weeks. PRDX6 was upregulated in the epididymis due to t-BHP; PRDX1 and catalase, although not significant, followed similar trend of increase. Testis of treated rats did not show signs of oxidative stress nor upregulation of antioxidant enzymes. We concluded that t-BHP-dependent oxidative stress promoted long-term changes in the epididymis and maturing spermatozoa that result in the impairment of sperm quality.

Keywords: reactive oxygen species; testis; antioxidant enzymes; peroxiredoxins; sperm maturation

---

1. Introduction

Infertility is a concerning pathophysiological condition that affects about 16% of couples worldwide, and approximately half of the cases are attributable to male factors [1]. Even though the cause for the majority of the male infertile cases is unknown, oxidative stress caused by a high amount of reactive oxygen species (ROS) has been observed in 30–80% of infertile patients [2,3]. Oxidative stress is a common feature associated to environmental pollutants, chemotherapeutic agents and other drugs, smoke, toxins, radiation, and diseases such as many types of cancer that have negative effects on fertility [4–8]. Even though low ROS levels are required for the acquisition of fertilization ability by the spermatozoon [9], oxidative stress damages spermatozoa by reducing motility, and increasing levels of DNA, protein oxidation and lipid peroxidation [10–12]. As occurring in somatic cells, this oxidative stress observed in spermatozoa is the result of an imbalance between the antioxidant defense system and the endogenous generation of ROS.

Peroxiredoxins (PRDXs) are a family of antioxidant enzymes that are highly expressed from yeast to humans. They are peroxidases that do not require co-factors such as heme group or selenium and contains one or two cysteine (Cys) residues in their active site which are essential for their antioxidant function [13,14]. PRDXs isoforms are divided into 2-Cys PRDXs (PRDX1-4), atypical PRDX (PRDX5) and 1-Cys PRDX (PRDX6). These enzymes are important antioxidants in spermatozoa that regulate the level of ROS such as peroxides (H_2O_2 and organic hydroperoxides) and peroxinitrite ($ONOO^-$), to avoid cellular toxicity [14,15]. Studies using the PRDX6 knockout mouse model have shown sub-fertility associated with severe impairment of sperm motility and high levels of lipid peroxidation and sperm DNA damage. Interestingly, this abnormal reproductive phenotype worsens with aging [16,17]. Other antioxidant enzymes that fight against oxidative stress in testis and epididymis are catalase, glutathione peroxidases, and thioredoxins [18–20].

Epididymal maturation is a crucial step in the formation of viable and healthy spermatozoa in humans and other mammals [21]. After spermatogenesis, fully formed yet immature and immotile spermatozoa enter into the epididymis, and by the time of exit, they acquire the ability to move and morphological features that optimize their fertilization capacity [21,22]. In addition to sperm maturation, the epididymis also provides essential proteins to spermatozoa via epididymosomes to maintain their cellular functions and to protect them from potential damages such as oxidative stress-dependent injuries [23].

We previously reported that spermatozoa from rats challenged with an in vivo oxidative stress using tert-butyl hydroperoxide (t-BHP), and collected 24 h after the treatment, have higher levels of DNA oxidation and lipid peroxidation and displayed poor motility compared to untreated controls [18]. There was a differential expression of PRDX1 and PRDX6 in a different segment of the epididymis of the treated rats. Interestingly, these spermatozoa contain high levels of PRDXs, as an attempt of the epididymis to fight against the oxidative stress established by the treatment [18].

In this study, we aimed to elucidate the long-term effect of t-BHP induced oxidative stress on rat reproductive system by assessing the oxidative damage and the expression of significant antioxidants enzymes that fight against hydroperoxides in spermatozoa, epididymis and testis.

2. Materials and Methods

2.1. Materials

Tert-butyl hydroperoxide (t-BHP), sodium dodecyl sulfate (SDS), phosphotungstic acid, buthylated hydroxytoluene, 2-thiobarbituric acid and malonaldehyde bis(dimethyl acetal), the Bicinchoninic protein determination assay and the anti-α-tubulin were purchased from Sigma-Aldrich Chemical Co. (St. Louis, MO, USA). The following were purchased from Abcam Inc., (Cambridge, MA, USA): rabbit polyclonal anti-PRDX1, mouse monoclonal anti-PRDX6, mouse monoclonal anti-4-Hydroxynonenal (4-HNE), rabbit polyclonal anti-catalase, and 8-hydroxy-deoxyguanosine (8-OHdG). Anti-thioredoxin 1 antibody was purchased from Cell Singaling Thecnology (Danvers, MA, USA), Polyvinylidene fluoride (PVDF) membranes (0.22 μm pore size; Osmonics Inc., Minnetonka, MN, USA), donkey anti-rabbit IgG and goat anti-mouse IgG, both conjugated to horseradish peroxidase (Cedarlane Laboratories Ltd., Hornby, ON, Canada), an enhanced chemiluminescence kit (Lumi-Light; Roche Molecular Biochemicals, Laval, QC, Canada) and radiographic films (Denville Scientific, Inc., Saint-Laurent, QC, Canada) were also used for immunodetection of blotted proteins. Other chemicals were used at the reagent level.

2.2. Animals and Treatment

Adult male Sprague Dawley rats ($n = 24$) were randomly distributed in t-BHP and control groups and were treated with 300 μmoles tert-BHP/kg b.w. or saline (control) once a day intraperitoneally for 15 days, respectively as done previously [18]. Treatment with tert-BHP showed to have no effects on the health of rats [18,24]. Animals were euthanized at 3, 6, and 9 weeks post-treatment. These end points correspond to late, middle and early spermatogenesis, respectively [25]. At each given end time, reproductive organs were collected, weighted and kept at −80 °C until use. For sperm motility determination, cauda epididymes were cut one time at the base with a surgical blade and placed in phosphate-buffered saline (PBS; 1 mM KH_2PO_4, 10 mM Na_2HPO_4, 137 mM NaCl, 2.7 mM KCl, pH 7.4) at 37 °C. Spermatozoa were allowed to swim-out for 10 min and were collected in clean tubes. All procedures with rats (handling, euthanasia, collections of tissues, etc.) were carried out following the regulations of the Canadian Council for Animal Care (CACC) and according the protocol #2009-5656 approved by the Facility Animal Care Committee (FACC) of the Research Institute, McGill University Health Centre.

2.3. Testes and Epididymes Homogenates Preparation

Control and t-BHP treated adult male Sprague-Dawley rats' frozen testis, caput and cauda epididymis were thawed, weighed and homogenized in a glass potter in RIPA buffer containing protease inhibitors. The samples were then sonicated for 20 s at 30% amplitude twice with 20 s intervals with a Sonic Vibracell (Sonics and Materials, Inc., Newtown, CT, USA). The samples were centrifuged at 21,000× g for 20 min at 4 °C. The supernatant was extracted, aliquoted and stored at −80 °C.

2.4. Sperm Motility and DNA Oxidation Determinations

Sperm motility was assessed by the same observer (CO) using the Olympus BH-2 microscope at 100 magnification with a thermal plate at 37 °C. At least 200 spermatozoa per duplicate were analyzed to determine percentage of total motility in each sample [18]. Sperm DNA oxidation was determined by immunohistochemistry using the anti-8-OHdG antibody as done previously [18]. Briefly, sperm samples were centrifuged at 2000× g for 5 min to remove the PBS medium and resuspended in 20 mM phosphate buffer (pH 6.0) with 1 mM EDTA for 5 min. Samples were then centrifuged and resuspended in 50 mM Tris-HCl (pH 7.4), 1% SDS and 40 mM dithiothreitol for 30 min. Final centrifugation of 5 min to replace the mixture with PBS was performed. The sperm PBS solution was smeared on Superfrost Plus slides (Fischer Scientific, Ottawa, ON, Canada) and they were fixed with 100% methanol at 20 °C for 30 min. Slides were incubated with 5% horse serum for 30 min at room temperature, then washed with PBS-T for 5 min and incubated with anti-8-OHdG antibody (1:100) (SMC-155D, StressMarq Biosciences Inc., Victoria, BC, Canada) diluted in 1% horse serum overnight at 4 °C. After a wash with PBS, the samples were incubated with biotinylated horse anti-mouse antibody in 1% BSA and PBS-T for 1 h, washed and finally incubated with Alexa Fluor 555-streptavidin (1:500 in PBS) for 45 min at 20 °C. ProLong Gold antifade with DAPI was added and smears sealed. Slides were analyzed with Zeiss Axiophot fluorescence microscopy (Carl Zeiss, Oberkochen, Germany). Two hundred spermatozoa per slide were counted in duplicate. A positive control was done by incubating spermatozoa with 2 mM H_2O_2 for 1 h at 37 °C. The specificity of the antibody was confirmed previously [18].

2.5. Testes Histological Analysis, and Sperm Count

Testes were dissected, weighed, and fixed immediately with Bouin fixative for 24 h before processing and embedding in paraffin blocks, and tissues were sectioned (5 µm) and stained with hematoxylin-eosin as previously described [26]. Spermatozoa heads from testis homogenates were counted in a hematocytometer as previously described [27].

2.6. SDS-PAGE and Immunoblotting

The bicinchoninic acid assay was performed to determine protein concentration in each tissue homogenate sample. Testis and epididymis tissue samples were mixed in electrophoresis sample buffer supplemented with 100 mM DTT, incubated at 95 °C for 5 min, and then centrifuged at 21,000× g for 5 min. Proteins in the supernatant were electrophoresed on 12% polyacrylamide gels and electrotransferred to polyvinylidene difluoride membranes. Then, the membranes were incubated in a solution of skim milk (5%, *w/v*) in Tween-containing Tris-buffered saline (TTBS; 20 mM Tris, 0.1% *v/v* Tween, pH 7.8) for 30 min followed by the incubation in anti-PRDX-1 (1:10,000), anti-PRDX-6 (1:10,000), anti-catalase (1:1000), anti-4-hydroxynonenal (4HNE) (1:100), anti-Thioredoxin1 (TRX1) (1:500) primary antibodies overnight. To test the specificity, 0.4 µg/mL of anti-PRDX1 was incubated with 2 µg/mL of its antigenic peptide in TBS-T supplemented with 3% BSA for 2 h at room temperature [18]. The absence of non-specific binding was confirmed by the incubation of tissue samples with the secondary antibody (goat anti-mouse or donkey anti-rabbit IgG) only. After being washed with TTBS, the membranes were incubated with goat anti-mouse or donkey anti-rabbit IgG conjugated with horseradish peroxidase (diluted 1:2000 in TTBS) for 45 min at room temperature and washed again with TTBS. The immunoreactive bands were detected using Lumi-Light chemiluminescence kit. Then, the membranes were stripped and re-blotted with an anti-tubulin antibody to determine equal loading. Silver staining was used to determine equal loading in samples under non-reducing conditions. The membrane detection was done by using both Amersham Imager 600 (Thermo Fisher Scientific, Inc., Toronto, ON, Canada) and autoradiography films. The digital images were analyzed using Image J win-64 software (University of Wisconsin-Madison, Madison, WI, USA). The band intensities of the protein were normalized to that of the tubulin to compare the level of expression of the protein of interest.

2.7. Statistical Analysis

All data were presented as mean ± SEM. Normality of the data and homogeneity of variances were determined by the Shapiro–Wilk and Bartlett tests, respectively. Because we euthanized different rats at each time point, statistical differences between groups were determined using Two-Way ANOVA and Bonferroni test (to assess treatment and time-specific changes) using GraphPad Prism 5 (GraphPad Software, Inc., San Diego, CA, USA). The Mann–Whitney test was used to determine statistical differences in sperm motility and DNA oxidation among groups. Differences with a *p*-value of ≤0.05 were considered significant.

3. Results

3.1. Low Sperm Motility and High DNA Oxidation Suggest Compromised Sperm Quality Due to Oxidative Stress

Spermatozoa from cauda epididymis were collected and analyzed 3 weeks, 6 weeks and 9 weeks after the end of the t-BHP treatment to determine whether there were damages due to the treatment during the late, middle and early spermatogenesis, respectively. Spermatozoa from t-BHP-treated animals showed a significant reduction of their motility at all times compared to those sperm from the control group. Moreover, we observed significantly higher DNA oxidation levels in spermatozoa from treated rats compared to controls (Figure 1).

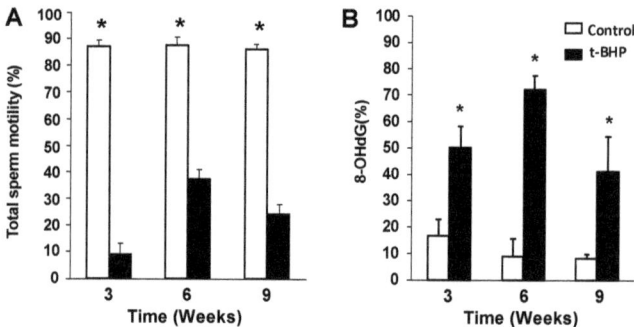

Figure 1. Impairment of sperm quality due to t-BHP treatment in male rats. (**A**) Sperm motility and (**B**) sperm DNA oxidation determined by 8-deoxyguanosine (8-OHdG) levels. The results are expressed as mean ± SEM. * Means higher than the other group at the same time point ($p \leq 0.05$; Mann–Whitney test, $n = 4$).

3.2. Lipid Peroxidation Increased in Caput and Cauda Epididymis of t-BHP Treated Male Rats

We determined oxidative damage on lipids by detecting 4-hydroxynonenal (4-HNE), a known marker of lipid peroxidation. We observed multiple bands detected by the anti-4HNE antibody, suggesting that different proteins contain the 4-HNE adduct in the tissues analyzed (Figure 2). A significant increase of 4-HNE levels was observed in both the caput and cauda epididymis of treated rats at 3 weeks and 6 weeks after the end of the t-BHP treatment compared to controls. Interestingly, the levels of 4-HNE at 9 weeks after the end of treatment were similar in the caput epididymis when comparing treated and control rats but were significantly higher than those seen in the control group at 6 weeks. In cauda epididymis, the 4-HNE levels returned to control values at 9 weeks after treatment. Noteworthily, the levels of lipid peroxidation at 9 weeks was higher in caput compared to cauda epididymis, suggesting an early dysregulation of the antioxidant response in the caput epididymis.

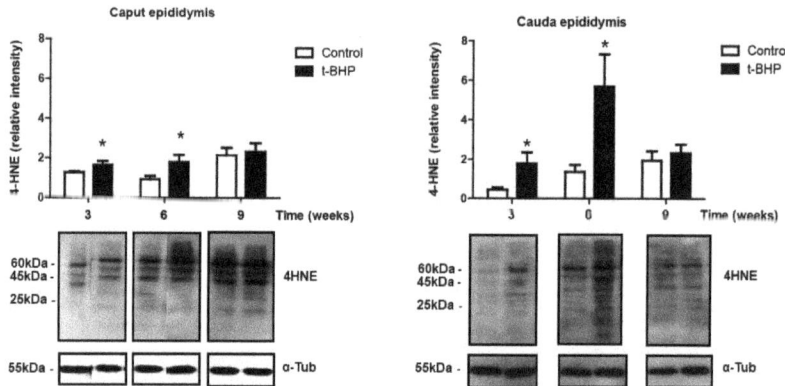

Figure 2. Lipid peroxidation (determined by 4-HNE levels) increased in caput and cauda epididymis in t-BHP compared to control male rats. The results of relative intensities (upper panels) are expressed as mean ± SEM. The blots presented are representative of experiments with 4 different rats. Some lanes showing protein bands have been pasted but belong to the same blot and have the same film exposure. * Means higher than the other group at the same time point ($p \leq 0.05$; Two-way ANOVA and Bonferroni post-hoc test, $n = 4$).

3.3. PRDX1 and PRDX6 Are Differentially Upregulated in Caput and Cauda Epididymis at Different Time Points

We assessed the expression levels of PRDX1 and PRDX6 in rat epididymis, and we observed a significant increase of PRDX6 in caput epididymis at 3 weeks and in cauda epididymis at 3 and 6 weeks post-treatment (Figure 3). We also found a trend of increase in PRDX1 expression levels in caput epididymis at week 3 and 9, and in cauda epididymis at week 9, but these increases were not significant (Figure 4).

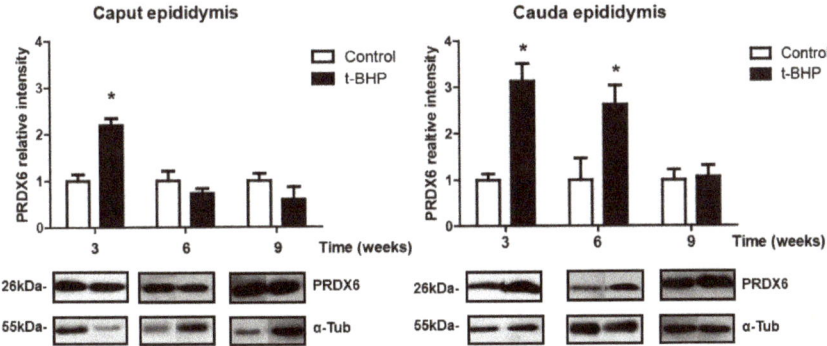

Figure 3. Peroxiredoxin 6 (PRDX6) expression in caput and cauda epididymis of control and t-BPH-treated male rats. The results of relative intensities (upper panels) are expressed as mean ± SEM. The blots presented are representative of experiments with 4 different rats. Some lanes showing protein bands have been pasted but belong to the same blot and have the same film exposure. * Means higher than the other group at the same time point ($p \leq 0.05$; Two-way ANOVA, $n = 4$).

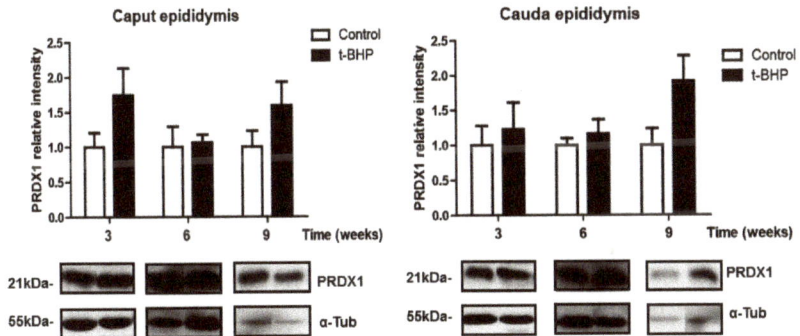

Figure 4. Peroxiredoxin 1 (PRDX1) expression in caput and cauda epididymis of control and t-BPH-treated male rats. The results of relative intensities (upper panels) are expressed as mean ± SEM, ($p > 0.05$; Two-way ANOVA and Bonferroni post-test, $n = 4$). The blots presented are representative of experiments with 4 different rats. Some lanes showing protein bands have been pasted but belong to the same blot and have the same film exposure.

3.4. Catalase Expression Shows Trends of Increase, with Significant Individual Variation in the Epididymis

Catalase expression levels were determined in rat epididymis from control and treated rats. Although not significant, we observed a trend of increase in treated rats compared to control at 3 and 6 weeks in cauda epididymis (Figure 5). Caput epididymis did not show upregulation of catalase at any time point.

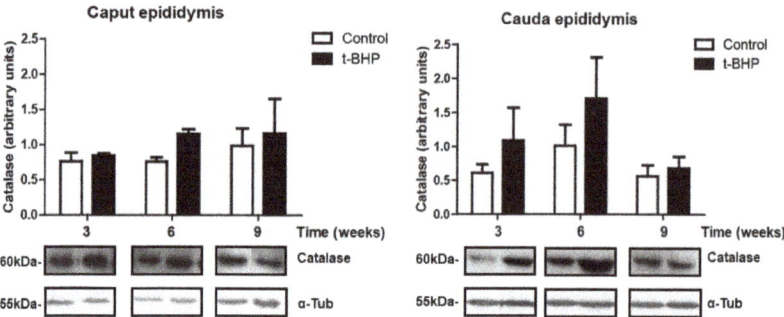

Figure 5. Catalase expression in caput and cauda epididymis of control and t-BPH-treated male rats. The results of relative intensities (upper panels) are expressed as mean ± SEM, ($p > 0.05$; Two-way ANOVA and Bonferroni post-test, $n = 4$). The blots presented are representative of experiments with 4 different rats. Some lanes showing protein bands have been pasted but belong to the same blot and have the same film exposure.

3.5. PRDXs, Catalase and Thioredoxin Expression Levels and Lipid Peroxidation Are Similar in Testis Despite the t-BHP Treatment

No significant difference was observed between the control and treated groups at any time points for PRDX1, PRDX6, catalase or TRX-1 (Figures 6 and 7).

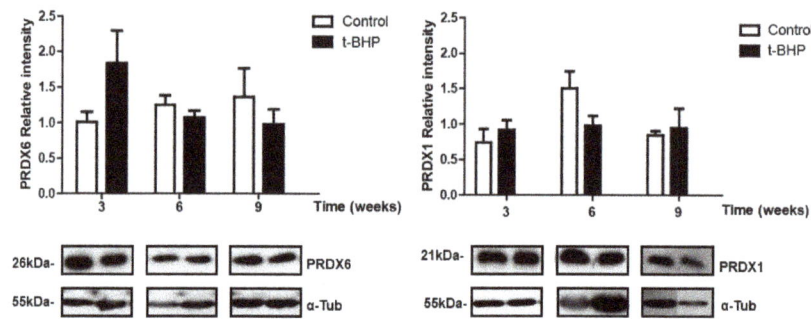

Figure 6. Peroxiredoxin 1 (PRDX1) and 6 (PRDX6) expression in testis of control and t-BPH-treated male rats. The results of relative intensities (upper panels) are expressed as mean ± SEM, ($p > 0.05$; Two-way ANOVA and Bonferroni post-test, $n = 4$). The blots presented are representative of experiments with 4 different rats. Some lanes showing protein bands have been pasted but belong to the same blot and have the same film exposure.

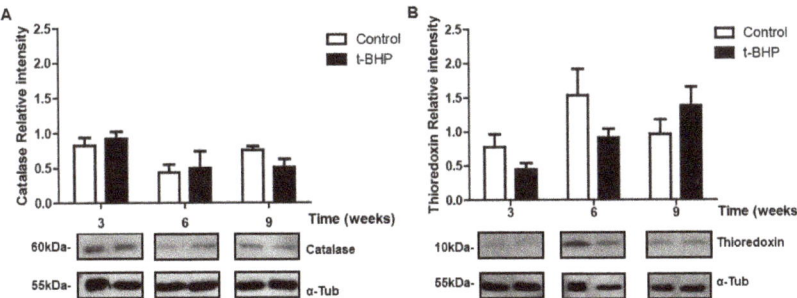

Figure 7. Catalase (**A**) and thioredoxin (**B**) expression in testis of control and t-BPH-treated male rats. The results of relative intensities (upper panels) are expressed as mean ± SEM, ($p > 0.05$; Two-way ANOVA and Bonferroni post-test, $n = 4$). The blots presented are representative of experiments with 4 different rats. Some lanes showing protein bands have been pasted but belong to the same blot and have the same film exposure.

The levels of 4-HNE in testis were similar in control and treated rats, suggesting that the oxidative stress generated by the treatment was well tolerated by the testis (Figure 8). Noteworthy, the levels of 4-HNE increased in testis from both control and treated rats at 6 and 9 weeks, suggesting that there is time dependency in the levels of lipid peroxidation in this organ.

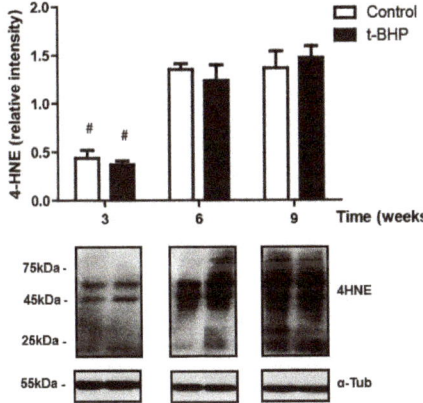

Figure 8. Lipid peroxidation in the testis of control and t-BPH-treated male rats. The results of relative intensities (upper panels) are expressed as mean ± SEM. The blots presented are representative of experiments with 4 different rats. Lanes showing protein bands have been pasted but belong to the same blot and have the same film exposure. # Means lower than all other groups, ($p \leq 0.05$; Two-way ANOVA and Bonferroni post-test, $n = 4$).

3.6. Reproductive Organs Weight, Spermatogenesis, and Sperm Production Were Not Affected by the t-BHP Treatment

To determine whether the damages observed in epididymal spermatozoa may have originated due to impairment of spermatogenesis by t-BHP treatment, we analyzed testis sections from control and treated groups to identify the different stages of spermatogenesis, compared the reproductive organs weight and sperm production in the experimental groups. We did not observe differences in body and organ weights between the treated and control rats at any time point (Figure 9 and Supplementary Materials Table S1).

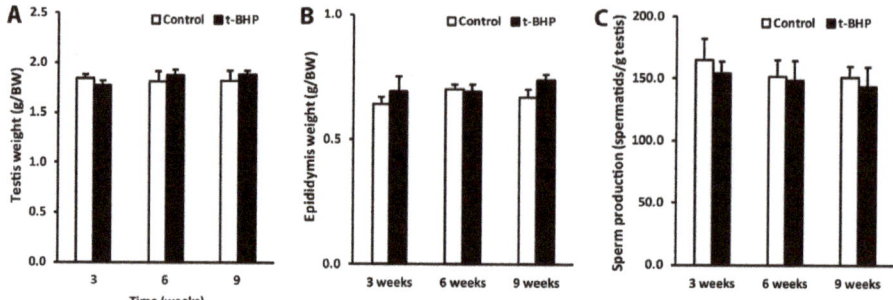

Figure 9. Testis and epididymis weight (A and B) and sperm production (C). ($n = 4$, Two-way ANOVA, $p > 0.05$).

The analysis of testis sections revealed that spermatogenesis proceeded normally in both control and treated rats during the 9 weeks after the end of treatment (Figure 10) that spermatogonia are transformed into spermatozoa [25]. We identified all the stages of the spermatogenesis in the rat, including the stages VII and VIII that contained elongating spermatids and fully formed spermatozoa, respectively in the luminal edge of the seminiferous epithelium, indicating active sperm production by the testes. As shown in Figure 10, some seminiferous tubules contain spermatozoa in the lumen (stage VIII), be ready to be spermiated to enter the epididymis. In addition, Sertoli and Leydig cells were morphologically normal.

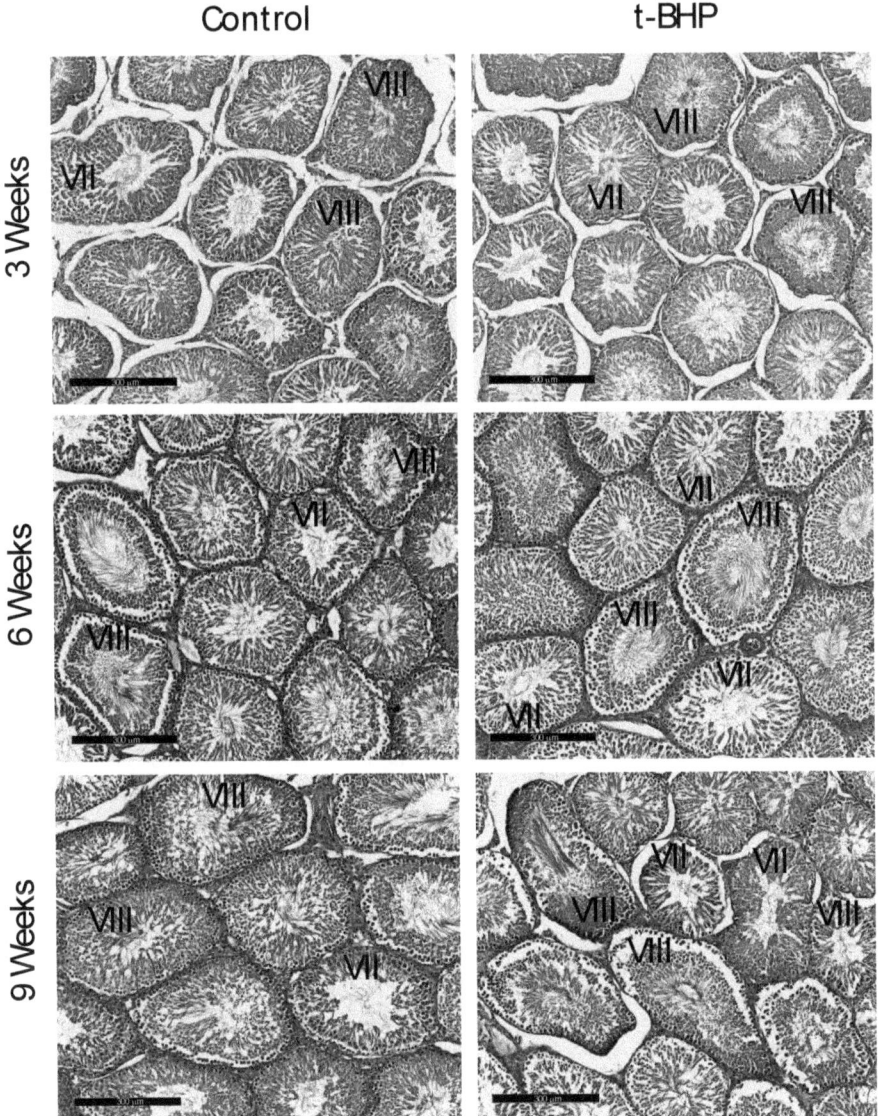

Figure 10. Histological analysis of testes from control and t-BHP-treated male rats. Testis sections were stained with hematoxylin and eosin to evaluate spermatogenesis. All testis sections displayed normal spermatogenesis (Stages VII and VIII showing elongating spermatids and spermatozoa in the lumen of the seminiferous tubules, respectively), $n = 4$. Bar = 300 µm.

4. Discussion

The present study shows for the first time the long-term adverse effect of t-BHP induced in vivo oxidative stress on rat spermatozoa and epididymis and the differential expression of antioxidant enzymes that fight against hydroperoxides in epididymis and testis. The reduced motility and high DNA damage observed in rat cauda epididymal spermatozoa collected at 3 weeks, 6 weeks and 9 weeks after the end of t-BHP treatment indicates the long-lasting adverse effects of oxidative stress on sperm quality. This result was unexpected since after one cycle of spermatogenesis (9 weeks in the rat) [25], new spermatozoa were produced from spermatogonia and these spermatozoa were those present in the epididymis at the time of collection [28]. Indeed, these collected spermatozoa showed significant DNA oxidative and impaired motility (Figure 1). These findings suggest that the balance between ROS and the antioxidant system has been compromised in the testis and or epididymis after the t-BHP-induced oxidative stress.

Based on the time of the collections and the length and stages of rat spermatogenesis, the collected rat spermatozoa were spermatids, spermatocytes or spermatogonia at the time that the rat was exposed to the t-BHP treatment [28]. When spermatozoa leave the testis, they enter into the epididymis to undergo their maturation before ejaculation [21]. Since the treatment has struck both testis and epididymis, the detrimental effects observed in the spermatozoa retrieved form the cauda epididymis could be a consequence of a direct impact of oxidative stress on the germ cells during spermatogenesis or due to the detrimental and persistent effects of high levels of ROS in the epididymal epithelium that impairs the proper epididymal maturation of spermatozoa.

In the epididymis, we expected that the increased expression of antioxidant enzymes would reduce the oxidative damage in spermatozoa because of the transfer of antioxidant enzymes from the epididymis to spermatozoa through the secretion of epididymosomes [23]. However, high 4-HNE levels in caput and cauda epididymis are an indication of developing lipid peroxidation (Figure 2) and suggest that the epididymal epithelium itself was damaged by the oxidative stress and therefore unable to protect spermatozoa during their maturation. The upregulation of PRDX6 and the trend of increase of PRDX1 and catalase expression in epididymis were not sufficient to scavenge the excessive ROS and restore the healthy cellular environment for a normal sperm epididymal maturation. We observed a similar response by the epididymis of rats treated for two weeks with t-BHP, with an upregulation of PRDX1 and PRDX6 but not of catalase [18].

Contrarily to what was observed in the epididymis, there were no significant differences in the levels of antioxidant enzyme expression and lipid peroxidation in the testes of treated rats compared to controls. Furthermore, the histological analysis and the testis weight and spermatid count indicated that there was no disruption of spermatogenesis. These findings suggest that there was no evidence of oxidative stress during spermatogenesis that could damage the collected spermatozoa at the different end points. Testicular spermatozoa have lower levels of DNA damage compared to the ejaculated counterparts [29,30]. Moreover, the fact that testicular spermatozoa from infertile men with obstructive azoospermia have low levels of DNA oxidation that do not interfere with the formation of an embryo by intracytoplasmic sperm injection suggests that the level of DNA oxidation in testicular sperm is not detrimental for male fertility [31]. It is plausible that the testicular spermatozoa are more resistant to oxidative stress compared to epididymal spermatozoa that flow freely in the lumen because the developing spermatozoa are guarded by the Sertoli cells that provide nutrients [32,33] and antioxidant protection through SOD, GSTs, GPXs, and PRDXs [34–36].

During epididymal maturation, the sperm chromatin is further compacted and could be exposed to oxidative stress generated by different conditions. Thus, it is of paramount importance that the epididymal epithelium protect the maturing spermatozoon against oxidative stress. The finding that PRDX6 expression levels are high when lipid peroxidation (measured by 4-HNE levels) are increased in cauda epididymis of treated rats, collected at week 3 and week 6, while these levels return to those of controls at week 9, indicate that PRDX6 is an essential component of the antioxidant response in the epididymis.

In a previous study, we challenged rats with the same t-BHP treatment and found that caudal epididymal spermatozoa collected 24 h after the end of the treatment had increased DNA oxidation, and reduced motility [18]. These findings indicated the negative effect of in vivo oxidative stress exclusively on epididymal maturation. We observed similar damages in mouse lacking PRDX6, a condition that generates an in vivo oxidative stress and is associated with male infertility [16]. The $Prdx6^{-/-}$ spermatozoa have low motility and high levels of DNA oxidation and lipid peroxidation.

The *Prdx6*^{−/−} spermatozoa also had higher percentages of cytoplasmic droplet retention compared to wild-type cells [16], an indication of abnormal epididymal maturation. During the epididymal transit, spermatozoa shed the residual cytoplasm; thus, an increase in spermatozoa carrying cytoplasmic droplets is an indication of abnormal epididymal maturation [37].

PRDX6 is a unique antioxidant enzyme as it is the only antioxidant enzyme known to date, with calcium-independent phospholipase A_2 (iPLA$_2$) [38] and lysophosphatidylcholine acyl transferase activities (LCAT) [39]. Both PRDX6 iPLA$_2$ and LCTAT activities are essential to remove and replace the oxidized phospholipids with newly synthesized phospholipids [39,40]. The epididymis increases PRDX6 in response to the oxidative damage caused by ROS to try to repair oxidized membrane lipids. A recovery of lipid peroxidation to the control level was observed in epididymis and spermatozoa at week 9, suggesting that PRDX6 repaired the damaged lipid membranes in this organ. This tendency was found in cauda epididymis but not in caput epididymis. We previously indicated that PRDX6 plays a crucial role in protecting both the epithelium and the spermatozoa in the cauda epididymis segment specifically [18]. Noteworthily, the higher levels of 4-HNE found in caput compared to cauda epididymis at week 9 suggest a differential capacity of antioxidant response in the different parts of the epididymis.

Although some repair of oxidative damage such as lipid peroxidation was observed in the caput and cauda epididymis at 6 and 9 weeks (Figure 2) and in spermatozoa at 9 weeks (Figure 1), epididymal spermatozoa had significant DNA oxidation at all time points. There is a possibility that the oxidative damage sperm DNA is a consequence of the impact of the treatment on the testis. Treated rats had similar sperm production than controls; their testes did not show long-lasting oxidative damage as the epididymis and were morphologically similar to control testis. Thus, it is less likely that the sperm DNA oxidation is due to problems during spermatogenesis. However, we cannot exclude the possibility that some of the damage observed in the sperm DNA may occur during the formation of spermatozoa in the testis. The oxidative stress generated by t-BHP altered the expression of miRNAs involved in the antioxidant response and spermatogenesis in mouse testis [41]. Although we did not see significant changes in spermatogenesis and the antioxidant response appears to be intact in rat testis, there is a possibility for the disruption of molecular mechanisms driven by miRNAs or epigenetic changes that can be associated with the permanent sperm DNA damage observed in this study. Further studies are required to rule out these possibilities.

While we presented evidence that the antioxidant response of the rat epididymis against oxidative stress is altered and may explain the poor quality of spermatozoa observed in this study, multiple factors could contribute to the persistent DNA oxidation observed in spermatozoa from t-BHP-treated rats. Sperm chromatin is a highly organized structure that differs from that of the somatic cells. Protamines replace histones during spermatogenesis, allowing the chromatin to tightly compact [42,43]. During epididymal transit, protamines become thiol oxidized and make disulfide bridges among them, thus, making the sperm chromatin more compacted [43]. Low mature protamine to protamine precursor ratio has been found in infertile patients and is correlated with high DNA damage, suggesting that chromatin compaction is critical for the protection of sperm DNA [43,44]. In our study, the persistent DNA damage observed in rat spermatozoa could be due to alterations of the sperm chromatin structure that interfered with normal sperm chromatin compaction, thus making sperm DNA more susceptible to the oxidative stress seen after 9 weeks of the end of treatment. Permanent oxidative stress in the male reproductive system as the one observed in *Prdx6*^{−/−} male mice leads to changes in the sperm chromatin with increased DNA oxidation and lower protamination (amounts of protamines) and DNA compaction compared to the wild-types controls [16]. Exposure of male rats to the chemotherapeutic agent cyclophosphamide, known to produce ROS as part of the mechanism of action, decreased the level of protamination and subsequently increased DNA damage of rat spermatozoa [45].

Similar long-lasting damages as those found in this study were observed in testicular cancer survivors who underwent chemotherapy with cisplatin and bleomycin, two drugs that generate high levels of ROS in cells exposed to them [46,47]. Cancer patients treated with polychemotherapy, including ROS-generating compounds, have high levels of lipid peroxides in blood, indicating the establishment of oxidative stress due to the treatment [48]. Spermatozoa from testicular cancer survivors displayed high DNA damage and low DNA compaction up to two and one years, respectively, after the end of chemotherapy [49,50]. With this significant clinical relevance, the present study provides insight into the understanding of the long-term effect of oxidative stress, a condition often seen in male infertility [2,3].

The molecular mechanism behind the long-term, lasting oxidative stress observed in this study is yet to be determined. A potential candidate is the dysregulation of mitochondria due to high levels of 4-HNE. We hypothesize that the high oxidative stress due to t-BHP promoted significant 4-HNE levels that impaired mitochondrial proteins leading to dysregulation of this organelle. We observed that the inhibition of PRDX6 iPLA$_2$ activity by MJ33 increased the levels of 4-HNE and impaired the sperm mitochondrial membrane potential, leading to the generation of oxidative stress and the oxidation of the DNA in human spermatozoa [15]. 4-HNE is capable of inducing mutations of the mitochondrial DNA and form adducts with mitochondrial proteins that lead to mitochondrial dysfunction [51]. Further studies are needed to elucidate the molecular mechanisms behind the alterations in spermatozoa and epididymis observed in the present study.

5. Conclusions

In conclusion, we reported the unexpected long-term effects of t-BHP treatment in the male rat reproductive system that impairs sperm quality after one complete cycle of spermatogenesis. The epididymis, in contrast to the testis, was primarily affected by the treatment displaying markers of oxidative stress such as high levels of 4-HNE up to 9 weeks after the end of the treatment. An antioxidant response by the upregulation of PRDX6 and possibly PRDX1 and catalase attempt to correct the oxidative stress in the epididymis results in the decrease of lipid peroxidation at 9 weeks in cauda epididymis, but it appears not to be sufficient to repair the oxidative damage observed in spermatozoa. Further studies will be needed to elucidate the molecular mechanism that generates long-term oxidative stress that impairs sperm quality and fertility. These studies are relevant since many conditions such as diseases (i.e., cancer, diabetes), drugs and even lifestyles (i.e., smoking) generate chronic oxidative stress that impacts male fertility.

Supplementary Materials: The following are available online at http://www.mdpi.com/2076-3921/9/2/170/s1, Table S1: Body and reproductive organs weight and sperm production.

Author Contributions: C.O. designed the experiments; P.Y.W. and E.S. performed the experiments; P.Y.W., E.S. and C.O. analyzed the results. P.Y.W. wrote the original draft; P.Y.W., E.S. and C.O. revised and edited the final version of the manuscript. All authors have read and agreed to the published version of the manuscript.

Funding: This research was funded by the Canadian Institutes of Health Research (CIHR), operating grant # MOP133661 to CO.

Acknowledgments: We thank Yannan Liu and Jeremie Desrosiers for their technical assistance.

Conflicts of Interest: The authors declare no conflict of interest.

References

1. World Health Organization. Towards more objectivity in diagnosis and management of male fertility. *Int. J. Androl.* **1997**, *7*, 1–53.
2. Iwasaki, A.; Gagnon, C. Formation of reactive oxygen species in spermatozoa of infertile patients. *Fertil. Steril.* **1992**, *57*, 409–416. [CrossRef]
3. Tremellen, K. Oxidative stress and male infertility: A clinical perspective. *Hum. Reprod. Update* **2008**, *14*, 243–258. [CrossRef] [PubMed]
4. Anderson, B.J.; Williamson, R.C. Testicular torsion in Bristol: A 25-year review. *Br. J. Surg.* **1988**, *75*, 988–992. [CrossRef]
5. Hasegawa, M.; Wilson, G.; Russell, L.D.; Meistrich, M.L. Radiation-induced cell death in the mouse testis: Relationship to apoptosis. *Radiat. Res.* **1997**, *147*, 457–467. [CrossRef]

6. Brennemann, W.; Stoffel-Wagner, B.; Helmers, A.; Mezger, J.; Jager, N.; Klingmuller, D. Gonadal function of patients treated with cisplatin based chemotherapy for germ cell cancer. *J. Urol.* **1997**, *158*, 844–850. [CrossRef]
7. Smith, R.; Kaune, H.; Parodi, D.; Madariaga, M.; Ríos, R.; Morales, I.; Castro, A. Increased sperm DNA damage in patients with varicocele: Relationship with seminal oxidative stress. *Hum. Reprod.* **2006**, *21*, 986–993. [CrossRef]
8. Turner, T.T. The study of varicocele through the use of animal models. *Hum. Reprod. Update* **2001**, *7*, 78–84. [CrossRef]
9. de Lamirande, E.; O'Flaherty, C. Sperm Capacitation as an Oxidative Event. In *Studies on Men's Health and Fertility, Oxidative Stress in Applied Basic Research and Clinical Practice*; Aitken, J., Alvarez, J., Agawarl, A., Eds.; Springer: Berlin/Heidelber, Germany, 2012; pp. 57–94.
10. Storey, B.T. Biochemistry of the induction and prevention of lipoperoxidative damage in human spermatozoa. *Mol. Hum. Reprod.* **1997**, *3*, 203–213. [CrossRef]
11. Aitken, R.J.; Gordon, E.; Harkiss, D.; Twigg, J.P.; Milne, P.; Jennings, Z.; Irvine, D.S. Relative impact of oxidative stress on the functional competence and genomic integrity of human spermatozoa. *Biol. Reprod.* **1998**, *59*, 1037–1046. [CrossRef]
12. Sikka, S.C.; Rajasekaran, M.; Hellstrom, W.J. Role of oxidative stress and antioxidants in male infertility. *J. Androl.* **1995**, *16*, 464–468. [PubMed]
13. O'Flaherty, C. The Enzymatic Antioxidant System of Human Spermatozoa. *Adv. Androl.* **2014**, *2014*, 1–15.
14. O'Flaherty, C. Peroxiredoxins: Hidden players in the antioxidant defence of human spermatozoa. *Basic Clin. Androl.* **2014**, *24*, 4. [CrossRef] [PubMed]
15. Fernandez, M.C.; O'Flaherty, C. Peroxiredoxin 6 activates maintenance of viability and DNA integrity in human spermatozoa. *Hum. Reprod.* **2018**, *33*, 1394–1407. [CrossRef]
16. Ozkosem, B.; Feinstein, S.I.; Fisher, A.B.; O'Flaherty, C. Absence of Peroxiredoxin 6 Amplifies the Effect of Oxidant Stress on Mobility and SCSA/CMA3 Defined Chromatin Quality and Impairs Fertilizing Ability of Mouse Spermatozoa. *Biol. Reprod.* **2016**, *94*, 1–10. [CrossRef]
17. Ozkosem, B.; Feinstein, S.I.; Fisher, A.B.; O'Flaherty, C. Advancing age increases sperm chromatin damage and impairs fertility in peroxiredoxin 6 null mice. *Redox Biol.* **2015**, *5*, 15–23. [CrossRef]
18. Liu, Y.; O'Flaherty, C. In vivo oxidative stress alters thiol redox status of peroxiredoxin 1 and 6 and impairs rat sperm quality. *Asian J. Androl.* **2017**, *19*, 73–79.
19. Miranda-Vizuete, A.; Sadek, C.M.; Jiménez, A.; Krause, W.J.; Sutovsky, P.; Oko, R. The Mammalian Testis-Specific Thioredoxin System. *Antioxid. Redox Signal.* **2004**, *6*, 25–40. [CrossRef]
20. Chabory, E.; Damon, C.; Lenoir, A.; Henry-Berger, J.; Vernet, P.; Cadet, R.; Drevet, J.R. Mammalian glutathione peroxidases control acquisition and maintenance of spermatozoa integrity. *J. Anim. Sci.* **2009**, *88*, 1321–1331. [CrossRef]
21. Robaire, B.; Hinton, B.T. Orgebin-Crist, The Epididymis. In *Knobil and Neill's Physiology of Reproduction*, 3rd ed.; Academic Press: St. Louis, MO, USA, 2006; pp. 1071–1148.
22. Sommer, R.J.; Ippolito, D.L.; Peterson, R.E. In Uteroand Lactational Exposure of the Male Holtzman Rat to 2,3,7,8-Tetrachlorodibenzo-p-dioxin: Decreased Epididymal and Ejaculated Sperm Numbers without Alterations in Sperm Transit Rate. *Toxicol. Appl. Pharmacol.* **1996**, *140*, 146–153. [CrossRef]
23. Sullivan, R.; Frenette, G.; Girouard, J. Epididymosomes are involved in the acquisition of new sperm proteins during epididymal transit. *Asian J. Androl.* **2007**, *9*, 483–491. [CrossRef] [PubMed]
24. Kumar, T.R. Induction of Oxidative Stress by Organic Hydroperoxides in Testis and Epididymal Sperm of Rats In Vivo. *J. Androl.* **2007**, *28*, 77–85. [CrossRef] [PubMed]
25. Clermont, Y. Kinetics of spermatogenesis in mammals: Seminiferous epithelium cycle and spermatogonial renewal. *Physiol. Rev.* **1972**, *52*, 198–236. [CrossRef] [PubMed]
26. Russell, L.E.R.; Sinha Hikim, A.; Clegg, E. *Histological and Histopathological Evaluation of the Testis*; Cache River Press: St. Louis, MO, USA, 1990.
27. Robb, G.W.; Amann, R.P.; Killian, G.J. Daily sperm production and epididymal sperm reserves of pubertal and adult rats. *Reproduction* **1978**, *54*, 103–107. [CrossRef] [PubMed]
28. Valli, H.; Phillips, B.T.; Shetty, G.; Byrne, J.A.; Clark, A.T.; Meistrich, M.L.; Orwig, K.E. Germline stem cells: Toward the regeneration of spermatogenesis. *Fertil. Steril.* **2014**, *101*, 3–13. [CrossRef] [PubMed]

29. Moskovtsev, S.I.; Alladin, N.; Lo, K.C.; Jarvi, K.; Mullen, J.B.M.; Librach, C.L. A comparison of ejaculated and testicular spermatozoa aneuploidy rates in patients with high sperm DNA damage. *Syst. Biol. Reprod. Med.* **2012**, *58*, 142–148. [CrossRef]
30. Greco, E.; Scarselli, F.; Iacobelli, M.; Rienzi, L.; Ubaldi, F.; Ferrero, S.; Tesarik, J. Efficient treatment of infertility due to sperm DNA damage by ICSI with testicular spermatozoa. *Hum. Reprod.* **2005**, *20*, 226–230. [CrossRef]
31. Aguilar, C.; Meseguer, M.; García-Herrero, S.; Gil-Salom, M.; O'Connor, J.E.; Garrido, N. Relevance of testicular sperm DNA oxidation for the outcome of ovum donation cycles. *Fertil. Steril.* **2010**, *94*, 979–988. [CrossRef]
32. Russell, L.D.; Peterson, R.N. Sertoli cell junctions: Morphological and functional correlates. *Int. Rev. Cytol.* **1985**, *94*, 177–211.
33. Griswold, M.D.; McLean, D. The Sertoli Cell. In *Knobil and Neill's Physiology of Reproduction*, 3rd ed.; Academic Press: St. Louis, MO, USA, 2006; pp. 949–975.
34. Yoganathan, T.; Eskild, W.; Hansson, V. Investigation of detoxification capacity of rat testicular germ cells and sertoli cells. *Free Radic. Biol. Med.* **1989**, *7*, 355–359. [CrossRef]
35. O'Flaherty, C.; Boisvert, A.; Manku, G.; Culty, M. Protective Role of Peroxiredoxins against Reactive Oxygen Species in Neonatal Rat Testicular Gonocytes. *Antioxidants* **2019**, *9*, 32. [CrossRef] [PubMed]
36. Bauché, F.; Fouchard, M.-H.; Jégou, B. Antioxidant system in rat testicular cells. *FEBS Lett.* **1994**, *349*, 392–396. [CrossRef]
37. Syntin, P.; Robaire, B. Sperm Structural and Motility Changes During Aging in the Brown Norway Rat. *J. Androl.* **2001**, *22*, 235–244.
38. Fisher, A.B. Peroxiredoxin 6 in the repair of peroxidized cell membranes and cell signaling. *Arch. Biochem. Biophys.* **2017**, *617*, 68–83. [CrossRef]
39. Fisher, A.B.; Dodia, C.; Sorokina, E.M.; Li, H.; Zhou, S.; Raabe, T.; Feinstein, S.I. A novel LysoPhosphatidylcholine Acyl Transferase Activity is Expressed by Peroxiredoxin 6. *J. Lipid Res.* **2016**, *31*, 292–303. [CrossRef]
40. Fisher, A.B.; Vasquez-Medina, J.P.; Dodia, C.; Sorokina, E.M.; Tao, J.Q.; Feinstein, S.I. Peroxiredoxin 6 phospholipid hydroperoxidase activity in the repair of peroxidized cell membranes. *Redox Biol.* **2018**, *14*, 41–46. [CrossRef]
41. Fatemi, N.; Sanati, M.H.; Shamsara, M.; Moayer, F.; Zavarehei, M.J.; Pouya, A.; Gourabi, H. TBHP-induced oxidative stress alters microRNAs expression in mouse testis. *J. Assist. Reprod. Genet.* **2014**, *31*, 1287–1293. [CrossRef]
42. Wykes, S.M.; Krawetz, S.A. The structural organization of sperm chromatin. *J. Biol. Chem.* **2003**, *278*, 29471–29477. [CrossRef]
43. Bedford, J.M.; Bent, M.J.; Calvin, H. Variations in the structural character and stability of the nuclear chromatin in morphologically normal human spermatozoa. *J. Reprod. Fertil.* **1973**, *33*, 19–29. [CrossRef]
44. Castillo, J.; Simon, L.; de Mateo, S.; Lewis, S.; Oliva, R. Protamine/DNA ratios and DNA damage in native and density gradient centrifuged sperm from infertile patients. *J. Androl.* **2011**, *32*, 324–332. [CrossRef]
45. Codrington, A.M.; Hales, B.F.; Robaire, B. Exposure of male rats to cyclophosphamide alters the chromatin structure and basic proteome in spermatozoa. *Hum. Reprod.* **2007**, *22*, 1431–1442. [CrossRef] [PubMed]
46. Weijl, N.I.; Hopman, G.D.; Wipkink-Bakker, A.; Lentjes EG, W.M.; Berger, H.M.; Cleton, F.J.; Osanto, S. Cisplatin combination chemotherapy induces a fall in plasma antioxidants of cancer patients. *Ann. Oncol.* **1998**, *9*, 1331–1337. [CrossRef] [PubMed]
47. Caporossi, D.; Ciafrè, S.A.; Pittaluga, M.; Savini, I.; Farace, M.G. Cellular responses to H2O2 and bleomycin-induced oxidative stress in L6C5 rat myoblasts. *Free Radic. Biol. Med.* **2003**, *35*, 1355–1364. [CrossRef] [PubMed]
48. Look, M.P.; Musch, E. Lipid Peroxides in the Polychemotherapy of Cancer Patients. *Chemotherapy* **1994**, *40*, 8–15. [CrossRef]
49. O'flaherty, C.M.; Chan, P.T.; Hales, B.F.; Robaire, B. Sperm Chromatin Structure Components Are Differentially Repaired in Cancer Survivors. *J. Androl.* **2012**, *33*, 629–636. [CrossRef]

50. O'Flaherty, C.; Hales, B.F.; Chan, P.; Robaire, B. Impact of chemotherapeutics and advanced testicular cancer or Hodgkin lymphoma on sperm deoxyribonucleic acid integrity. *Fertil. Steril.* **2010**, *94*, 1374–1379. [CrossRef]
51. Xiao, M.; Zhong, H.; Xia, L.; Tao, Y.; Yin, H. Pathophysiology of mitochondrial lipid oxidation: Role of 4-hydroxynonenal (4-HNE) and other bioactive lipids in mitochondria. *Free Radic. Biol. Med.* **2017**, *111*, 316–327. [CrossRef]

© 2020 by the authors. Licensee MDPI, Basel, Switzerland. This article is an open access article distributed under the terms and conditions of the Creative Commons Attribution (CC BY) license (http://creativecommons.org/licenses/by/4.0/).

Review

Molecular Changes Induced by Oxidative Stress that Impair Human Sperm Motility

Karolina Nowicka-Bauer [1],* and Brett Nixon [2,3]

1. Institute of Human Genetics, Polish Academy of Sciences, 60-479 Poznan, Poland
2. Priority Research Centre for Reproductive Science, School of Environmental and Life Sciences, Discipline of Biological Sciences, University of Newcastle, Callaghan, Newcastle, NSW 2308, Australia; brett.nixon@newcastle.edu.au
3. Hunter Medical Research Institute, Pregnancy and Reproduction Program, New Lambton Heights, Newcastle, NSW 2305, Australia
* Correspondence: karolina.nowicka-bauer@amu.edu.pl

Received: 21 January 2020; Accepted: 31 January 2020; Published: 4 February 2020

Abstract: A state of oxidative stress (OS) and the presence of reactive oxygen species (ROS) in the male reproductive tract are strongly correlated with infertility. While physiological levels of ROS are necessary for normal sperm functioning, elevated ROS production can overwhelm the cell's limited antioxidant defenses leading to dysfunction and loss of fertilizing potential. Among the deleterious pleiotropic impacts arising from OS, sperm motility appears to be particularly vulnerable. Here, we present a mechanistic account for how OS contributes to altered sperm motility profiles. In our model, it is suggested that the abundant polyunsaturated fatty acids (PUFAs) residing in the sperm membrane serve to sensitize the male germ cell to ROS attack by virtue of their ability to act as substrates for lipid peroxidation (LPO) cascades. Upon initiation, LPO leads to dramatic remodeling of the composition and biophysical properties of sperm membranes and, in the case of the mitochondria, this manifests in a dissipation of membrane potential, electron leakage, increased ROS production and reduced capacity for energy production. This situation is exacerbated by the production of cytotoxic LPO byproducts such as 4-hydroxynonenal, which dysregulate molecules associated with sperm bioenergetic pathways as well as the structural and signaling components of the motility apparatus. The impact of ROS also extends to lesions in the paternal genome, as is commonly seen in the defective spermatozoa of asthenozoospermic males. Concluding, the presence of OS in the male reproductive tract is strongly and positively correlated with reduced sperm motility and fertilizing potential, thus providing a rational target for the development of new therapeutic interventions.

Keywords: 4-hydroxynonenal (4HNE); infertility; lipid peroxidation; male germ cells; oxidative stress; reactive oxygen species; spermatozoa; sperm capacitation; sperm motility

1. Introduction

Male infertility accounts for approximately 40% of all cases of infertility [1] and affects approximately 7% of all men worldwide [2]. Recently, considerable attention has focused on the role of oxidative stress (OS) in the pathophysiology of male infertility. OS is associated with the excessive generation of free radicals, such as reactive oxygen species (ROS), and/or decreased efficacy of antioxidant defenses. In humans, the pervasive impact of OS has been linked to the development of a variety of diseases as diverse as Alzheimer's disease [3], cancer [4], heart failure [5] and obesity [6]. OS is also a prevalent biomarker associated with the semen of approximately 35% of infertile men [7] and the presence of elevated levels of seminal ROS has been reported arising from male reproductive tract pathologies such as varicocele [8], inflammatory [9] and prostate cancer [10]. Owing to their unique architecture, featuring an abundance of oxidizable substrates and limited intracellular antioxidant defenses, the male germ cell is particularly vulnerable to elevated levels of ROS. The resultant OS commonly manifests in a spectrum of adverse sequelae, which drive germ cell dysfunction and culminate in their apoptotic demise [11]. Accordingly, patients with increased ROS levels in their seminal plasma [12] often present with reduced sperm count (oligozoospermia); a condition attributed, at least in part, to apoptosis within the developing germ cell population [13,14]. However, in addition to the loss of sperm viability, OS has also been causally linked to lesions in the motility profile of mature spermatozoa [15].

Progressive sperm movement is required for delivery of the male gamete to the site of fertilization within the ampulla of the Fallopian tubes, as well as the resultant syngamy that facilitates transfer of paternal genetic and epigenetic information to an oocyte during natural conception. Effective sperm motility also plays a crucial role during the assisted reproduction technology (ART) procedures of intrauterine insemination and in vitro fertilization. According to World Health Organization (WHO) criteria [16], the presence of less than 32% of spermatozoa with progressive motility in an ejaculate is defined as asthenozoospermia. It is estimated that asthenozoospermia accounts for as much as 30% of all cases of male infertility [17] and, in the absence of genetic defects, bacterial infection or abnormal semen liquefaction, this condition is often directly linked to the presence of OS in the male reproductive tract and semen [18]. Indeed, elevated levels of seminal ROS have repeatedly been documented in studies of astheno- and oligoasthenozoospermic individuals [18–21]. In terms of the mechanistic basis of sperm dysfunction leading to asthenozoospermia, defective mitochondria and a concomitant reduction in the production of energy required to support normal movement have been identified as a common etiology [18]. However, structural changes in the motility apparatus housed within the sperm flagellum and dysregulation of motility-associated signaling pathways have also been reported in response to OS [22], thus complicating the diagnosis of sperm motility defects. Here, we survey the literature pertaining to the generation of ROS in the male reproductive tract and the deleterious influence of OS on the biochemical pathways and structural features of the sperm cell responsible for modulating their motility. We also highlight the role of antioxidants in combating the burden of OS associated infertility.

2. Sources of ROS in Semen

Seminal ROS originate from a variety of different endogenous and exogenous sources. In addition to generation by spermatozoa themselves, other cellular contaminants such as immature round germ cells, leukocytes and epithelial cells, can also have a direct bearing on the levels of ROS within an ejaculate. Additionally, a number of environmental and lifestyle factors can exert direct and indirect influence over the levels of OS encountered within the male reproductive tract.

2.1. Endogenous Sources of ROS

2.1.1. Spermatozoa

At least two distinct pathways have been implicated in ROS generation in mature spermatozoa. One such pathway is localized within the sperm plasma membrane and is linked to the activity of the nicotinamide adenine dinucleotide phosphate (NADPH) oxidase system, whilst the second is associated with electron leakage from the mitochondrial electron transport chain (ETC) [23]. The latter of these pathways represents the main source of ROS production in spermatozoa and occurs as a result of the premature exit of electrons from the respiratory chain. This leakage prevents the reduction of oxygen to water at cytochrome c oxidase, with the escaped electrons instead reacting with molecular oxygen (O_2) to form the superoxide radical ($O_2\bullet^-$). Basal levels of electron leakage can potentially occur from multiple sites within the ETC (reviewed in [24]), and, unlike the $O_2\bullet^-$ generated at the level of the plasma membrane that supports sperm capacitation [25], mitochondrial derived $O_2\bullet^-$ is generally associated with pathological damage. Indeed, when superoxide anion generation exceeds the limited antioxidant capacity of the sperm cell, it has the ability to propagate the formation of non-radicals, including hydrogen peroxide (H_2O_2), and in the presence of Fe^{3+}, the formation of alternative radicals such the hydroxyl anion (OH^-) via Haber-Weiss and Fenton reactions [26]. Additional ROS containing nitrogen atoms (such as NO, NO_3^-, NO^-, N_2O, and HNO_3) are also able to be formed [27,28] and collectively, these powerful oxidants have the potential to trigger the peroxidation of membrane lipids and a concomitant loss of sperm function [29]. In studies conducted in rats it has been shown that the presence of dead spermatozoa can also promote higher than normal levels of H_2O_2 [30]. This response appears to be related to the apoptosis cascade during which the disruption of mitochondrial membranes leads to the release of cytochrome c and elevated ROS production [31].

2.1.2. Immature Germ Cells

In addition to mature spermatozoa, ejaculated semen samples also frequently contain variable amounts of contaminating immature sperm cells, which have failed to complete normal morphological differentiation during spermatogenesis [32]. Depending on the timing of such errors, they may result in the presence of either round cells or seemingly mature spermatozoa that retain a considerable portion of their cytoplasm. The latter cells originate from defects encountered in the final phase of spermatogenesis (i.e., spermiogenesis), during which the majority of the cell's cytoplasm would otherwise be shed to create the highly streamlined profile of the sperm head [33]. The residual cytoplasm retained in these immature sperm cells tends to accumulate in the vicinity of the mid-piece and contains high levels of glucose-6-phosphate dehydrogenase; an enzyme that catalyzes NADPH production via the pentose phosphate pathway [34]. NADPH, in turn, is capable of acting as a substrate to fuel ROS-generating NADPH oxidases. Thus, the presence of immature germ cells harboring excess cytoplasm has the potential to elevate endogenous ROS levels within an ejaculate and deleteriously affect the function of their otherwise normal counterparts [34].

2.1.3. Leukocytes

A state of infection and/or inflammation in the male reproductive tract (i.e., testis, epididymis, seminal vesicles and/or prostate) can result in increased infiltration of leukocytes bringing with them an attendant risk of elevated ROS within an ejaculate [35]. Indeed, when appropriately stimulated, phagocytic leukocytes are capable of metabolizing oxygen to produce copious quantities of ROS in a process often referred to as a respiratory burst [36]. This microbicidal defense response is mediated by the NADPH-oxidase complex, and enhanced by the presence of cytokines, which themselves are released during inflammation [9,37]. It follows that the presence of leukocytes, and predominantly polymorphonuclear neutrophils, is recognized as a major source of ROS in human semen [37,38].

2.1.4. Varicocele

Varicocele is a pathology associated with an abnormal enlargement of the pampiniform venous plexus surrounding the spermatic cord in the scrotum [39]. This clinical condition occurs in approximately 15% of males and accordingly represents the most common cause of primary and secondary infertility in men; accounting for as much as 40% of all cases of male infertility [40]. Consistent with this clear association between varicocele and male infertility, a considerable body of evidence supports OS as a key causative agent in the pathophysiology of varicocele [41]. Indeed, both infertile patients and fertile men with varicocele frequently present with higher levels of ROS, NO, and lipid peroxidation products in their reproductive tract than men without varicocele [42–46]. The most widely accepted model to account for these findings suggests that varicoceles lead to an increase in scrotal temperature owing to the reflux and accumulation of warmed abdominal blood within the pampiniform plexus. The resultant heat stress has, in turn, been postulated to enhance ETC electron leakage via thermal inhibition of mitochondrial complexes, thus accentuating mitochondrial ROS production (reviewed in [47]).

3. Exogenous Sources of ROS

In addition to endogenous influences, modern lifestyle and environment factors have also been increasingly linked with a range of adverse health sequelae, including poorer quality semen parameters. Below, we give brief consideration to physical and chemical factors that have been associated with heightened levels of seminal ROS.

3.1. Physical Factors

One of the contemporary issues of the modern lifestyle is excess heating of testes. In scrotal animals such as our own, the testicular environment has evolved to operate optimally at temperatures that generally lie at least 1–2 °C below that of core body temperature (reviewed in [48]). Accordingly, the spermatogenic process can be adversely impacted by scrotal hyperthermia as occurs in certain professions such as those that involve extended periods of seated activity (e.g., at a desk or in a vehicle) or those that involve direct exposure to high ambient temperatures (e.g., steel fabrication and welding). The occupational practice of using a laptop computer in such a way that it covers the testes can also cause localized heat stress, as does the wearing of tight clothing and in particular, underwear. Through mechanisms described in the previous section, the heat-stressed testes have been shown to produce excessive ROS, which is linked to the impairment of sperm function [49,50]. It should be noted that alternative lifestyle habits such as frequent sauna or warm bath exposures also represent potential sources of heat stress that may impact on male infertility [51–53]. In support of this conclusion, a recent study by Houston et al. [54] demonstrated that the exposure of mice to elevated ambient temperature led to increased sperm mitochondrial ROS generation and OS-induced molecular changes during germ cell development.

Similarly, increasing attention has been focused on the potential impact of non-ionizing radiation, such as the electromagnetic energy (EME) emitted by mobile technologies and other forms of microwave equipment, on the male germ line [55]. In this context, the human body has the potential to behave in a manner analogous to that of an antenna that receives EME [56]. The exposure of human tissues to EME can have various biological effects including localized elevation of the temperature in the affected tissue, including the testes [56]. EME can also alter cellular membrane potential and impact molecular bonding, with the polar side chains of amino acids being particularly affected by EME exposure [57]. Such changes not only have the potential to influence protein structure, and hence interfere with enzymatic activity, but can also perturb the transmembrane transport of ions [56,57]. As an extension of these findings, multiple studies have now demonstrated that supraphysiological levels of EME can negatively affect mitochondrial membrane characteristics and/or overall functioning [58–60]. Indeed, the exposure of isolated cells to EME can lead to increased activation of mitochondrial function and an attendant elevation of ROS production associated with complexes I and III of ETC, independent of changes in mitochondrial membrane potential [56,58]. In model cells such as those of the human amniotic epithelial lineage, magnetic fields can induce mitochondrial permeability transition and cytochrome-c release together with increased intracellular ROS generation, via a pathway that is dependent on glycogen synthase kinase-3β [58]. As proof of concept, it has also been shown that human spermatozoa exposed to EME at frequencies designed to simulate that emitted by mobile phones experience reduced motility and vitality; defects that were associated with increased mitochondrial ROS production and numerous molecular alterations that are synonymous with OS [59].

3.2. Chemical Factors

Aside from the physical factors discussed above, male fertility is also sensitive to a variety of toxicants, such as those arising from industrial processes or from common everyday materials, which accumulate in the human body. Indeed, it is well established that the accumulation of the heavy metals lead and cadmium can impair multiple semen parameters, including sperm motility [61]. Similarly, male rats treated with lithium display OS in their testes and experience reduced sperm count and motility [62]. The induction of OS in the male reproductive tract has also been cited as a causal agent responsible for elevated levels of apoptosis among developing germ cells, defects in sperm morphology and impaired sperm function in mice treated with industrial contaminants used in the production of plastics, such as bisphenol-A [63], mono-butyl phthalate [64] and other related compounds [65]. This spectrum of deleterious OS-related effects extends to other forms of chemical exposure including those associated with excessive alcohol consumption or cigarette smoking. In this context, chemicals contained within cigarette smoke have been shown to cause local inflammation, an attendant 48% increase in seminal leukocytes, and a 107% increase in seminal ROS levels [66]. It follows that the semen of cigarette smokers is not only characterized by increased ROS levels, but also extensive molecular changes in the spermatozoa and reduced overall semen quality [67]. Likewise, excessive alcohol consumption can lead to ethanol-induced cell membrane destruction, increased production of ROS and impaired sperm function [68,69].

From the preceding discussion, it is apparent that acute and/or chronic exposure to a variety of external or internal factors can trigger the overproduction of ROS and reduce antioxidant defenses within the male reproductive tract, thus propagating an OS cascade and resulting in LPO (summarized in Figure 1).

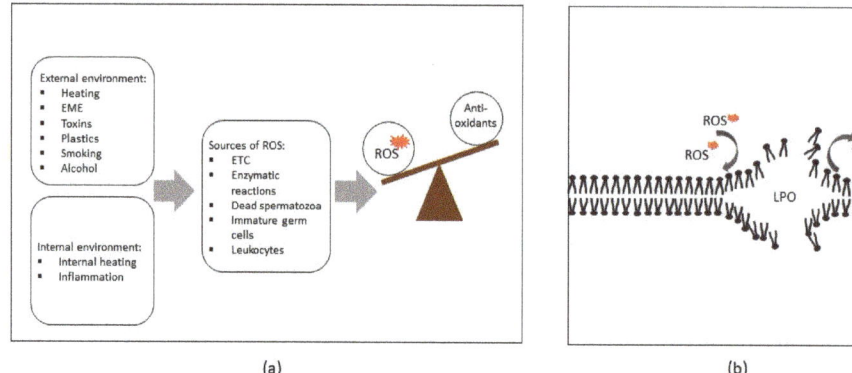

Figure 1. (a) Sources of reactive oxygen species (ROS) in human spermatozoa and the relationship between the rate of their production and antioxidant defenses during oxidative stress. (b) ROS are capable of attacking polyunsaturated fatty acids (PUFAs) within cellular membranes, initiating lipid peroxidation cascades (LPO) and resulting in the production of cytotoxic lipid aldehydes such as 4-hydroxynonenal (4HNE). Abbreviations: EME, electromagnetic energy; ETC, electron transport chain.

4. ROS-Induced Lesions Detected in Low-Motility Spermatozoa

With the diverse range of factors that can amplify the levels of ROS in semen, attention has naturally focused on the impact of these powerful oxidants on sperm function. Through decades of research we have come to realize that the highly specialized sperm cell is exceptionally vulnerable to disturbance in ROS levels owing to the presence of modest antioxidant defenses and conversely, a myriad of oxidizable substrates [70]. Not the least of these are the polyunsaturated fatty acids (PUFAs; such as linolenic, arachidonic and decohexaenoic acids) that dominate the lipid architecture of the sperm plasma membrane. In the case of human spermatozoa, the predominant PUFA is decohexaenoic acid (DHA), a lipid that accounts for >50% of all membrane PUFAs [70,71]. PUFAs such as DHA not only play a major role in the regulation of sperm membrane fluidity, but owing to the presence of multiple carbon-carbon double bonds, they also serve as prime substrates for ROS attack. The resultant cascade of lipid peroxidation (LPO) reactions catalyze the formation of numerous breakdown products including a suite of highly reactive lipid aldehydes [e.g., malondialdehyde (MDA), 4-hydroxynonenal (4HNE) and acrolein] [11] (see Figure 1b).

Under physiological conditions, aldehyde-metabolizing enzymes function to detoxify and prevent the accumulation of these advanced end products of LPO [72]. However, excessive OS can promote the accrual of lipid aldehydes and, owing to their inherent stability (relative to that of free radicals), these electrophiles can elicit widespread cellular damage and pathological dysfunction in human spermatozoa [73,74]. In the case of 4HNE, which ranks among the most abundant and cytotoxic of the lipid aldehydes, the chemical structure contains three reactive functional groups: a C2=C3 double bond, a C1=O carbonyl group and a hydroxyl group on C4 [75]. These structural elements render the 4HNE electrophile highly reactive toward nucleophilic groups, enabling the formation of both the Michael addition of thiol or amino compounds and Schiff bases with primary amines. Thus, 4HNE has the ability to react with proteins (principally those containing histidine, cysteine and lysine residues), lipids, and nucleic acids (mostly with the guanosine moiety of DNA) [76–78]. In spermatozoa, the creation of 4HNE adducts has been linked with compromised membrane integrity, motility defects and reduced ability to participate in oocyte interactions [22,73,79]; thus reducing overall fertility. Moreover, studies by Bromfield and colleagues [73,79] have shown that the impact of 4HNE can vary depending on the timing and of the insult. Thus, 4HNE can drive post-meiotic round spermatids towards a ferroptotic cell death pathway, whereas an equivalent exposure of mature spermatozoa can elicit functional lesions, which compromise their fertilization potential (e.g., dysregulation of the molecular chaperone Heat Shock Protein A2 and an accompanying disruption of oocyte recognition), but does not overtly impact their viability [73,79]. Such differential pathogenesis may be attributed to the highly specialized architecture of the male germ cell, which depending on their stage of differentiation, features an abundance of substrates for 4HNE attack, minimal antioxidant defense enzymes, and limited capacity for self-repair when 4HNE-mediated damage is sustained [11]. However, excess ROS production can also directly impact sperm function via increases in redox driven protein modifications [80]. Thus, OS has been shown to promote an increase in S-glutathionylation and tyrosine nitration of sperm proteins, both of which adversely impact motility [80,81]. Similar alterations have been documented by Vignini et al. [82], who demonstrated an increase in $ONOO^-$ concentrations and tyrosine nitration in human asthenozoospermic sperm samples.

4.1. Compromised Sperm Membrane Integrity

The peroxidation of membrane lipids leads to the catabolism of phospholipids and liberation of PUFAs, thus directly contributing to increased membrane fluidity and permeability to ions, which can elicit downstream effects in terms of inactivating membrane enzymes and receptors [20,79]. The loss of sperm membrane integrity is also tightly correlated with reduced sperm motility [83]. Indeed, the compromise of both of these parameters has been well documented in the context of cryopreservation [84–87] in which post-thaw sperm samples typically experience a burst of ROS and high levels of OS [85,86,88]. Sperm motility is also highly sensitive to pH and ion concentration (reviewed in [89]). The disruption of sperm membrane integrity alters the diffusion of ions across the membrane and dysregulates the function of ion pumps and channels, especially as those ion channels that are regulated by PUFAs (reviewed in [90]). Although there is currently limited direct evidence linking sperm membrane integrity and ion channel function, the peroxidation of lipids and the downregulation of some ion channels in low-motility spermatozoa has been documented [83,91–94]. In this context, Baker et al. [22] reported two sperm ion channels, voltage-dependent anion-selective channel protein 2 (VDAC2) and sodium/potassium-transporting ATPase subunit alpha-4 (ATP1A4) that are susceptible to the formation of potentially deleterious adducts with 4HNE. Similarly, decreased VDAC and ATP1A4 abundance has been reported in the spermatozoa of males with asthenozoospermia [95] and in patients with unilateral varicocele (a pathology correlated with OS) [96], respectively.

The deleterious impact of ROS extends beyond the plasma membrane to also include those membranes surrounding subcellular organelles such as the mitochondria. In contrast to somatic cells, sperm mitochondria are localized in one specific region of the sperm cell—the mid-piece of the flagellum. Within this domain, the mitochondria are organized end to end to form long spirals that wrap tightly around the axoneme [97,98]. The correct functioning of mitochondria is especially important for sperm cells in terms of supplying the energy needed for efficient movement. The elevated production of mitochondrial ROS and the resultant propagation of sperm plasma membrane LPO can, in turn, initiate secondary LPO in mitochondrial membranes. Similar to the plasma membrane, the peroxidation of lipids residing in the mitochondrial membranes also changes the fluidity of these structures, thus resulting in the upregulation of proton and electron leakage through the inner membrane [99,100]. Such a situation leads to the loss of mitochondrial membrane potential (ΔΨ), compromises the efficiency of mitochondrial ATP generation, and triggers a cycle of elevated electron leakage and the generation of additional mitochondrial ROS, which collectively exacerbates the impact of OS on sperm function [18,101,102].

4.2. Dysregualtion of Sperm Metabolic Enzymes

In human sperm cells, the majority of the energy needed to support motility appears to originate from glycolysis, which takes place in the sperm tail. However, it has been argued that mitochondrial oxidative phosphorylation (OXPHOS) also plays a secondary role in the bioenergetic pathways associated with motility. In this context, a key substrate for OXPHOS are the endogenous lipids present in the sperm membrane [103], but other metabolic substrates and their associated pathways have also been implicated [18,103,104]. Indeed, depending on the prevailing environmental conditions that sperm encounter on route to the site of fertilization, these cells may be able to utilize different metabolic pathways.

As previously mentioned, the cytotoxic aldehydes generated during LPO can form adducts with multiple elements of the sperm proteome. Curiously, however, not all sperm proteins appear to display equivalent susceptibility to lipid aldehyde carbonylation reactions [105]. By way of example, the application of affinity-based isolation techniques coupled with mass spectrometry has revealed that 4HNE preferentially adducts to a relatively small number of putative targets in human spermatozoa [22]. Notably, many of these 4HNE targets serve as metabolic enzymes responsible for the production of energy needed to support sperm motility (depicted in Figure 2), including several glycolytic enzymes [phosphofructokinase (PFKP); aldolase A, fructose-bisphosphate (ALDOA); phosphoglycerate kinase (PGK), pyruvate kinase (PKM); lactate dehydrogenase C chain (LDHC)], enzymes involved in the TCA cycle [malate dehydrogenase 2, NAD (mitochondrial) (MDH2)] [22], electron transport chain [succinate dehydrogenase complex, subunit A, flavoprotein (SDHA) [106]; ubiquinol-cytochrome c reductase, Rieske iron-sulfur polypeptide 1 (UQCRFS1)] [22] and beta-oxidation [fatty acid amide hydrolase (FAAH); acyl-CoA synthetase long-chain family member 1 (ACSL1); hydroxyacyl-CoA dehydrogenase trifunctional multienzyme complex subunit beta (HADHB); acetyl-Coenzyme A acetyltransferase 1 (ACAT1)] [22]. Moreover, 4HNE has also been found to adduct with solute carrier family 25 (mitochondrial carrier; adenine nucleotide translocator), member 31 (SLC25A31), a mitochondrial carrier protein involved in the exchange of cytoplasmic ADP with mitochondrial ATP [22]. Although the consequences of damage caused by the insertion of bulky 4HNE (C-9) carbonyl adducts has yet to be investigated across all targeted sperm proteins, it is reasonable to suspect that these modifications could elicit protein mis-folding, poor substrate recognition, and/or degradation of the protein itself [107,108]. Indeed, recent work by Aitken et al. [106] has demonstrated that 4HNE adduction to SDHA can activate mitochondrial electron leakage and disrupt ΔΨ in human spermatozoa. These data raise the prospect that 4HNE may primarily perturb the bioenergetic pathways that sustain sperm motility and thus provide a mechanistic model to account for the dysregulation of sperm function commonly reported in cells burdened by excessive ROS production [68]. Notably, the self-perpetuating nature of this pathway may also account for the ability of 4HNE to drive human spermatozoa towards an intrinsic apoptotic pathway.

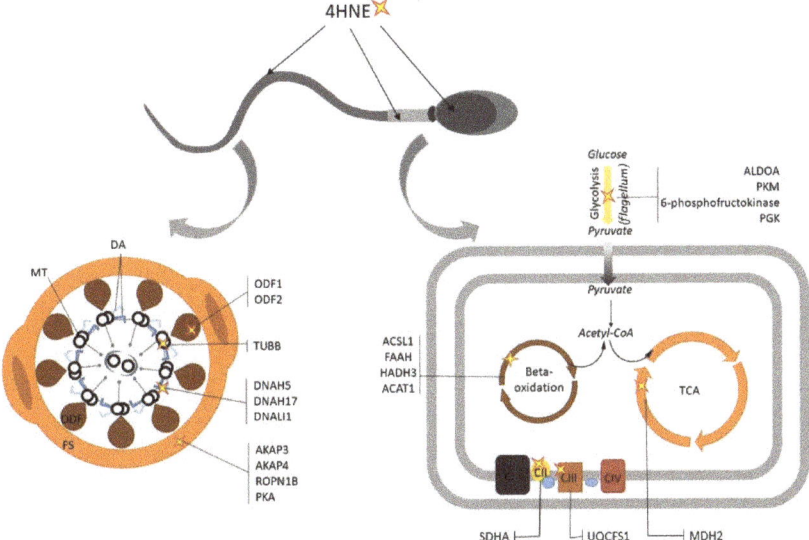

Figure 2. The influence of 4-hydroxynonenol (4HNE) on sperm function. The lipid aldehyde 4HNE has repeatedly been shown to form adducts with sperm flagellum proteins associated with the motility apparatus, signaling pathways and metabolism. In the context of the mitochondria, 4HNE adduction has been linked to adverse effects on enzymes of beta-oxidation, the tricarboxylic acid (TCA) cycle and the electron transport chain (ETC), thus attenuating the energy production. Abbreviations: CI, ETC Complex I; CII, ETC Complex II; CIII, ETC Complex III; CIV, ETC Complex IV; DA, dynein arms; FS, fibrous sheath; MT, microtubules; ODF, outer dense fibers.

Such a model takes on added significance in view of proteomic data emerging from studies of the spermatozoa of asthenozoospermic patients. Whilst not universal, a common theme to emerge from this work is that proteins involved in energy production are generally downregulated in the defective spermatozoa of asthenozoospermic individuals [18,105,109]. Indeed, the majority of proteins so affected in asthenozoospermia are linked to glycolysis, pyruvate metabolism, tricarboxylic acid cycle (TCA), OXPHOS, beta-oxidation or alternative metabolic pathways [18,109,110]. By way of example, the glycolytic enzyme, glucose-6-phosphate isomerase (GPI), has been shown to be under-represented in asthenozoospermic individuals, as has the sperm-specific glyceraldehyde-3-phosphate dehydrogenase (GAPDHS) [110–113]; the activity of which is also attenuated after treatment of spermatozoa with ROS [112]. Moreover, several mitochondrial enzymes involved in the downstream conversion of pyruvate to acetyl-CoA, TCA cycle, ETC and a plethora of ATP synthase subunits have also been shown to be dysregulated. Similar reductions in protein abundance also extend to several enzymes involved in beta-oxidation of fatty acids [18,95,105,109,110]. Alternative proteomic studies of asthenozoospermia have revealed downregulation of one of the subunits of NADH dehydrogenase (NDUFA13), a key constituent of complex I of the ETC [114]. The knockdown of this protein in a mouse spermatocyte cell line led to the loss of $\Delta\Psi$, increased ROS production and apoptosis [114]. These data draw interesting parallels with the study of asthenozoospermic individuals by Nowicka-Bauer et al. [18], which clearly showed the downregulation of sperm mitochondrial metabolic pathways and $\Delta\Psi$; defects that were accompanied by increased production of mitochondrial ROS in the spermatozoa of asthenozoospermic patients. These data reinforce the notion that reduced sperm motility is predominantly associated with dysfunction of sperm mitochondria leading to elevated levels of ROS and a reciprocal reduction in energy production. However, it remains to be established what factor(s) are responsible for the under-representation of metabolic enzymes in the spermatozoa of asthenozoospermic individuals and whether this is in any way linked to localized ROS generation within the vicinity of the developing germ cells.

4.3. Defects in the Sperm Motility Apparatus and Signaling Pathways

The flagellum, which is responsible for the propagation of sperm motility, is constructed around a cytoskeletal structure known as the axoneme. The axoneme, in turn, consists of tubulin microfilaments arranged in a highly conserved 9 + 2 scheme (whereby 9 doublets surround one central pair). To each doublet of microtubules are attached two dynein arms (outer and inner), which act as motor proteins capable of 'walking' along the adjacent microtubules and effecting the sliding of microtubules relative to each other (reviewed in [115]). This creates an undulatory wave that is propagated along the length of the flagellum to generate the rhythmic beating patterns responsible for propelling sperm forward [116]. Aside from the core elements described above, the sperm axoneme is also supported by a family of accessory proteins known as outer dense fibers (ODF1–ODF4), which uniquely among mammalian spermatozoa, are localized around the microtubules and provide additional elasticity and stability to flagellum movement [117]. Given the fundamental importance of motility in terms of delivering spermatozoa to the site of fertilization, it follows that any defects in the axonemal structure, or the ODF accessory proteins, can compromise fertility [118]. Accordingly, several studies on asthenozoospermia have linked this condition with an attendant change in the levels of structural proteins residing in the flagellum. Among the most frequently reported of these are the dynein motor proteins, ODFs and those belonging to the tektin family of microtubule-stabilizing proteins [18,109,117].

In a similar context, studies on sperm protein adducts arising from direct 4HNE challenge have demonstrated that this aldehyde can readily bind to tubulin (TUBB), several members of the dynein family (DNAH5, DNAH17, DNALI1), and the ODF1 and ODF2 proteins [22] (see Figure 2). Notably, these findings align well with the proteomic deficits identified in the spermatozoa of asthenozoospermic individuals [18,109,110]. As an extension of this work, however, it has also been shown that core elements of the sperm fibrous sheath (a unique cytoskeletal structure that surrounds the axoneme and outer dense fibers and regulates the flexibility and shape of the flagellar beat [119]) are highly sensitive to 4HNE adduction. In particular, the major fibrous sheath components of A-Kinase Anchoring Protein (AKAP4 and AKAP3) and Rhophilin Associated Tail Protein 1B (ROPN1B), which contribute to the structural organization of the fibrous sheath [119,120], have been identified as dominant 4HNE targets [22,99,101]. Indeed, AKAP4 has recently been validated as a conserved target for 4HNE adduction in primary cultures of post-meiotic male germ cells (round spermatids) and in mature mouse and human spermatozoa [121]. Through the application of an exogenous 4HNE treatment regimen, we further demonstrated that 4HNE modification resulted in a substantial reduction in the levels of AKAP4 detected in round spermatids and mature spermatozoa alike. Moreover, reduced AKAP4 levels were correlated with dysregulation of the capacitation-associated signaling framework assembled around the AKAP4 scaffold.

These data accord with the demonstration that the spermatozoa of male mice lacking AKAP4 display reduced progressive movement and are infertile [122]. Similarly, reduced levels of AKAP4 have also been implicated as a biomarker of human asthenozoospermic males [95]. In addition to AKAP4, alternative cAMP-responsive elements such as protein kinase A (PKA), have also been validated as primary targets of 4HNE-mediated modification [22], whilst seminal OS has been correlated with altered levels of several proteins mapping to the MAPK/ERK pathways [123]. Such defects have been correlated with a commensurate decrease in global levels of tyrosine phosphorylation in human spermatozoa [22]; a form of post-translational modification that underpins sperm motility, and is particularly important in the induction of hyperactivation [124]. Aside from the targets mentioned above, 4HNE also has the potential to modify sperm phosphatase activity, and hence motility, via the targeting of serine/threonine-protein phosphatase 2B catalytic subunit gamma isoform (PPP3CC) [22]. Indeed, male mice lacking PPP3CC display an infertility phenotype linked to a reduction in sperm motility [125]. Thus, whilst metabolic-related proteins are clearly over represented among those targeted for ROS and lipid aldehyde attack [22], numerous other proteins implicated in sperm motility also appear sensitive to OS.

Recent evidence suggests that mammalian spermatozoa may possess an unfolded protein response (UPR); a conserved cellular pathway that is activated upon accumulation of unfolded or misfolded proteins during stress conditions [126]. The activation of the UPR triggers the expression of chaperones to assist protein refolding (reviewed in [127]), whilst also blocking the synthesis of other proteins via the phosphorylation of eukaryotic translation initiation factor-2 (eIF2α) [128]. Santiago et al. [126] have reported increased levels of proteins involved in the UPR (heat shock proteins; HSF1, HSP90, HSPD1, HSP27; and eIF2α) and reduced motility in human spermatozoa exposed to H_2O_2. Despite this, it is possible that prolonged OS within the testes can compromise the UPR owing to 4HNE adduction of key elements of this pathway including the 60 kDa Heat Shock Protein, Mitochondrial (HSPD1) [22]. If this were to be the case, then the inactivation of UPR could contribute to the downregulation of some proteins associated with asthenozoospermia. In any case, such broad-spectrum effects make it extremely challenging to determine whether individual sperm proteins play a dominant role in the pathophysiological responses to OS or whether this phenomenon is instead attributed to dysregulation of multiple targets. This situation is further complicated when considering that 4HNE and other forms of ROS have the potential to damage not only sperm proteins, but also the DNA comprising the paternal and mitochondrial genomes.

4.4. Sperm DNA Modifications

At the genomic level, the fidelity of mitochondrial ATP production is controlled by the interplay of mitochondrial (mtDNA) and nuclear (nDNA) DNA, which encode the various components necessary for the proper assembly and function of the mitochondrial complexes of OXPHOS. Specifically, the mtDNA encodes a subset of the protein subunits of Complex I, Complex IV, cytochrome b and ATP synthase, while the nDNA encodes the remainder of the enzymatic components. It follows that the integrity of the mitochondrial and nuclear genomes within the developing germ cell are crucial for the subsequent establishment of normal sperm motility profiles and conversely, that ROS can elicit a deleterious impact on this aspect of sperm function via the induction of DNA damage (reviewed in [129]).

Sperm DNA is highly vulnerable to ROS damage owing to the progressive silencing of the germ cells transcriptional machinery during the latter phases of their development and the attendant reduction in their capacity to mount an effective DNA repair response. Indeed, the replacement of sperm histones with protamines during spermiogenesis promotes extreme compaction of the nDNA; a phenomenon that protects sperm chromatin against ROS-mediated, and other sources, of damage. However, if the integrity of the protamination process is compromised such that portions of poorly compacted DNA remain, these genomic regions are placed at heightened risk of oxidative attack. Accordingly, recent genome wide analyses have demonstrated that oxidative sperm DNA damage occurs predominantly on specific chromatin regions with lower compaction associated with histones and inter-linker domains attached to the nuclear matrix [130,131]. Thus, ROS such as H_2O_2, $O_2\bullet^-$ or $\bullet OH$ are all capable of directly damaging DNA integrity by way of base modifications, induction of single- and double-strand DNA brakes, chromatin cross-linking and/or deletions [129]. Additionally, 4HNE has also been shown to promote the formation of DNA adducts such as 8-oxoguanine (8-oxoG), 1,N6-ethenoadenosine and 1,N2-ethenoguanosine in human sperm cells [132]. Accordingly, elevated levels of oxidative DNA damage, and in particular 8-oxoG lesions, are frequently encountered in the spermatozoa of male infertility patients [129,132]; emphasizing the importance of evaluating sperm DNA fragmentation in individuals considering assisted reproduction treatments.

In males with asthenozoospermia, the levels of sperm DNA fragmentation have been found to be significantly higher than that of fertile controls [133]. Bonanno et al. [20] have also reported that asthenozoospermic males had elevated ROS levels in their semen, which was correlated with decreased mtDNA integrity in their spermatozoa. By comparison, nDNA fragmentation was only detected in less than one-fifth of the patients analyzed in this study [20]. This phenomenon may reflect less efficient packaging of mtDNA, rendering it more susceptible to ROS and 4HNE attack [134–136]. Alternatively, in spermatozoa the mtDNA is placed in much closer proximity to the main source of ROS generation than that of the nDNA. Aside from changes in mtDNA integrity, it has also been reported that spermatozoa with low-motility possess elevated levels of mtDNA copy number; suggesting that these cells incorporate more mtDNA content during spermatogenesis [20,137–139]. By contrast, measurement of cell-free mtDNA copy number in the seminal plasma of asthenozoospermic and oligozoospermic males has identified a reciprocal relationship, whereby lower levels of cell-free mtDNA are associated with increased levels of ROS [21]. A testicular origin for these lesions is supported by the presence of mtDNA deletions in the spermatozoa of infertile males [140,141]. Further, increases in both mtDNA copy number and mtDNA deletions have been recorded in the same patient samples, wherein they were strongly associated with the presence of OS [142].

One of the best-described large-scale mtDNA deletions is the specific 4977 bp deletion (mtDNA^{4977}), which occurs between nucleotides 8470 and 13447 and eliminates seven genes encoding four subunits of Complex I (ND3, ND4, ND4L, partial ND5), one subunit of Complex IV (COIII) and two subunits of ATP-synthase (ATP6 and partial ATP8); all of which are crucial for OXPHOS (depicted in Figure 3). Accordingly, mtDNA^{4977} has been linked to a spectrum of disorders, including heart disease, different forms of cancer and mitochondrial diseases [143–146]. Notably, this mutation has also been reported to accumulate in different human tissues as a consequence of natural aging [147]. If the mtDNA^{4977} deletion occurs during spermatogenesis and/or spermiogenesis, the mature spermatozoa are endowed with dysfunctional mitochondria, which can act as a source of ROS. Interestingly, mtDNA^{4977} has been recorded in the spermatozoa of infertile males with asthenozoospermia and oligoasthenozoospermia [148,149]. An alternative large-scale mtDNA deletion comprising 7436 bp (mtDNA^{7436}) has also been reported in low-motility human spermatozoa [150]. In addition to the genes excised by mtDNA^{4977}, mtDNA^{7436} also eliminates genes encoding subunit 6 of Complex I (ND6) and cytochrome b from the mtDNA genome (see Figure 3). Although positive correlations between mtDNA deletions and motility have not been universally established [151,152], Kumar et al. [19] have shown that mutations in sperm mtDNA (giving rise to nucleotide changes in subunits of: ATP6, ATP8, ND2, ND3, ND4 and ND5) do exist in males with oligoasthenozoospermia and that these defects are associated with elevated ROS levels in their spermatozoa. In any case, the crucial role of mitochondria in sperm bioenergetics, as well as the potential for defective mitochondria to generate excessive ROS, motivates a better understanding of the factors responsible for perturbation of mtDNA in developing germ cells.

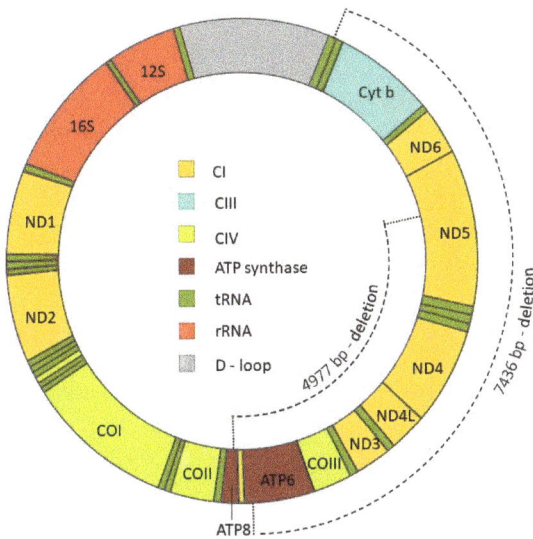

Figure 3. Mitochondrial DNA and location of the mtDNA 4977 bp and 7436 bp deletions. The mtDNA^{4977} deletion includes genes encoding two subunits of ATP-synthase (ATP6 and ATP8), cytochrome oxidase III (COIII), NADH dehydrogenase subunit 3 (ND3), ND4, ND4 subunit L (ND4L), and ND5, whereas mtDNA^{7436} includes ATP6, COIII, ND3, ND4, ND4L, ND5, ND6 and the entire cytochrome b (Cyt b). As a consequence, spermatozoa harboring mtDNA4977 and mtDNA7436 deletions lack several essential OXPHOS genes, fail to assemble functional ETC complexes, and experience compromised energy production. Abbreviations: CI, ETC Complex I; CIII, ETC Complex III; CIV, ETC Complex IV.

5. Antioxidant Systems in the Male Reproductive Tract

The male reproductive tract produces a wide range of antioxidant scavengers capable of defending spermatozoa from ROS attack. Given that mature spermatozoa are translationally inactive and carry with them minimal endogenous antioxidant defenses, they are highly dependent on these exogenous sources of enzymes during spermatogenesis and their residence in the male reproductive tract. In semen, the most abundant antioxidant enzymes are those belonging to the glutathione peroxidase (GPX) and peroxiredoxin (PRDX) families (reviewed in [153]). The GPX4 isoform is predominantly synthesized in the testis and in mature spermatozoa is abundantly localized in the midpiece. Transgenic mice lacking GPX4 display impaired sperm quality, including deficits in sperm motility and structural abnormalities in the midpiece of the flagellum [154]. An alternative GPX isoform, GPX5, is highly expressed in the caput epididymis and has been implicated in the protection of sperm DNA based on the demonstration that the spermatozoa of *Gpx5*-null mice display lower levels of DNA compaction and higher levels of 8-oxoG than their wildtype counterparts [155]. In addition to GPXs, PRDXs are also highly expressed in the caput and cauda epididymis and have been documented within seminal plasma and in virtually all domains of the human spermatozoon [156]. PRDXs are a highly-conserved family of thiol-dependent peroxidases that regulate antioxidant defense systems by virtue of their ability to reduce H_2O_2, peroxynitrite ($ONOO^-$) and hydroperoxides (ROOH); themselves becoming inactivated upon oxidization [156,157]. PRDX1 and PRDX6 have been shown to be abundantly expressed in rat epididymal spermatozoa and to become highly oxidized after the induction of OS with substrates such as tert-butyl hydroperoxide (tert-BHP) [158]. Due to its catalase and calcium-independent phospholipase A2 (Ca^{2+}-iPLA2) activities, PRDX6 has also been implicated in the prevention of LPO and the repair of oxidized membranes [159]. Accordingly, the inhibition of PRDX6 Ca^{2+}-iPLA2 activity in human spermatozoa has recently been shown to promote extensive oxidative damage (including high levels of LPO, DNA oxidation and reduced ΔΨ), which contributes to a phenotype of reduced sperm motility [160]. Aside from GPXs and PRDXs, alternative enzymatic antioxidants such as superoxide dismutase (SOD) and catalase (CAT) collaborate to protect spermatozoa held within male reproductive tract. Such enzymatic defenses can be supplemented by dietary derived non-enzymatic antioxidants, including vitamins (C, E and B), carotenoids, glutathione, inositol, carnitines, cysteines, hyaluronan, serum albumin, and zinc (reviewed in [161]).

Highlighting the physiological importance of these collective antioxidant defenses, patients with poor sperm motility parameters commonly have an attendant deficit in their levels of seminal plasma antioxidants. By way of example, studies of asthenozoospermic individuals with idiopathic or varicocele-related background have shown these lesions are commonly accompanied by relatively low levels of PRDX6 and PRDX1 enzymes within their seminal plasma and spermatozoa, respectively [162]. Similarly, the levels of lactoferrin (LTF; an iron-binding glycoprotein secreted by the mammalian epididymis [163,164] that possesses antioxidant properties and is capable of binding receptors on the sperm head and midpiece [165]) are reduced in spermatozoa from asthenozoospermic males [166]. Whilst others have reported the opposing trend; i.e., higher levels of LTF in spermatozoa from males with asthenozoospermia [18], such discrepancies may reflect the alternative defects that give rise to this pathology and/or the integrity of sperm membrane domains wherein LTF receptors are localized. Building on this body of evidence, the levels of vitamin C, vitamin E and the reduced form of glutathione (GSH; an endogenous antioxidant that can function synergistically with vitamin C [167]), have each been found to be diminished in the semen of asthenozoospermic males compared to normozoospermic controls [168,169]. In contrast to this general trend, data from meta-analyses have failed to document significant changes in the levels of alternative seminal plasma antioxidants such as SOD between males with different forms of infertility (including oligoasthenozoospermia) and that of healthy controls [170], suggesting that not all antioxidants are of equivalent importance in terms of maintaining sperm function. One possible explanation is that the attenuation of antioxidant capacity in infertile individuals may be due, at least in part, to redox driven modifications of specific antioxidant enzymes. Illustrative of this, PRDXs, which serve as primary antioxidants in ejaculated human spermatozoa, are especially vulnerable to ROS-induced modifications such as S-nitrosylation, tyrosine nitration, and S-glutathionylation (reviewed in [171]). Oxidative damage to specific antioxidants may therefore be among the first steps in the cascade of events that contribute to OS-mediated male infertility.

Taking into account the potential of antioxidants to ameliorate OS related pathologies, it is perhaps not surprising that different cocktails of enzymatic and non-enzymatic antioxidants have found therapeutic application during assisted reproductive interventions [172]. Among the most popular of these are vitamin C, zinc and L-carnitine [173–175], although it must be acknowledged the outcomes of these trials are far from consistent. In this context, zinc supplementation has been reported to improve sperm motility by way of reducing OS, apoptosis and DNA fragmentation; but notably, these results were only achieved in the presence of vitamins C and E [174]. Similarly, in a study by Garolla et al. [175], L-carnitine was shown to improve sperm motility but only in samples wherein normal levels of GPX were maintained. This caveat could explain the opposing data obtained by Sigman et al. [176], who reported that L-carnitine had no effect on sperm motility. In view of these dichotomous results, there is a clear imperative to continue basic research to advance our understanding of the interplay of ROS and sperm biology in order to inform the development of effective therapies to combat OS-mediated lesions.

6. Conclusions

Overall, this review highlights mechanisms contributing to an OS-induced decline in sperm motility associated with conditions such as asthenozoospermia. A clear consensus to emerge from the reviewed literature is that the presence of OS in the male reproductive tract is strongly and positively correlated with reduced sperm motility. This state of OS can be evoked not only by intrinsic factors but also by a diversity of environmental agents commonly encountered during modern life. Thus, the challenges presented to the male reproductive tract in terms of mounting an effective and prolonged defense against ROS can result in an attenuation of antioxidant capacity and a concomitant acceleration of OS-induced sperm damage. One particular pathway that appears to contribute to much of the OS response is that of LPO, which is responsible for the generation of highly reactive aldehyde species. These electrophiles are able to adversely impact sperm motility via the adduction and dysregulation of proteins involved in sperm bioenergetic pathways as well as the structural and signaling components of the motility apparatus. Given that these lesions go hand in hand with oxidative DNA damage, it is possible that they may serve a physiological role in terms of reducing the likelihood of such sperm from participating in fertilization and thus transmitting an altered paternal genome to the next generation. This possibility emphasizes the need for the development of novel therapeutic interventions to address the burden of OS-mediated dysfunction in the male germline.

Author Contributions: Conceptualization, K.N.-B. and B.N.; writing—original draft preparation, K.N.-B.; writing—review and editing, B.N. All authors have read and agreed to the published version of the manuscript.

Funding: Support was provided by a National Health and Medical Research Council of Australia (NHMRC) Project Grant (APP1163319) awarded to BN. BN is the recipient of a NHMRC Senior Research Fellowship.

Conflicts of Interest: The authors declare no conflict of interest.

References

1. Brugh, V.M.; Lipshultz, L.I. Male factor infertility: Evaluation and management. *Clin. North Am.* **2004**, *88*, 367–385. [CrossRef]
2. Lotti, F.; Maggi, M. Ultrasound of the male genital tract in relation to male reproductive health. *Hum. Reprod. Update* **2015**, *21*, 56–83. [CrossRef]
3. Vergallo, A.; Giampietri, L.; Baldacci, F.; Volpi, L.; Chico, L.; Pagni, C.; Giorgi, F.S.; Ceravolo, R.; Tognoni, G.; Siciliano, G.; et al. Oxidative Stress Assessment in Alzheimer's Disease: A Clinic Setting Study. *Am. J. Alzheimers Dis. Other Demen.* **2018**, *33*, 35–41. [CrossRef] [PubMed]
4. Ma-On, C.; Sanpavat, A.; Whongsiri, P.; Suwannasin, S.; Hirankarn, N.; Tangkijvanich, P.; Boonla, C. Oxidative stress indicated by elevated expression of Nrf2 and 8-OHdG promotes hepatocellular carcinoma progression. *Med. Oncol.* **2017**, *34*, 57. [CrossRef] [PubMed]
5. Li, B.; Chi, R.F.; Qin, F.Z.; Guo, X.F. Distinct changes of myocyte autophagy during myocardial hypertrophy and heart failure: Association with oxidative stress. *Exp. Physiol.* **2016**, *101*, 1050–1063. [CrossRef] [PubMed]
6. Dursun, E.; Akalin, F.A.; Genc, T.; Cinar, N.; Erel, O.; Yildiz, B.O. Oxidative Stress and Periodontal Disease in Obesity. *Medicine (Baltimore)* **2016**, *95*, e3136. [CrossRef]
7. Lanzafame, F.M.; La Vignera, S.; Vicari, E.; Calogero, A.E. Oxidative stress and medical antioxidant treatment in male infertility. *Reprod. Biomed. Online* **2009**, *19*, 638–659. [CrossRef]
8. Agarwal, A.; Prabakaran, S.; Allamaneni, S.S. Relationship between oxidative stress, varicocele and infertility: A meta-analysis. *Reprod. Biomed. Online* **2006**, *12*, 630–633. [CrossRef]
9. Fraczek, M.; Sanocka, D.; Kamieniczna, M.; Kurpisz, M. Proinflammatory cytokines as an intermediate factor enhancing lipid sperm membrane peroxidation in in vitro conditions. *J. Androl.* **2008**, *29*, 85–92. [CrossRef]
10. Kurfurstova, D.; Bartkova, J.; Vrtel, R.; Mickova, A.; Burdova, A.; Majera, D.; Mistrik, M.; Kral, M.; Santer, F.R.; Bouchal, J.; et al. DNA damage signalling barrier, oxidative stress and treatment-relevant DNA repair factor alterations during progression of human prostate cancer. *Mol. Oncol.* **2016**, *10*, 879–894. [CrossRef]
11. Walters, J.L.H.; De Iuliis, G.N.; Nixon, B.; Bromfield, E.G. Oxidative Stress in the Male Germline: A Review of Novel Strategies to Reduce 4-Hydroxynonenal Production. *Antioxidants* **2018**, *7*, 132. [CrossRef] [PubMed]
12. Agarwal, A.; Mulgund, A.; Sharma, R.; Sabanegh, E. Mechanisms of oligozoospermia: An oxidative stress perspective. *Syst. Biol. Reprod. Med.* **2014**, *60*, 206–216. [CrossRef] [PubMed]
13. Muratori, M.; Tamburrino, L.; Marchiani, S.; Cambi, M.; Olivito, B.; Azzari, C.; Forti, G.; Baldi, E. Investigation on the Origin of Sperm DNA Fragmentation: Role of Apoptosis, Immaturity and Oxidative Stress. *Mol. Med.* **2015**, *21*, 109–122. [CrossRef] [PubMed]
14. Koppers, A.J.; Mitchell, L.A.; Wang, P.; Lin, M.; Aitken, R.J. Phosphoinositide 3-kinase signalling pathway involvement in a truncated apoptotic cascade associated with motility loss and oxidative DNA damage in human spermatozoa. *Biochem J.* **2011**, *436*, 687–698. [CrossRef]

15. Aitken, R.J.; Smith, T.B.; Jobling, M.S.; Baker, M.A.; De Iuliis, G.N. Oxidative stress and male reproductive health. *Asian J. Androl.* **2014**, *16*, 31–38. [CrossRef]
16. World Health Organization, Department of Reproductive Health and Research. *WHO Laboratory Manual for the Examination and Processing of Human Semen*, 5th ed.; World Health Organization: Geneva, Switzerland, 2010; Volume 1, p. 224.
17. Liu, F.J.; Liu, X.; Han, J.L.; Wang, Y.W.; Jin, S.H.; Liu, X.X.; Liu, J.; Wang, W.T.; Wang, W.J. Aged men share the sperm protein PATE1 defect with young asthenozoospermia patients. *Hum. Reprod.* **2015**, *30*, 861–869. [CrossRef]
18. Nowicka-Bauer, K.; Lepczynski, A.; Ozgo, M.; Kamieniczna, M.; Fraczek, M.; Stanski, L.; Olszewska, M.; Malcher, A.; Skrzypczak, W.; Kurpisz, M.K. Sperm mitochondrial dysfunction and oxidative stress as possible reasons for isolated asthenozoospermia. *J. Physiol. Pharmacol.* **2018**, *69*, 403–417. [CrossRef]
19. Kumar, R.; Venkatesh, S.; Kumar, M.; Tanwar, M.; Shasmsi, M.B.; Kumar, R.; Gupta, N.P.; Sharma, R.K.; Talwar, P.; Dada, R. Oxidative stress and sperm mitochondrial DNA mutation in idiopathic oligoasthenozoospermic men. *Indian J. Biochem. Biophys.* **2009**, *46*, 172–177.
20. Bonanno, O.; Romeo, G.; Asero, P.; Pezzino, F.M.; Castiglione, R.; Burrello, N.; Sidoti, G.; Frajese, G.V.; Vicari, E.; D'Agata, R. Sperm of patients with severe asthenozoospermia show biochemical, molecular and genomic alterations. *Reproduction* **2016**, *152*, 695–704. [CrossRef]
21. Chen, Y.; Liao, T.; Zhu, L.; Lin, X.; Wu, R.; Jin, L. Seminal plasma cell-free mitochondrial DNA copy number is associated with human semen quality. *Eur. J. Obstet. Gynecol. Reprod. Biol.* **2018**, *231*, 164–168. [CrossRef]
22. Baker, M.A.; Weinberg, A.; Hetherington, L.; Villaverde, A.I.; Velkov, T.; Baell, J.; Gordon, C.P. Defining the mechanisms by which the reactive oxygen species by-product, 4-hydroxynonenal, affects human sperm cell function. *Biol. Reprod.* **2015**, *92*, 108. [CrossRef] [PubMed]
23. Gavella, M.; Lipovac, V. NADH-dependent oxidoreductase (diaphorase) activity and isozyme pattern of sperm in infertile men. *Arch. Androl.* **1992**, *28*, 135–141. [CrossRef] [PubMed]
24. Jastroch, M.; Divakaruni, A.S.; Mookerjee, S.; Treberg, J.R.; Brand, M.D. Mitochondrial proton and electron leaks. *Essays Biochem.* **2010**, *47*, 53–67. [CrossRef] [PubMed]
25. O'Flaherty, C.M.; Beorlegui, N.B.; Beconi, M.T. Reactive oxygen species requirements for bovine sperm capacitation and acrosome reaction. *Theriogenology* **1999**, *52*, 289–301. [CrossRef]
26. Koppenol, W.H. The Haber-Weiss cycle—70 years later. *Redox Rep.* **2001**, *6*, 229–234. [CrossRef]
27. Buzadzic, B.; Vucetic, M.; Jankovic, A.; Stancic, A.; Korac, A.; Korac, B.; Otasevic, V. New insights into male (in)fertility: The importance of NO. *Br. J. Pharmacol.* **2015**, *172*, 1455–1467. [CrossRef]
28. Otasevic, V.; Stancic, A.; Korac, A.; Jankovic, A.; Korac, B. Reactive oxygen, nitrogen, and sulfur species in human male fertility. A crossroad of cellular signaling and pathology. *Biofactors* **2019**. [CrossRef]
29. Chen, S.J.; Allam, J.P.; Duan, Y.G.; Haidl, G. Influence of reactive oxygen species on human sperm functions and fertilizing capacity including therapeutical approaches. *Arch. Gynecol. Obstet.* **2013**, *288*, 191–199. [CrossRef]
30. Alomar, M.; Alzoabi, M.; Zarkawi, M. Kinetics of hydrogen peroxide generated from live and dead ram spermatozoa and the effects of catalase and oxidase substrates addition. *Czech. J. Anim. Sci.* **2016**, *61*, 1–7. [CrossRef]
31. Gao, W.; Pu, Y.; Luo, K.Q.; Chang, D.C. Temporal relationship between cytochrome c release and mitochondrial swelling during UV-induced apoptosis in living HeLa cells. *J. Cell Sci.* **2001**, *114*, 2855–2862.
32. Patil, P.S.; Humbarwadi, R.S.; Patil, A.D.; Gune, A.R. Immature germ cells in semen-correlation with total sperm count and sperm motility. *J. Cytol.* **2013**, *30*, 185–189. [CrossRef]
33. Hermo, L.; Pelletier, R.M.; Cyr, D.G.; Smith, C.E. Surfing the wave, cycle, life history, and genes/proteins expressed by testicular germ cells. Part 1: Background to spermatogenesis, spermatogonia, and spermatocytes. *Microsc. Res. Tech.* **2010**, *73*, 241–278. [CrossRef]
34. Gomez, E.; Buckingham, D.W.; Brindle, J.; Lanzafame, F.; Irvine, D.S.; Aitken, R.J. Development of an image analysis system to monitor the retention of residual cytoplasm by human spermatozoa: Correlation with biochemical markers of the cytoplasmic space, oxidative stress, and sperm function. *J. Androl.* **1996**, *17*, 276–287.
35. Agarwal, A.; Rana, M.; Qiu, E.; AlBunni, H.; Bui, A.D.; Henkel, R. Role of oxidative stress, infection and inflammation in male infertility. *Andrologia* **2018**, *50*, e13126. [CrossRef]

36. Plante, M.; de Lamirande, E.; Gagnon, C. Reactive oxygen species released by activated neutrophils, but not by deficient spermatozoa, are sufficient to affect normal sperm motility. *Fertil. Steril.* **1994**, *62*, 387–393. [CrossRef]
37. Wang, A.; Fanning, L.; Anderson, D.J.; Loughlin, K.R. Generation of reactive oxygen species by leukocytes and sperm following exposure to urogenital tract infection. *Arch. Androl.* **1997**, *39*, 11–17. [CrossRef]
38. Tamura, M.; Tamura, T.; Tyagi, S.R.; Lambeth, J.D. The superoxide-generating respiratory burst oxidase of human neutrophil plasma membrane. Phosphatidylserine as an effector of the activated enzyme. *J. Biol. Chem.* **1988**, *263*, 17621–17626. [PubMed]
39. Bostwick, D.G.; Ma, J. Spermatic Cord and Testicular Adnexa. In *Urologic Surgical Pathology*, 4th ed.; Elsevier—Health Sciences Division: Philadelphia, PA, USA, 2020; pp. 834–852.
40. Practice Committee of the American Society for Reproductive Medicine; Society for Male Reproduction and Urology. Report on varicocele and infertility: A committee opinion. *Fertil. Steril.* **2014**, *102*, 1556–1560. [CrossRef]
41. Hendin, B.N.; Kolettis, P.N.; Sharma, R.K.; Thomas, A.J., Jr.; Agarwal, A. Varicocele is associated with elevated spermatozoal reactive oxygen species production and diminished seminal plasma antioxidant capacity. *J. Urol.* **1999**, *161*, 1831–1834. [CrossRef]
42. Sakamoto, Y.; Ishikawa, T.; Kondo, Y.; Yamaguchi, K.; Fujisawa, M. The assessment of oxidative stress in infertile patients with varicocele. *BJU Int.* **2008**, *101*, 1547–1552. [CrossRef] [PubMed]
43. Agarwal, A.; Hamada, A.; Esteves, S.C. Insight into oxidative stress in varicocele-associated male infertility: Part 1. *Nat. Rev. Urol.* **2012**, *9*, 678–690. [CrossRef] [PubMed]
44. Mostafa, T.; Rashed, L.; Taymour, M. Seminal cyclooxygenase relationship with oxidative stress in infertile oligoasthenoteratozoospermic men with varicocele. *Andrologia* **2016**, *48*, 137–142. [CrossRef] [PubMed]
45. Gul, M.; Bugday, M.S.; Erel, O. Thiol-disulphide homoeostasis as an oxidative stress marker in men with varicocele. *Andrologia* **2018**. [CrossRef] [PubMed]
46. Erfani Majd, N.; Sadeghi, N.; Tavalaee, M.; Tabandeh, M.R.; Nasr-Esfahani, M.H. Evaluation of Oxidative Stress in Testis and Sperm of Rat Following Induced Varicocele. *Urol. J.* **2019**, *16*, 300–306. [CrossRef]
47. Cho, C.L.; Esteves, S.C.; Agarwal, A. Novel insights into the pathophysiology of varicocele and its association with reactive oxygen species and sperm DNA fragmentation. *Asian J. Androl.* **2016**, *18*, 186–193. [CrossRef]
48. Bedford, J.M. Human spermatozoa and temperature: The elephant in the room. *Biol. Reprod.* **2015**, *93*, 97. [CrossRef]
49. Ikeda, M.; Kodama, H.; Fukuda, J.; Shimizu, Y.; Murata, M.; Kumagai, J.; Tanaka, T. Role of radical oxygen species in rat testicular germ cell apoptosis induced by heat stress. *Biol. Reprod.* **1999**, *61*, 393–399. [CrossRef]
50. Paul, C.; Murray, A.A.; Spears, N.; Saunders, P.T. A single, mild, transient scrotal heat stress causes DNA damage, subfertility and impairs formation of blastocysts in mice. *Reproduction* **2008**, *136*, 73–84. [CrossRef]
51. Saikhun, J.; Kitiyanant, Y.; Vanadurongwan, V.; Pavasuthipaisit, K. Effects of sauna on sperm movement characteristics of normal men measured by computer-assisted sperm analysis. *Int. J. Androl.* **1998**, *21*, 358–363. [CrossRef]
52. Shefi, S.; Tarapore, P.E.; Walsh, T.J.; Croughan, M.; Turek, P.J. Wet heat exposure: A potentially reversible cause of low semen quality in infertile men. *Int. Braz. J. Urol.* **2007**, *33*, 50–56. [CrossRef]
53. Garolla, A.; Torino, M.; Sartini, B.; Cosci, I.; Patassini, C.; Carraro, U.; Foresta, C. Seminal and molecular evidence that sauna exposure affects human spermatogenesis. *Hum. Reprod.* **2013**, *28*, 877–885. [CrossRef]
54. Houston, B.J.; Nixon, B.; Martin, J.H.; De Iuliis, G.N.; Trigg, N.A.; Bromfield, E.G.; McEwan, K.E.; Aitken, R.J. Heat exposure induces oxidative stress and DNA damage in the male germ line. *Biol. Reprod.* **2018**, *98*, 593–606. [CrossRef]
55. Houston, B.J.; Nixon, B.; King, B.V.; De Iuliis, G.N.; Aitken, R.J. The effects of radiofrequency electromagnetic radiation on sperm function. *Reproduction* **2016**, *152*, R263–R276. [CrossRef]
56. Agarwal, A.; Singh, A.; Hamada, A.; Kesari, K. Cell phones and male infertility: A review of recent innovations in technology and consequences. *Int. Braz. J. Urol.* **2011**, *37*, 432–454. [CrossRef] [PubMed]
57. Wdowiak, A.; Mazurek, P.A.; Wdowiak, A.; Bojar, I. Effect of electromagnetic waves on human reproduction. *Ann. Agric. Environ. Med.* **2017**, *24*, 13–18. [CrossRef]
58. Koppers, A.J.; De Iuliis, G.N.; Finnie, J.M.; McLaughlin, E.A.; Aitken, R.J. Significance of mitochondrial reactive oxygen species in the generation of oxidative stress in spermatozoa. *J. Clin. Endocrinol. Metab.* **2008**, *93*, 3199–3207. [CrossRef] [PubMed]

59. De Iuliis, G.N.; Newey, R.J.; King, B.V.; Aitken, R.J. Mobile phone radiation induces reactive oxygen species production and DNA damage in human spermatozoa in vitro. *PLoS ONE* **2009**, *4*, e6446. [CrossRef] [PubMed]
60. Feng, B.; Qiu, L.; Ye, C.; Chen, L.; Fu, Y.; Sun, W. Exposure to a 50-Hz magnetic field induced mitochondrial permeability transition through the ROS/GSK-3β signaling pathway. *Int. J. Radiat. Biol.* **2016**, *92*, 148–155. [CrossRef] [PubMed]
61. Wijesekara, G.U.; Fernando, D.M.; Wijerathna, S.; Bandara, N. Environmental and occupational exposures as a cause of male infertility. *Ceylon Med. J.* **2015**, *60*, 52–56. [CrossRef] [PubMed]
62. Saad, A.B.; Rjeibi, I.; Alimi, H.; Ncib, S.; Smida, A.; Zouari, N.; Zourgui, L. Lithium induced, oxidative stress and related damages in testes and heart in male rats: The protective effects of Malva sylvestris extract. *Biomed. Pharmacother.* **2017**, *86*, 127–135. [CrossRef]
63. Kaur, S.; Saluja, M.; Bansal, M.P. Bisphenol A induced oxidative stress and apoptosis in mice testes: Modulation by selenium. *Andrologia* **2018**, *50*. [CrossRef] [PubMed]
64. Du, J.; Xiong, D.; Zhang, Q.; Li, X.; Liu, X.; You, H.; Ding, S.; Yang, X.; Yuan, J. Mono-butyl phthalate-induced mouse testis injury is associated with oxidative stress and down-regulated expression of Sox9 and *Dazl*. *J. Toxicol. Sci.* **2017**, *42*, 319–328. [CrossRef] [PubMed]
65. Wang, Y.X.; Zeng, Q.; Sun, Y.; You, L.; Wang, P.; Li, M.; Yang, P.; Li, J.; Huang, Z.; Wang, C.; et al. Phthalate exposure in association with serum hormone levels, sperm DNA damage and spermatozoa apoptosis: A cross-sectional study in China. *Environ. Res.* **2016**, *150*, 557–565. [CrossRef] [PubMed]
66. Saleh, R.A.; Agarwal, A.; Sharma, R.K.; Nelson, D.R.; Thomas, A.J., Jr. Effect of cigarette smoking on levels of seminal oxidative stress in infertile men: A prospective study. *Fertil. Steril.* **2002**, *78*, 491–499. [CrossRef]
67. Hamad, M.F.; Shelko, N.; Kartarius, S.; Montenarh, M.; Hammadeh, M.E. Impact of cigarette smoking on histone (H2B) to protamine ratio in human spermatozoa and its relation to sperm parameters. *Andrology* **2014**, *2*, 666–677. [CrossRef]
68. Oh, S.I.; Lee, M.S.; Kim, C.I.; Song, K.Y.; Park, S.C. Aspartate modulates the ethanol-induced oxidative stress and glutathione utilizing enzymes in rat testes. *Exp. Mol. Med.* **2002**, *34*, 47–52. [CrossRef]
69. Siervo, G.E.; Vieira, H.R.; Ogo, F.M.; Fernandez, C.D.; Gonçalves, G.D.; Mesquita, S.F.; Anselmo-Franci, J.A.; Cecchini, R.; Guarnier, F.A.; Fernandes, G.S. Spermatic and testicular damages in rats exposed to ethanol: Influence of lipid peroxidation but not testosterone. *Toxicology* **2015**, *330*, 1–8. [CrossRef]
70. Aitken, R.J.; Baker, M.A.; De Iuliis, G.N.; Nixon, B. New Insights into Sperm Physiology and Pathology. In *Fertility Control. Handbook of Experimental Pharmacology*; Habenicht, U.F., Aitken, R., Eds.; Springer: Berlin/Heidelberg, Germany, 2010; pp. 99–115.
71. Lenzi, A.; Gandini, L.; Picardo, M.; Tramer, F.; Sandri, G.; Panfili, E. Lipoperoxidation damage of spermatozoa polyunsaturated fatty acids (PUFA): Scavenger mechanisms and possible scavenger therapies. *Front. Biosci.* **2000**, *5*, E1–E15.
72. Hauck, A.K.; Bernlohr, D.A. Oxidative stress and lipotoxicity. *J. Lipid Res.* **2016**, *57*, 1976–1986. [CrossRef]
73. Bromfield, E.G.; Aitken, R.J.; McLaughlin, E.A.; Nixon, B. Proteolytic degradation of heat shock protein A2 occurs in response to oxidative stress in male germ cells of the mouse. *Mol. Hum. Reprod.* **2017**, *23*, 91–105. [CrossRef]
74. Badouard, C.; Ménézo, Y.; Panteix, G.; Ravanat, J.L.; Douki, T.; Cadet, J.; Favier, A. Determination of new types of DNA lesions in human sperm. *Zygote* **2008**, *16*, 9–13. [CrossRef] [PubMed]
75. Dalleau, S.; Baradat, M.; Guéraud, F.; Huc, L. Cell death and diseases related to oxidative stress: 4-hydroxynonenal (HNE) in the balance. *Cell Death Differ.* **2013**, *20*, 1615–1630. [CrossRef] [PubMed]
76. LoPachin, R.M.; Gavin, T.; Petersen, D.R.; Barber, D.S. Molecular mechanisms of 4-hydroxy-2-nonenal and acrolein toxicity: Nucleophilic targets and adduct formation. *Chem. Res. Toxicol.* **2009**, *22*, 1499–1508. [CrossRef] [PubMed]
77. Schaur, R.J.; Siems, W.; Bresgen, N.; Eckl, P.M. 4-Hydroxy-nonenal-A Bioactive Lipid Peroxidation Product. *Biomolecules* **2015**, *5*, 2247–2337. [CrossRef]
78. Doorn, J.A.; Petersen, D.R. Covalent adduction of nucleophilic amino acids by 4-hydroxynonenal and 4-oxononenal. *Chem. Biol. Interact.* **2003**, *143*, 93–100. [CrossRef]
79. Bromfield, E.G.; Aitken, R.J.; Anderson, A.L.; McLaughlin, E.A.; Nixon, B. The impact of oxidative stress on chaperone-mediated human sperm-egg interaction. *Hum. Reprod.* **2015**, *30*, 2597–2613. [CrossRef]
80. Morielli, T.; O'Flaherty, C. Oxidative stress impairs function and increases redox protein modifications in human spermatozoa. *Reproduction* **2015**, *149*, 113–123. [CrossRef]

81. Salvolini, E.; Buldreghini, E.; Lucarini, G.; Vignini, A.; Di Primio, R.; Balercia, G. Nitric oxide synthase and tyrosine nitration in idiopathic asthenozoospermia: An immunohistochemical study. *Fertil. Steril.* **2012**, *97*, 554–560. [CrossRef]
82. Vignini, A.; Nanetti, L.; Buldreghini, E.; Moroni, C.; Ricciardo-Lamonica, G.; Mantero, F.; Boscaro, M.; Mazzanti, L.; Balercia, G. The production of peroxynitrite by human spermatozoa may affect sperm motility through the formation of protein nitrotyrosine. *Fertil. Steril.* **2006**, *85*, 947–953. [CrossRef]
83. Ramirez, J.P.; Carreras, A.; Mendoza, C. Sperm plasma membrane integrity in fertile and infertile men. *Andrologia* **1992**, *24*, 141–144. [CrossRef]
84. Karimfar, M.H.; Niazvand, F.; Haghani, K.; Ghafourian, S.; Shirazi, R.; Bakhtiyari, S. The protective effects of melatonin against cryopreservation-induced oxidative stress in human sperm. *Int. J. Immunopathol. Pharmacol.* **2015**, *28*, 69–76. [CrossRef] [PubMed]
85. Ghorbani, M.; Vatannejad, A.; Khodadadi, I.; Amiri, I.; Tavilani, H. Protective effects of glutathione supplementation against oxidative stress during cryopreservation of human spermatozoa. *Cryo. Letters* **2016**, *37*, 34–40. [PubMed]
86. Najafi, A.; Daghigh Kia, H.; Mehdipour, M.; Shamsollahi, M.; Miller, D.J. Does fennel extract ameliorate oxidative stress frozen-thawed ram sperm? *Cryobiology* **2019**, *87*, 47–51. [CrossRef] [PubMed]
87. Varela, E.; Rojas, M.; Restrepo, G. Membrane stability and mitochondrial activity of bovine sperm frozen with low-density lipoproteins and trehalose. *Reprod. Domest. Anim.* **2019**. [CrossRef] [PubMed]
88. Dashtestani, F.; Ghourchian, H.; Najafi, A. Silver-gold-apoferritin nanozyme for suppressing oxidative stress during cryopreservation. *Mater. Sci. Eng. C. Mater. Biol. Appl.* **2019**, *94*, 831–840. [CrossRef]
89. Mishra, A.K.; Kumar, A.; Swain, D.K.; Yadav, S.; Nigam, R. Insights into pH regulatory mechanisms in mediating spermatozoa functions. *Vet. World* **2018**, *11*, 852–858. [CrossRef] [PubMed]
90. Elinder, F.; Liin, S.I. Actions and Mechanisms of Polyunsaturated Fatty Acids on Voltage-Gated Ion Channels. *Front. Physiol.* **2017**, *8*, 43. [CrossRef]
91. Hosseinzadeh Colagar, A.; Karimi, F.; Jorsaraei, S.G. Correlation of sperm parameters with semen lipid peroxidation and total antioxidants levels in astheno- and oligoasheno- teratospermic men. *Iran. Red. Crescent. Med. J.* **2013**, *15*, 780–785. [CrossRef]
92. Tamburrino, L.; Marchiani, S.; Vicini, E.; Muciaccia, B.; Cambi, M.; Pellegrini, S.; Forti, G.; Muratori, M.; Baldi, E. Quantification of CatSper1 expression in human spermatozoa and relation to functional parameters. *Hum. Reprod.* **2015**, *30*, 1532–1544. [CrossRef]
93. Liu, S.W.; Li, Y.; Zou, L.L.; Guan, Y.T.; Peng, S.; Zheng, L.X.; Deng, S.M.; Zhu, L.Y.; Wang, L.W.; Chen, L.X. Chloride channels are involved in sperm motility and are downregulated in spermatozoa from patients with asthenozoospermia. *Asian J. Androl.* **2017**, *19*, 418–424. [CrossRef]
94. Sinha, A.; Singh, V.; Singh, S.; Yadav, S. Proteomic analyses reveal lower expression of TEX40 and ATP6V0A2 proteins related to calcium ion entry and acrosomal acidification in asthenozoospermic males. *Life Sci.* **2019**, *218*, 81–88. [CrossRef]
95. Hashemitabar, M.; Sabbagh, S.; Orazizadeh, M.; Ghadiri, A.; Bahmanzadeh, M. A proteomic analysis on human sperm tail: Comparison between normozoospermia and asthenozoospermia. *J. Assist. Reprod. Genet.* **2015**, *32*, 853–863. [CrossRef]
96. Agarwal, A.; Sharma, R.; Durairajanayagam, D.; Ayaz, A.; Cui, Z.; Willard, B.; Gopalan, B.; Sabanegh, E. Major protein alterations in spermatozoa from infertile men with unilateral varicocele. *Reprod. Biol. Endocrinol.* **2015**, *13*, 8. [CrossRef] [PubMed]
97. Ursini, F.; Heim, S.; Kiess, M.; Maiorino, M.; Roveri, A.; Wissing, J.; Flohe, L. Dual function of the selenoprotein PHGPx during sperm maturation. *Science* **1999**, *285*, 1393–1396. [CrossRef] [PubMed]
98. Bahr, G.F.; Engler, W.F. Considerations of volume, mass, DNA, and arrangement of mitochondria in the midpiece of bull spermatozoa. *Exp. Cell Res.* **1970**, *60*, 338–340. [CrossRef]
99. Parker, N.; Vidal-Puig, A.; Brand, M.D. Stimulation of mitochondrial proton conductance by hydroxynonenal requires a high membrane potential. *Biosci. Rep.* **2008**, *28*, 83–88. [CrossRef]
100. Hanukoglu, I.; Rapoport, R.; Weiner, L.; Sklan, D. Electron leakage from the mitochondrial NADPH–adrenodoxin reductase–adrenodoxin–P450scc (cholesterol side chain cleavage) system. *Arch. Biochem. Biophys.* **1993**, *305*, 489–498. [CrossRef]
101. Amaral, A.; Ramalho-Santos, J. Assessment of mitochondrial potential: Implications for the correct monitoring of human sperm function. *Int. J. Androl.* **2010**, *33*, 180–186. [CrossRef]

102. Zhang, W.D.; Zhang, Z.; Jia, L.T.; Zhang, L.L.; Fu, T.; Li, Y.S.; Wang, P.; Sun, L.; Shi, Y.; Zhang, H.Z. Oxygen free radicals and mitochondrial signaling in oligospermia and asthenospermia. *Mol. Med. Rep.* **2014**, *10*, 1875–1880. [CrossRef]
103. Amaral, A.; Castillo, J.; Estanyol, J.M.; Ballesca, J.L.; Ramalho-Santos, J.; Oliva, R. Human sperm tail proteome suggests new endogenous metabolic pathways. *Mol. Cell Proteom.* **2013**, *12*, 330–342. [CrossRef]
104. Amaral, A.; Castillo, J.; Ramalho-Santos, J.; Oliva, R. The combined human sperm proteome: Cellular pathways and implications for basic and clinical science. *Hu. Reprod. Update* **2014**, *20*, 40–62. [CrossRef] [PubMed]
105. Moscatelli, N.; Lunetti, P.; Braccia, C.; Armirotti, A.; Pisanello, F.; De Vittorio, M.; Zara, V.; Ferramosca, A. Comparative Proteomic Analysis of Proteins Involved in Bioenergetics Pathways Associated with Human Sperm Motility. *Int. J. Mol. Sci.* **2019**, *20*, 3000. [CrossRef] [PubMed]
106. Aitken, R.J.; Whiting, S.; De Iuliis, G.N.; McClymont, S.; Mitchell, L.A.; Baker, M.A. Electrophilic aldehydes generated by sperm metabolism activate mitochondrial reactive oxygen species generation and apoptosis by targeting succinate dehydrogenase. *J. Biol. Chem.* **2012**, *287*, 33048–33060. [CrossRef] [PubMed]
107. Carbone, D.L.; Doorn, J.A.; Kiebler, Z.; Sampey, B.P.; Petersen, D.R. Inhibition of Hsp72-mediated protein refolding by 4-hydroxy-2-nonenal. *Chem. Res. Toxicol.* **2004**, *17*, 1459–1467. [CrossRef]
108. Carbone, D.L.; Doorn, J.A.; Petersen, D.R. 4-Hydroxynonenal regulates 26S proteasomal degradation of alcohol dehydrogenase. *Free Radic. Biol. Med.* **2004**, *37*, 1430–1439. [CrossRef]
109. Amaral, A.; Paiva, C.; Attardo Parrinello, C.; Estanyol, J.M.; Ballescà, J.L.; Ramalho-Santos, J.; Oliva, R. Identification of proteins involved in human sperm motility using highthroughput differential proteomics. *J. Proteome Res.* **2014**, *13*, 5670–5684. [CrossRef]
110. Guo, Y.; Jiang, W.; Yu, W.; Niu, X.; Liu, F.; Zhou, T.; Zhang, H.; Li, Y.; Zhu, H.; Zhou, Z.; et al. Proteomics analysis of asthenozoospermia and identification of glucose-6-phosphate isomerase as an important enzyme for sperm motility. *J. Proteom.* **2019**, *208*, 103478. [CrossRef]
111. Miki, K.; Qu, W.; Goulding, E.H.; Willis, W.D.; Bunch, D.O.; Strader, L.F.; Perreault, S.D.; Eddy, E.M.; O'Brien, D.A. Glyceraldehyde 3-phosphate dehydrogenase-S, a sperm-specific glycolytic enzyme, is required for sperm motility and male fertility. *Proc. Natl. Acad. Sci. USA* **2004**, *101*, 16501–16506. [CrossRef]
112. Elkina, Y.L.; Atroshchenko, M.M.; Bragina, E.E.; Muronetz, V.I.; Schmalhausen, E.V. Oxidation of glyceraldehyde-3-phosphate dehydrogenase decreases sperm motility. *Biochemistry* **2011**, *76*, 268–272. [CrossRef]
113. Liu, J.; Wang, Y.; Gong, L.; Sun, C. Oxidation of glyceraldehyde-3-phosphate dehydrogenase decreases sperm motility in diabetes mellitus. *Biochem. Biophys. Res. Commun.* **2015**, *465*, 245–248. [CrossRef]
114. Yang, Y.; Cheng, L.; Wang, Y.; Han, Y.; Liu, J.; Deng, X.; Chao, L. Expression of NDUFA13 in asthenozoospermia and possible pathogenesis. *Reprod. Biomed. Online* **2017**, *34*, 66–74. [CrossRef] [PubMed]
115. Roberts, A.J.; Kon, T.; Knight, P.J.; Sutoh, K.; Burgess, S.A. Functions and mechanics of dynein motor proteins. *Nat. Rev. Mol. Cell Biol.* **2013**, *14*, 713–726. [CrossRef] [PubMed]
116. Brokaw, C.J. Flagellar movement: A sliding filament model. *Science* **1972**, *178*, 455–462. [CrossRef] [PubMed]
117. Zhao, W.; Li, Z.; Ping, P.; Wang, G.; Yuan, X.; Sun, F. Outer dense fibers stabilize the axoneme to maintain sperm motility. *J. Cell Mol. Med.* **2018**, *22*, 1755–1768. [CrossRef]
118. Linck, R.W.; Chemes, H.; Albertini, D.F. The axoneme: The propulsive engine of spermatozoa and cilia and associated ciliopathies leading to infertility. *J. Assist. Reprod. Genet.* **2016**, *33*, 141–156. [CrossRef]
119. Eddy, E.M.; Toshimori, K.; O'Brien, D.A. Fibrous sheath of mammalian spermatozoa. *Microsc. Res. Tech.* **2003**, *61*, 103–115. [CrossRef]
120. Brown, P.R.; Miki, K.; Harper, D.B.; Eddy, E.M. A-kinase anchoring protein 4 binding proteins in the fibrous sheath of the sperm flagellum. *Biol. Reprod.* **2003**, *68*, 2241–2248. [CrossRef]
121. Nixon, B.; Bernstein, I.; Cafe, S.L.; Delehedde, M.; Sergeant, N.; Eamens, A.L.; Lord, T.; Dun, M.D.; De Iuliis, G.N.; Bromfield, E.G. A Kinase Anchor Protein 4 is vulnerable to oxidative adduction in male germ cells. *Front. Cell Dev. Biol.* **2019**. In press. [CrossRef]
122. Miki, K.; Willis, W.D.; Brown, P.R.; Goulding, E.H.; Fulcher, K.D.; Eddy, E.M. Targeted disruption of the Akap4 gene causes defects in sperm flagellum and motility. *Dev. Biol.* **2002**, *248*, 331–342. [CrossRef]
123. Ayaz, A.; Agarwal, A.; Sharma, R.; Kothandaraman, N.; Cakar, Z.; Sikka, S. Proteomic analysis of sperm proteins in infertile men with high levels of reactive oxygen species. *Andrologia* **2018**, *50*, e13015. [CrossRef]

124. Nassar, A.; Mahony, M.; Morshedi, M.; Lin, M.H.; Srisombut, C.; Oehninger, S. Modulation of sperm tail protein tyrosine phosphorylation by pentoxifylline and its correlation with hyperactivated motility. *Fertil. Steril.* **1999**, *71*, 919–923. [CrossRef]
125. Miyata, H.; Satouh, Y.; Mashiko, D.; Muto, M.; Nozawa, K.; Shiba, K.; Fujihara, Y.; Isotani, A.; Inaba, K.; Ikawa, M. Sperm calcineurin inhibition prevents mouse fertility with implications for male contraceptive. *Science* **2015**, *350*, 442–445. [CrossRef] [PubMed]
126. Santiago, J.; Silva, J.V.; Fardilha, M. First Insights on the Presence of the Unfolded Protein Response in Human Spermatozoa. *Int. J. Mol. Sci.* **2019**, *20*, 5518. [CrossRef] [PubMed]
127. Jovaisaite, V.; Mouchiroud, L.; Auwerx, J. The mitochondrial unfolded protein response, a conserved stress response pathway with implications in health and disease. *J. Exp. Biol.* **2014**, *217*, 137–143. [CrossRef]
128. Zhang, G.; Ling, X.; Liu, K.; Wang, Z.; Zou, P.; Gao, J.; Cao, J.; Ao, L. The p-eIF2α/ATF4 pathway links endoplasmic reticulum stress to autophagy following the production of reactive oxygen species in mouse spermatocyte-derived cells exposed to dibutyl phthalate. *Free Radic. Res.* **2016**, *50*, 698–707. [CrossRef]
129. Cocuzza, M.; Sikka, S.C.; Athayde, K.S.; Agarwal, A. Clinical relevance of oxidative stress and sperm chromatin damage in male infertility: An evidence based analysis. *Int. Braz. J. Urol.* **2007**, *33*, 603–621. [CrossRef]
130. Xavier, M.J.; Nixon, B.; Roman, S.D.; Scott, R.J.; Drevet, J.R.; Aitken, R.J. Paternal impacts on development: Identification of genomic regions vulnerable to oxidative DNA damage in human spermatozoa. *Hum. Reprod.* **2019**, *34*, 1876–1890. [CrossRef]
131. Xavier, M.J.; Nixon, B.; Roman, S.D.; Aitken, R.J. Improved methods of DNA extraction from human spermatozoa that mitigate experimentally-induced oxidative DNA damage. *PLoS ONE* **2018**, *13*, e0195003. [CrossRef]
132. Kodama, H.; Yamaguchi, R.; Fukuda, J.; Kasai, H.; Tanaka, T. Increased oxidative deoxyribonucleic acid damage in the spermatozoa of infertile male patients. *Fertil. Steril.* **1997**, *68*, 519–524. [CrossRef]
133. Piasecka, M.; Gaczarzewicz, D.; Laszczyńska, M.; Starczewski, A.; Brodowska, A. Flow cytometry application in the assessment of sperm DNA integrity of men with asthenozoospermia. *Folia Histochem. Cytobiol.* **2007**, *45*, S127–S136.
134. Yakes, F.M.; Van Houten, B. Mitochondrial DNA damage is more extensive and persists longer than nuclear DNA damage in human cells following oxidative stress. *Proc. Natl. Acad. Sci. USA* **1997**, *94*, 514–519. [CrossRef] [PubMed]
135. Salazar, J.J.; Van Houten, B. Preferential mitochondrial DNA injury caused by glucose oxidase as a steady generator of hydrogen peroxide in human fibroblasts. *Mutat. Res.* **1997**, *385*, 139–149. [CrossRef]
136. Sawyer, D.E.; Mercer, B.G.; Wiklendt, A.M.; Aitken, R.J. Quantitative analysis of gene-specific DNA damage in human spermatozoa. *Mutat. Res.* **2003**, *529*, 21–34. [CrossRef]
137. Díez-Sánchez, C.; Ruiz-Pesini, E.; Lapeña, A.C.; Montoya, J.; Pérez-Martos, A.; Enríquez, J.A.; López-Pérez, M.J. Mitochondrial DNA content of human spermatozoa. *Biol. Reprod.* **2003**, *68*, 180–185. [CrossRef]
138. Amaral, A.; Ramalho-Santos, J.; St John, J.C. The expression of polymerase gamma and mitochondrial transcription factor A and the regulation of mitochondrial DNA content in mature human sperm. *Hum. Reprod.* **2007**, *22*, 1585–1596. [CrossRef]
139. Song, G.J.; Lewis, V. Mitochondrial DNA integrity and copy number in sperm from infertile men. *Fertil. Steril.* **2008**, *90*, 2238–2244. [CrossRef]
140. Kao, S.; Chao, H.T.; Wei, Y.H. Mitochondrial deoxyribonucleic acid 4977-bp deletion is associated with diminished fertility and motility of human sperm. *Biol. Reprod.* **1995**, *52*, 729–736. [CrossRef]
141. Gashti, N.G.; Salehi, Z.; Madani, A.H.; Dalivandan, S.T. 4977-bp mitochondrial DNA deletion in infertile patients with varicocele. *Andrologia* **2014**, *46*, 258–262. [CrossRef]
142. Lin, P.H.; Lee, S.H.; Su, C.P.; Wei, Y.H. Oxidative damage to mitochondrial DNA in atrial muscle of patients with atrial fibrillation. *Free Radic. Biol. Med.* **2003**, *35*, 1310–1318. [CrossRef]
143. Vecoli, C.; Borghini, A.; Pulignani, S.; Mercuri, A.; Turchi, S.; Carpeggiani, C.; Picano, E.; Andreassi, M.G. Prognostic value of mitochondrial DNA4977 deletion and mitochondrial DNA copy number in patients with stable coronary artery disease. *Atherosclerosis* **2018**, *276*, 91–97. [CrossRef]

144. Dimberg, J.; Hong, T.T.; Nguyen, L.T.T.; Skarstedt, M.; Löfgren, S.; Matussek, A. Common 4977 bp deletion and novel alterations in mitochondrial DNA in Vietnamese patients with breast cancer. *Springerplus* **2015**, *4*, 58. [CrossRef]
145. Guo, Z.S.; Jin, C.L.; Yao, Z.J.; Wang, Y.M.; Xu, B.T. Analysis of the Mitochondrial 4977 Bp Deletion in Patients with Hepatocellular Carcinoma. *Balkan J. Med. Genet.* **2017**, *20*, 81–86. [CrossRef]
146. Zhang, Y.; Ma, Y.; Bu, D.; Liu, H.; Xia, C.; Zhang, Y.; Zhu, S.; Pan, H.; Pei, P.; Zheng, X.; et al. Deletion of a 4977-bp Fragment in the Mitochondrial Genome Is Associated with Mitochondrial Disease Severity. *PLoS ONE* **2015**, *10*, e0128624. [CrossRef]
147. Lee, H.C.; Pang, C.Y.; Hsu, H.S.; Wei, Y.H. Differential accumulations of 4,977 bp deletion in mitochondrial DNA of various tissues in human ageing. *Biochim. Biophys. Acta.* **1994**, *1226*, 37–43. [CrossRef]
148. Ambulkar, P.S.; Chuadhari, A.R.; Pal, A.K. Association of large scale 4977-bp "common" deletions in sperm mitochondrial DNA with asthenozoospermia and oligoasthenoteratozoospermia. *J. Hum. Reprod Sci.* **2016**, *9*, 35–40. [CrossRef]
149. Bahrehmand Namaghi, I.; Vaziri, H. Sperm mitochondrial DNA deletion in Iranian infertiles with asthenozoospermia. *Andrologia* **2017**, *49*. [CrossRef]
150. Ambulkar, P.S.; Waghmare, J.E.; Chaudhari, A.R.; Wankhede, V.R.; Tarnekar, A.M.; Shende, M.R.; Pal, A.K. Large Scale 7436-bp Deletions in Human Sperm Mitochondrial DNA with Spermatozoa Dysfunction and Male Infertility. *J. Clin. Diagn. Res.* **2016**, *10*, GC09–GC12. [CrossRef]
151. Cummins, J.M.; Jequier, A.M.; Martin, R.; Mehmet, D.; Goldblatt, J. Semen levels of mitochondrial DNA deletions in men attending an infertility clinic do not correlate with phenotype. *Int. J. Androl.* **1998**, *21*, 47–52. [CrossRef]
152. St John, J.C.; Jokhi, R.P.; Barratt, C.L. Men with oligoasthenoteratozoospermia harbour higher numbers of multiple mitochondrial DNA deletions in their spermatozoa, but individual deletions are not indicative of overall aetiology. *Mol. Hum. Reprod.* **2001**, *7*, 103–111. [CrossRef]
153. O'Flaherty, C. Orchestrating the antioxidant defenses in the epididymis. *Andrology* **2019**, *7*, 662–668. [CrossRef]
154. Schneider, M.; Förster, H.; Boersma, A.; Seiler, A.; Wehnes, H.; Sinowatz, F.; Neumüller, C.; Deutsch, M.J.; Walch, A.; Hrabé de Angelis, M.; et al. Mitochondrial glutathione peroxidase 4 disruption causes male infertility. *FASEB J.* **2009**, *23*, 3233–3242. [CrossRef]
155. Chabory, E.; Damon, C.; Lenoir, A.; Kauselmann, G.; Kern, H.; Zevnik, B.; Garrel, C.; Saez, F.; Cadet, R.; Henry-Berger, J.; et al. Epididymis seleno-independent glutathione peroxidase 5 maintains sperm DNA integrity in mice. *J. Clin. Invest.* **2009**, *119*, 2074–2085. [CrossRef]
156. O'Flaherty, C.; de Souza, A.R. Hydrogen peroxide modifies human sperm peroxiredoxins in a dose-dependent manner. *Biol. Reprod.* **2011**, *84*, 238–247. [CrossRef]
157. Dubuisson, M.; Vander Stricht, D.; Clippe, A.; Etienne, F.; Nauser, T.; Kissner, R.; Koppenol, W.H.; Rees, J.F.; Knoops, B. Human peroxiredoxin 5 is a peroxynitrite reductase. *FEBS Lett.* **2004**, *571*, 161–165. [CrossRef]
158. Liu, Y.; O'Flaherty, C. In vivo oxidative stress alters thiol redox status of peroxiredoxin 1 and 6 and impairs rat sperm quality. *Asian. J. Androl.* **2017**, *19*, 73–79. [CrossRef]
159. Fisher, A.B. Peroxiredoxin 6 in the repair of peroxidized cell membranes and cell signaling. *Arch. Biochem. Biophys.* **2017**, *617*, 68–83. [CrossRef]
160. Fernandez, M.C.; O'Flaherty, C. Peroxiredoxin 6 is the primary antioxidant enzyme for the maintenance of viability and DNA integrity in human spermatozoa. *Hum. Reprod.* **2018**, *33*, 1394–1407. [CrossRef]
161. Bansal, A.K.; Bilaspuri, G.S. Impacts of oxidative stress and antioxidants on semen functions. *Vet. Med. Int.* **2010**, *2010*, 686137. [CrossRef]
162. Gong, S.; San Gabriel, M.C.; Zini, A.; Chan, P.; O'Flaherty, C. Low amounts and high thiol oxidation of peroxiredoxins in spermatozoa from infertile men. *J. Androl.* **2012**, *33*, 1342–1351. [CrossRef]
163. Wichmann, L.; Vaalasti, A.; Vaalasti, T.; Tuohimaa, P. Localization of lactoferrin in the male reproductive tract. *Int. J. Androl.* **1989**, *12*, 179–186. [CrossRef]
164. Pearl, C.A.; Roser, J.F. Expression of lactoferrin in the boar epididymis: Effects of reduced estrogen. *Domest. Anim. Endocrinol.* **2008**, *34*, 153–159. [CrossRef] [PubMed]
165. Wang, P.; Liu, B.; Wang, Z.; Niu, X.; Su, S.; Zhang, W.; Wang, X. Characterization of lactoferrin receptor on human spermatozoa. *Reprod. Biomed. Online* **2011**, *22*, 155–161. [CrossRef] [PubMed]

166. Hamada, A.; Sharma, R.; du Plessis, S.S.; Willard, B.; Yadav, S.P.; Sabanegh, E.; Agarwal, A. Two-dimensional differential in-gel electrophoresis-based proteomics of male gametes in relation to oxidative stress. *Fertil. Steril.* **2013**, *99*, 1216–1226. [CrossRef]
167. Montecinos, V.; Guzmán, P.; Barra, V.; Villagrán, M.; Muñoz-Montesino, C.; Sotomayor, K.; Escobar, E.; Godoy, A.; Mardones, L.; Sotomayor, P.; et al. Vitamin C is an essential antioxidant that enhances survival of oxidatively stressed human vascular endothelial cells in the presence of a vast molar excess of glutathione. *J. Biol. Chem.* **2007**, *282*, 15506–15515. [CrossRef]
168. Nouri, M.; Ghasemzadeh, A.; Farzadi, L.; Shahnazi, V.; Ghaffari-Novin, M. Vitamins C, E and lipid peroxidation levels in sperm and seminal plasma of asthenoteratozoospermic and normozoospermic men. *Iran. J. Reprod. Med.* **2008**, *6*, 1–5.
169. Micheli, L.; Cerretani, D.; Collodel, G.; Menchiari, A.; Moltoni, L.; Fiaschi, A.I.; Moretti, E. Evaluation of enzymatic and non-enzymatic antioxidants in seminal plasma of men with genitourinary infections, varicocele and idiopathic infertility. *Andrology* **2016**, *4*, 456–464. [CrossRef]
170. Huang, C.; Cao, X.; Pang, D.; Li, C.; Luo, Q.; Zou, Y.; Feng, B.; Li, L.; Cheng, A.; Chen, Z. Is male infertility associated with increased oxidative stress in seminal plasma? A-meta analysis. *Oncotarget* **2018**, *9*, 24494–24513. [CrossRef]
171. O'Flaherty, C.; Matsushita-Fournier, D. Reactive oxygen species and protein modifications in spermatozoa. *Biol. Reprod.* **2017**, *97*, 577–585. [CrossRef]
172. Piomboni, P.; Gambera, L.; Serafini, F.; Campanella, G.; Morgante, G.; De Leo, V. Sperm quality improvement after natural anti-oxidant treatment of asthenoteratospermic men with leukocytospermia. *Asian J. Androl.* **2008**, *10*, 201–206. [CrossRef]
173. Akmal, M.; Qadri, J.Q.; Al-Waili, N.S.; Thangal, S.; Haq, A.; Saloom, K.Y. Improvement in human semen quality after oral supplementation of vitamin C. *J. Med. Food.* **2006**, *9*, 440–442. [CrossRef]
174. Omu, A.E.; Al-Azemi, M.K.; Kehinde, E.O.; Anim, J.T.; Oriowo, M.A.; Mathew, T.C. Indications of the mechanisms involved in improved sperm parameters by zinc therapy. *Med. Princ. Pract.* **2008**, *17*, 108–116. [CrossRef] [PubMed]
175. Garolla, A.; Maiorino, M.; Roverato, A.; Roveri, A.; Ursini, F.; Foresta, C. Oral carnitine supplementation increases sperm motility in asthenozoospermic men with normal sperm phospholipid hydroperoxide glutathione peroxidase levels. *Fertil. Steril.* **2005**, *83*, 355–361. [CrossRef] [PubMed]
176. Sigman, M.; Glass, S.; Campagnone, J.; Pryor, J.L. Carnitine for the treatment of idiopathic asthenospermia: A randomized, double-blind, placebo-controlled trial. *Fertil. Steril.* **2006**, *85*, 1409–1414. [CrossRef] [PubMed]

© 2020 by the authors. Licensee MDPI, Basel, Switzerland. This article is an open access article distributed under the terms and conditions of the Creative Commons Attribution (CC BY) license (http://creativecommons.org/licenses/by/4.0/).

Article

GSTO2 Isoforms Participate in the Oxidative Regulation of the Plasmalemma in Eutherian Spermatozoa during Capacitation

Lauren E. Hamilton [1], Michal Zigo [2], Jiude Mao [2], Wei Xu [1], Peter Sutovsky [2,3], Cristian O'Flaherty [4] and Richard Oko [1,*]

[1] Department of Biomedical and Molecular Sciences, Queen's University, Kingston, ON K7L 3N6, Canada; 9leh5@queensu.ca (L.E.H.); wx@queensu.ca (W.X.)
[2] Division of Animal Sciences, College of Food, Agriculture and Natural Resources, Columbia, MO 65211, USA; zigom@missouri.edu (M.Z.); maoj@missouri.edu (J.M.); sutovskyp@missouri.edu (P.S.)
[3] Division of Obstetrics, Gynecology and Women's Health, School of Medicine, University of Missouri, Columbia, MO 65211, USA
[4] Department of Surgery (Urology Division), Faculty of Medicine, McGill University, Montreal, QC H4A 3JI, Canada; cristian.oflaherty@mcgill.ca
* Correspondence: ro3@queensu.ca

Received: 17 October 2019; Accepted: 26 November 2019; Published: 29 November 2019

Abstract: In addition to perinuclear theca anchored glutathione-s-transferase omega 2 (GSTO2), whose function is to participate in sperm nuclear decondensation during fertilization (Biol Reprod. 2019, 101:368–376), we herein provide evidence that GSTO2 is acquired on the sperm plasmalemma during epididymal maturation. This novel membrane localization was reinforced by the isolation and identification of biotin-conjugated surface proteins from ejaculated and capacitated boar and mouse spermatozoa, prompting us to hypothesize that GSTO2 has an oxidative/reductive role in regulating sperm function during capacitation. Utilizing an inhibitor specific to the active site of GSTO2 in spermatozoa, inhibition of this enzyme led to a decrease in tyrosine phosphorylation late in the capacitation process, followed by an expected decrease in acrosome exocytosis and motility. These changes were accompanied by an increase in reactive oxygen species (ROS) levels and membrane lipid peroxidation and culminated in a significant decrease in the percentage of oocytes successfully penetrated by sperm during in vitro fertilization. We conclude that GSTO2 participates in the regulation of sperm function during capacitation, most likely through protection against oxidative stress on the sperm surface.

Keywords: glutathione-s-transferase omega 2; capacitation; fertilization; male fertility; oxidative regulation; spermatozoa; reactive oxygen species (ROS)

1. Introduction

The spermatozoa expelled from the seminiferous tubules at the end of spermatogenesis lack progressive motility and the ability to fertilize the oocyte. It is only through epididymal maturation and functional capacitation that spermatozoa undergo the necessary transformational changes needed to fertilize.

Capacitation is a highly orchestrated set of reactions that ultimately culminates in the acrosome exocytosis reaction and spermatozoa obtaining their fertilizing competency as they approach the oocyte in the oviduct [1]. Due to the lack of transcriptional and translational activity within mature spermatozoa, any changes that the cell undergoes must occur through post-translational modifications of pre-existing proteins. A key modulator of cell signaling and enzymatic function are the levels of reactive oxygen species (ROS) [2–8].

High sensitivity to membrane lipid peroxidation requires eutherian spermatozoa to maintain a fine balance between the production and regulation of ROS [9–12]. High concentrations of oxygen radicals and peroxidation by-products have been well documented as contributors to male infertility [9,13–17]. Ideally, when ROS is maintained at low concentrations, the ROS molecules are utilized by spermatozoa to modulate cellular functions such as the initiation of capacitation and hyperactivated motility [1,7,18–22]. Spermatozoa are, therefore, critically dependent on their network of regulatory and detoxifying enzymes to ensure that optimal concentrations of ROS are maintained.

Eutherian spermatozoa are largely devoid of the cytoplasmic reservoir of antioxidant enzymes seen in most somatic cells, and therefore have a large reliance on the detoxifying capacity of the fluids in their surroundings. Epididymal secretions and semen are amongst the most antioxidant rich fluids within the body, equipped with specialized ROS scavenging molecules and enzymes that aid in buffering the oxidative stress levels of sperm cells [23–27]. It is also through these secretions that surface-anchored detoxifying enzymes can be imparted to the plasmalemma of spermatozoa as they progress through the male reproductive system.

The proteomic analysis of epididymal secretions, semen and surface bound spermatozoon proteins helps to decode the detoxifying landscape of spermatozoa and identify the regulatory systems at play [28,29]. At the center of many detoxifying systems is the tripeptide thiol glutathione. Glutathione provides a recyclable source of reducing power and facilitates many groups of antioxidant enzymes residing within and on the sperm surface, such as glutathione reductases, glutaredoxins, thioredoxins, perioxiredoxins and glutathione-s-transferases.

Glutathione-s-transferases (GSTs) are a large super-family of phase II detoxification enzymes that are ubiquitously found throughout the body, and well represented within the male reproductive environment [30]. GSTs of the Mu, Pi, Theta, Alpha, Zeta and Omega classes have all been identified as components of the seminal plasma or as sperm resident proteins, with their own functionally distinct roles [31–34]. Furthermore, several classes of GST have recently been shown to have functional multimodality, acting in both sperm–egg interactions and redox regulation of the plasmalemma [31]. Harboring enzymes that can facilitate more than one cellular process may be a valuable asset for cells such as spermatozoa that have evolved to be streamlined and devoid of most cytosolic resources.

The Omega class of GSTs have also been shown to have multifunctionality [30,35,36]. Equipped with a cysteine residue at their activity site, GSTOs are not only able to facilitate glutathione reductase reactions but have also been shown to have dehydroascorbate reductase capabilities [30,35,36]. Moreover, GSTO2, one of only two functionally active enzymes within the Omega class, has been credited as having the highest levels of dehydroascorbate reductase functionality within mammals [37]. With the epididymis and seminal plasma having some of the highest concentrations of ascorbic acid (AA) found within the body, maintaining optimal concentrations of AA within the sperm may prove vital to its fitness [26]. Therefore, the enzymatic versatility of facilitating reactions in both glutathione and AA centered pathways may make GSTO2 a highly valuable surface-bound enzyme.

In addition to its previous characterization as a constituent of the postacrosomal sheath and perforatorium of the perinuclear theca [34,38], we demonstrate that GSTO2 is also present on the sperm surface in mouse and boar spermatozoa. Through surface protein isolation, indirect immunofluorescence, fluorescence immunohistochemistry, functional inhibition studies and computer assisted sperm analysis, we characterize the surface localization of GSTO2 isoforms and demonstrate the functional role of GSTO2 in facilitating sperm capacitation through oxidative regulation.

2. Materials and Methods

2.1. Animals

Retired breeder CD1 and C57BL/6 mice were purchased from Charles River Laboratories (Charles River, St-Constant, QC, Canada). All procedures in this study were conducted under the Animal Utilization Protocols approved by Queen's University Animal Care Committee (protocol # 2017-1742) and complied with the Guidelines of the Canadian Council on Animal Care. Fertile, non-transgenic boar semen samples were collected at the University of Missouri's National Swine Resource and Research Center and processed in their Division of Animal Sciences, College of Food, Agriculture and Natural Resources (Columbia, MO, USA), under the strict guidance of the University of Missouri's Animal Welfare Assurance Number and Animal Care and Use Committee (ACUC) protocol # A3394-01.

2.2. Sperm Extractions from Mice

Spermatozoa were obtained from the fresh cauda epididymis' of mature CD1 and C57BL/6 males. Cauda epididymis' were placed in approximately 0.5 mL of phosphate-buffered saline (PBS) and pierced with a $26\frac{1}{2}$ gauge needle to allow the sperm to diffuse out into the solution.

2.3. Antibodies and Reagents

The central antibody was a goat polyclonal anti-GSTO2 antibody (Y-12, Santa Cruz Biotechnology, Dallas, TX, USA), used at a concentration of 0.2 µg/mL for Western blot analysis, and 6.67 µg/mL for fluorescence immunocytochemistry. For the measure of protein tyrosine phosphorylation, the clone 4G10 anti-tyrosine phosphorylation antibody (Millipore-Sigma, 05-321, St. Louis, MO, USA) was used at a concentration of 0.1 µg/mL and standardized using an anti-alpha tubulin antibody (Sigma T6074, Burlington, MA, USA). For immunohistochemistry a rabbit-polyclonal anti-GSTO2 (Sigma Prestige, HPA048141, Burlington, MA, USA) was used at a concentration of 6.67 µg/mL. For Western blot analysis, a rabbit anti-goat IgG-HRP (horseradish peroxidase) (0.4 µg/mL, Santa Cruz Biotechnology, Santa Cruz, CA, USA) secondary antibody was used, and, for indirect immunofluorescence studies, a donkey anti-goat IgG-CFL (colorized fluorochrome) 555 (Santa Cruz, 2 µg/mL), or donkey-anti-rabbit-IgG-CFL 488 was used (2 µg/mL, Abcam, Cambridge, MA, USA). For the assessment of the acrosome exocytosis reaction lectin PNA (*Arachis hypogaea*) conjugated to the colorized fluorocrome 647 (Invitrogen, L32460, Waltham, MA, USA) was used at a concentration of 15 µg/mL. For peroxidation analysis, a BODIPY 581/591 C11 probe (4,4-difluoro-5-(4-phenyl-1,3-butadienyl)-4-bora-3a,4a-diaza-s-indacene-3-undecanoic acid) (Invitrogen Molecular Probes, D3861) was used at a final concentration of 5 µM. For evaluation of the total reactive oxygen species, the Cellular ROS Detection Assay kit (ab186029 Abcam, Cambridge, MA, USA) was used following the manufacturer's manual. For all enzymatic inhibition studies, a membrane permeable cell tracker probe (Invitrogen Molecular Probes C7025) was used. The probe has been shown to bind covalently and irreversibly to the active site of GSTO isozymes [39] and was used at a concentration of 100 µM in all our experiments. Any additional reagents were purchased from Millipore-Sigma (Burlington, MA, USA).

2.4. Fluorescence Electrophoresis and Western Blotting Analysis

All sperm samples were freshly extracted and subsequently incubated with our inhibitory probe for 25 min at 37 or 38 °C, depending on the species. The samples were washed twice and then solubilized in a non-reducing sample buffer (200 mM Tris pH 6.8, 4% SDS, 0.1% bromophenol blue, 40% glycerol, 5% β-mercaptoethanol). A BLUeye pre-stained protein ladder (GeneDirex) was loaded along with approximately 1–2 million cells per lane and resolved on 4% stacking and 12% separating polyacrylamide gels, as described by Laemmli [40]. The gel was run at 100 volts for 110 min before being placed in transfer buffer and imaged in a fluorescence biophotonic chamber. After transfer to a polyvinylidene fluoride (PVDF) membrane (Millipore) for 120 min in Tris-glycine transfer buffer on ice using a Hoefer Transfer apparatus (Hoefer Scientific Instruments), the membrane was also imaged in the same fluorescence chamber. The membrane was then blocked in a 10% skim milk and phosphate-buffered saline (PBS) solution with 0.05% Tween-20 (PBS-T) for 30 min to prevent non-specific binding. The membrane was incubated with primary antibody overnight at 4 °C with slight agitation. The next day, the membrane was washed in PBS-T six times, each for five minutes, before a two hour incubation with secondary antibody conjugated to horseradish peroxidase. The membrane was then washed extensively and subjected to an immunodetection reaction that was visualized using Clarity Western ECL Substrate (Bio Rad Laboratories, Hercules, CA, USA). The membrane was exposed to X-ray film for developing. For the evaluation of tyrosine phosphorylation (PY) during capacitation, relative intensities were calculate using Image J. Total PY was calculated for each sample and normalized using the intensity of tubulin in each sample.

2.5. Fluorescence Immunocytochemistry

Mouse and boar spermatozoa were mounted on poly-L lysine coated coverslips and fixed in 2% formaldehyde for 40 min. Non-specific binding was blocked using 5% bovine serum albumin (BSA) for 25 min before incubation in primary antibody overnight at 4 °C. The next day, the coverslips were washed extensively in 1% BSA-PBS before a 40 min incubation with secondary antibody conjugated to a fluorescent marker and DAPI (4′,6-Diamidino-2-Phenylindole, Dihydrochloride) at room temperature and hidden from light. The coverslips were washed again before being mounted on glass slides using VectaShield mounting medium (Vector Laboratories, Burlingame, CA, USA) and sealed with nail polish. Spermatozoa from mouse and boar were also incubated with our fluorescent inhibitor for 25 min, washed extensively and mounted onto glass slides to visualize the binding pattern. Fluorescence images were taken at the Queen's University Cancer Research Institute Imaging Centre, using a Quorum Wave Effects spinning disc confocal microscope or at the University of Missouri-Columbia, using a Nikon Eclipse 800 microscope with CoolSnap CCD camera (Nikon, Tokyo, Japan). All images were subsequently analyzed using MetaMorph Imaging Software (Molecular Devices, San Jose, CA, USA).

2.6. Boar Surface Protein Extractions

Boar surface sperm protein extractions were performed on approximately 3×10^8 cells of both ejaculated and in vitro capacitated samples, as previously described in [41] with the Pierce Surface Protein Biotinylation kit (ThermoFisher Scientific, A44390). Extracts were subsequently precipitated with the use of a 2-D Clean up Kit (GE Healthcare, Uppsala, Sweden) and resolubilized with Laemmli reducing sample buffer for Western Blot Analysis.

2.7. Fluorescent Immunohistochemistry

Paraformaldehyde fixed, and paraffin embedded boar and mouse epididymal sections were deparaffinized in xylene and hydrated through a graded series of methanol solutions. After hydration, the sections were treated to abolish autofluorescence and subjected to antigen retrieval by microwave irradiation in a 5% Urea, Tris-HCL solution, pH 9.5 [18]. Before primary antibody incubation, sections were blocked with 5% normal goat serum (NGS) diluted with phosphate-buffered saline (PBS). The primary antibody was made in 1% NGS-PBS and incubated at 4 °C overnight. Slides were washed with 1% NGS-PBS before incubation with fluorescent-tagged secondary antibody and DAPI. Once completed, slides were washed in PBS before being mounted to coverslips using Vectashield mounting Media (Vector Laboratories) and sealed with clear nail polish. Images were captured at the Queen's University Cancer Research Institute Imaging Centre, using a Quorum Wave Effects spinning disc confocal microscope and analyzed using MetaMorph imaging software.

2.8. Mouse In Vitro Capacitation and Acrosome Exocytosis Reaction

Spermatozoa were extracted from fresh cauda epididymis by piercing with a 26 $\frac{1}{2}$ gauge needle and allowing the sperm to diffuse out into a Whittens-Hepes Medium. Sperm were then incubated for

25 min at 37 °C, 5% CO_2 in Whittens-Hepes Medium with either the inhibitor, or dimethyl sulfoxide (DMSO, Control). After treatment, sperm were diluted to a final concentration of approximately two million cells/mL with either a capacitation medium (Modified Whittens-Hepes medium supplemented with 5mg/mL BSA and 20 mM $NaHCO_3$) or the non-capacitiation medium (Whittens-Hepes without supplementation) and left to incubate for up to 90 min at 37 °C, 5% CO_2. For the analysis of tyrosine phosphorylation, samples were collected every 45 min, washed twice and placed in sample buffer before being run on a Western blot. For the acrosome exocytosis reaction, sperm cells were left to incubate in either capacitating or non-capacitating media for 60 min and then progesterone was added to a final concentration of 10 µM and left for an additional 30 min. Cells were subsequently washed in PBS and allowed to air dry on glass slides. Cells were fixed with ice cold absolute methanol for 15 min, washed in PBS and stained with PNA (15µg/mL) and DAPI for 30 min. Slides were rinsed one final time and coverslips were mounted. In each treatment, 200 cells were randomly selected and the acrosome was scored as either intact or reacting/reacted. At least three replicates of each treatment were assessed and the reported values represent the average per group. The Western blotting results were analyzed using an ANOVA with a post-hoc Tukey test, comparing the mean intensities of three replicates. The acrosome exocytosis reaction results were compared using a t-test with Welch's correction, comparing the average percentage of reacted cells from three replicates.

2.9. Boar In Vitro Capacitation

Boar spermatozoa were collected from fresh ejaculate and were subsequently centrifuged and washed to remove seminal plasma. Sperm were counted using a hemocytometer (ThermoFisher Scientific, Waltham, MA), and incubated in a modified TL-HEPES medium with either the GSTO Inhibitor or DMSO (Control) for 25 min at 38 °C. Sperm cells were then placed in either capacitating (modified TL-HEPES supplemented with 5 mM sodium pyruvate, 11 mM D-glucose, 2 mM $CaCl_2$, 2 mM sodium bicarbonate, and 2% (*m/v*) bovine serum albumin) or non-capacitating medium (modified TL-HEPES medium without supplementation) to a concentration of two million cells/mL and incubated at 38 °C for four hours. Cells were removed every hour, washed, and placed in reducing sample buffer, for tyrosine phosphorylation analysis by Western blotting up until the four-hour time point. The results were analyzed using an ANOVA with a post-hoc Tukey test, comparing the mean intensities from three replicates.

2.10. Mouse In Vitro Fertilization

Mouse oocytes were obtained from super-ovulated 6 week old CD1 females. Females were given 10 UI of pregnant mare serum gonadotropin (PMSG) through intraperitoneal (IP) injection, followed 48 h later by 10 UI of human chorionic gonadotropin (hCG), also administered through IP injection. Oviducts were harvested from the sacrificed females 12–13 h after the last hCG injection into Advanced KSOM medium (MR-101-D, Millipore-Sigma), warmed to 37 °C. Cumulus-oocyte complexes were extracted from the oviducts into warmed Advanced KSOM medium, washed and rested at 37 °C, 5% CO_2 under mineral oil until fertilization droplets were prepared, to a maximum of 30 min. Spermatozoa were extracted from the cauda epididymis and into Whitten-Hepes Medium where they were either incubated with the GSTO inhibitor (100 µM) or DMSO for 25 min under mineral oil at 37 °C, 5% CO_2. Sperm were subsequently washed and placed into EmbryoMax human tubal fluid (HTF, MR-070-D, Millipore-Sigma) to capacitate for a minimum of 1 h. Sperm were then diluted with HTF into a 50 µl droplet with a concentration of approximately 1×10^5/mL sperm and between 20–25 oocytes were added to each droplet. Fertilization droplets were incubated at 37 °C, 5% CO_2 under mineral oil for 5 h before oocytes were removed, washed and further cultured in 50 µL Advanced KSOM droplets under mineral oil for approximately 8 h. Oocytes were then fixed in 2% formaldehyde, permeabilized in phosphate-buffered saline with 0.1% Triton-X-100 (PBS-Tx) and stained with DAPI to allow for the visualization of the sperm head or pronuclear formation, indicative of successful sperm penetration. Oocytes were then mounted onto glass slides using Vectashield mounting Medium (Vector

Laboratories) and sealed with clear nail polish. Images were captured at the Queen's University Cancer Research Institute Imaging Centre, using a Quorum Wave Effects spinning disc confocal microscope. Three replicates of each treatment were performed, all using different mice, and statistical analysis was performed using a t-test with Welch's correction.

2.11. Swine In Vitro Fertilization

Ovaries were obtained from a local slaughterhouse and aspirated to obtain oocytes from follicles of 3–6 mm in size. Oocyte with uniform ooplasm and compact cumulus cells were selected and in vitro matured at 38.5 °C, 5% CO_2, for 42 to 44 h. Cumulus cells from matured cumulus-oocyte complexes (COCs) were removed with 0.1% hyaluronidase in TL-HEPES-PVA and washed three times with TL-HEPES-PVA medium. Twenty-five to thirty oocytes were placed into 100 µL drops of the mTBM medium, while sperm were prepared. Boar semen was collected from sperm rich fraction the day before IVF. Sperm cells were incubated with the GSTO inhibitor for 25 min at 38.5 °C immediately after being isolated from the semen and before being placed in a short BTS extender (BTS, IMV Technologies, Maple Grove, MN, USA) until used. Mitochondria of the sperm tail were stained with a viable, mitochondrion-specific probe MitoTracker® Red CMXRos (Molecular Probes, Inc., Eugene, OR, USA) for 10 min at 38 °C in a warm incubator before the sperm solution was diluted to a concentration of 1×10^6 cells/mL. Co-incubation of the sperm and the oocytes was left for approximately 6 h before oocytes were removed and transferred to 100 µL drops of PZM-3 medium containing 0.4% BSA (A6003; Sigma, Burlington, MA, USA) for additional culture. Three replicates of each experiment were performed using three different ejaculates and presented as the average. Statistical analysis was performed using a t-test with Welch's correction.

2.12. Mouse Computer-Aided Sperm Analysis (CASA)

Spermatozoa were extracted from the cauda epididymis of mature C57BL/6 males and placed in Whittens-Hepes Medium. Sperm were subsequently placed with the GSTO inhibitor (100 µM) or DMSO (Control) for 25 min at 37 °C before being washed and diluted with additional Whittens-Hepes medium or capacitating medium (Whittens Hepes supplemented with 5mg/mL BSA and 20 mM $NaHCO_3$) to a final concentration of approximately 1–2 million cells/mL. Sperm were then placed back at 37 °C and left to capacitate for one hour. Samples were then gently spun down and resuspended for analysis. Analysis was done by Sperm Vision HR software version 1.01 (Minitube, Ingersoll, ON, Canada). At least 200 cells from each treatment were analyzed in each experimental trial. All cells that were not unequivocally identifiable as sperm were removed from the analysis by the technician. Four replicates were performed, each using a new mouse, and the data are presented as the average of all replicates. Multiple t-tests were performed to determine statistical significance.

2.13. Lipid peroxidation Intensity Analysis

Mouse spermatozoa were extracted from the cauda epididymis of mature C57BL/6 males and placed in Whitten-Hepes Medium. Sperm were diluted to a concentration of approximately 2×10^6/mL, and subsequently placed with the BODIPY 581/591 C11 (D3861, Invitrogen Molecular Probes, Eugene, OR, USA) to a final concentration of 5 µM for 20 min at 37 °C. A subset of sperm were not treated with the BODIPY C11 probe to act as a baseline for innate fluorescence. Sperm were washed twice at 650× g for 5 min before being placed with the GSTO inhibitor (100 µM) or DMSO (Control) for 25 min at 37 °C. Spermatozoa were washed again and placed in capacitating medium (Whittens-Hepes Medium supplemented with 5mg/mL BSA and 20 mM $NaHCO_3$) to a final concentration of approximately 1–2 million cells/mL and incubated for one hour at 37 °C, with agitation every 15 min. Spermatozoa were subsequently washed and placed on a glass slide, covered with a coverslip and sealed with clear nail polish. Three trials were performed for each treatment, and at least 200 cells per treatment were analyzed each trial by confocal microscopy. Images were captured at the Queen's University Cancer Research Institute Imaging Centre, using a Quorum Wave Effects spinning disc confocal microscope, and analysis was performed using the MetaMorph Imaging software. The intensities of both the red and green fluorescence were acquired for all treatments and controls. For GSTO Inhibited samples, the fluorescent intensity of the inhibitor itself was subtracted from the overall green intensity to ensure it did not impact the overall ratio of green and red fluorescence. The results were analyzed using a t-test with Welch's correction to determine statistical significance.

2.14. Cellular Reactive Oxygen Species Levels

Mouse spermatozoa were extracted from the cauda epididymis of mature CD1 males and placed in Whitten-Hepes Medium. The sperm were subsequently placed with the GSTO inhibitor (100 µM) or vehicle (DMSO) for 25 min at 37 °C. Spermatozoa were then diluted to a concentration 5×10^6 cells/mL and placed under capacitating conditions through supplementing the buffer with 5 mg/mL BSA and 20 mM $NaHCO_3$. Spermatozoa were left to incubate for 60 min, with agitation every 15 min at 37 °C/5% CO_2. Reactive oxygen species levels were then probed using the Cellular ROS Assay Kit (ab186029). Samples were subsequently washed and fixed in 2% formaldehyde before being analyzed using flow cytometry. Flow cytometry was performed at the Queen's University Cardiac Pulmonary Unit using the Sony SH800 Cell Sorter and least 10,000 events were recorded for each sample. Three experimental trails were done for each treatment, each using a different mouse, and the data presented include the average mean intensity over the three trials. Statistical analysis was done using a t-test with Welch's correction.

2.15. Statistical Analysis

Statistical analysis was performed using Prism 8 statistical software. An ANOVA with a Tukey post-hoc test was used for the statistical evaluation of the results obtained from in vitro capacitation experiments. Multiple t-tests were performed for the computer-aided sperm analysis and acrosome exocytosis reaction experiments comparing different treatment groups under the same testing conditions. A t-test with Welch's correction was performed in the lipid peroxidation, total cellular ROS levels and in vitro fertilization experiments. All experiments were performed at least three times and all results are the averages of all replicates. For each experiment, all replicates of mouse experiments were performed with a different mouse, whereas all experiments performed in boar used a different ejaculate.

3. Results

3.1. The presence of GSTO2 on the Plasmalemma of Mature Mouse and Boar Spermatozoa

Glutathione-S-Transferase Omega 2 was observed on the plasmalemma of non-permeabilized mouse and boar spermatozoa through the use of indirect immunofluorescence (Figure 1). This reactivity was inhibited when the antibody (Y-12) was pre-incubated with its corresponding peptide block, and was absent when samples were incubated in only the secondary antibody. Furthermore, when spermatozoa were incubated with a Glutathione-S-Transferase Omega (GSTO)-specific inhibitory binding molecule, the entirety of the sperm showed reactivity, suggesting enzymes of the GSTO family are present outside of the perinuclear theca, the sole localization of GSTO enzymes currently documented within eutherian spermatozoa [34].

To further investigate the surface localization of GSTO2, surface proteins of non-permeabilized fresh and capacitated boar spermatozoa were isolated through the use of the Pierce Cell Surface Biotinylation and Isolation Kit (A44390, ThermoFisher Scientific, Waltham, MA, USA), run on an SDS-PAGE gel and probed with an anti-GSTO2 antibody. The findings revealed that two GSTO2 isoforms were present on the surface of fresh boar spermatozoa (Figure 2, Lane 1) but that, following in vitro capacitation, only the higher isoform remained (Figure 2, Lane 2). This reactivity was inhibited when the antibody was pre-incubated with its corresponding peptide block (Figure 2, Lanes 3 and 4).

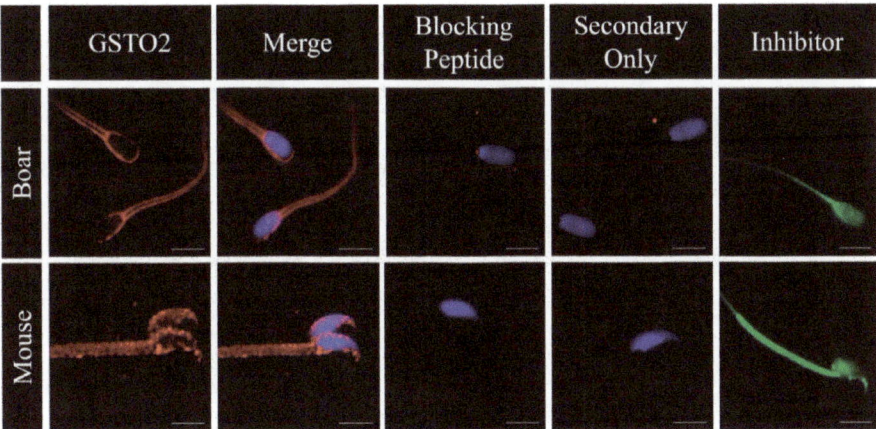

Figure 1. The surface reactivity of glutathione-s-transferase omega 2 (GSTO2) on non-permeabilized mouse and boar spermatozoa using indirect immunofluorescence. Non-permeabilized fresh boar and mouse spermatozoa were stained using a GSTO2 specific (Y-12) antibody (GSTO2), or a GSTO-specific fluorescent inhibitor (Inhibitor). Nuclei were labelled with DAPI. To confirm the specificity of the antibody, the anti-GSTO2 antibody was preincubated with its blocking agent (blocking peptide, the peptide used to raise the antibody) according to the manufacturer's instructions and a secondary-only control was done (Secondary Only). The GSTO inhibitor is a membrane permeable cell tracker probe (Invitrogen Molecular Probe C7025). The bar represents 10 μm.

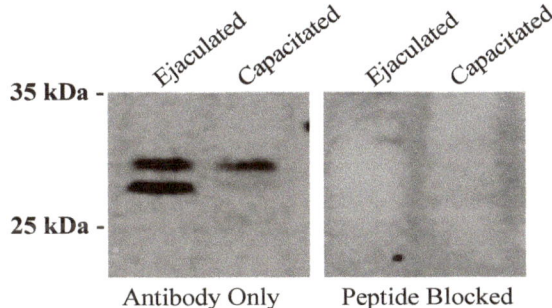

Figure 2. GSTO2 reactivity to biotin-isolated surface proteins of ejaculated and capacitated boar spermatozoa. Immunoblotting using an anti-GSTO2 (Y-12) antibody on biotin-isolated surface proteins from ejaculated boar sperm (Lane 1) and capacitated boar sperm (Lane 2). The antibody was pre-incubated with its corresponding blocking peptide to show specificity in both the ejaculated boar sperm (Lane 3) and capacitated boar sperm (Lane 4) samples.

Due to the spermatozoon's lack of transcriptional and translational capabilities after the round spermatid stage of spermiogenesis and the absence of GSTO2 membrane reactivity previously reported during spermatogenesis [34], we aimed to investigate if there was an external source of the enzyme that was secreted during epididymal transport. Fluorescence immunohistochemical staining of porcine and murine epididymal sections (Figure 3) revealed the presence of GSTO2 within the caput, corpus and caudal regions of the epididymis. The reactivity was concentrated at the luminal aspect of the epithelium, however, in the mouse caput, the GSTO2 reactivity was also seen in distinct regions, that spanned from the basal to the luminal aspect of the epithelium.

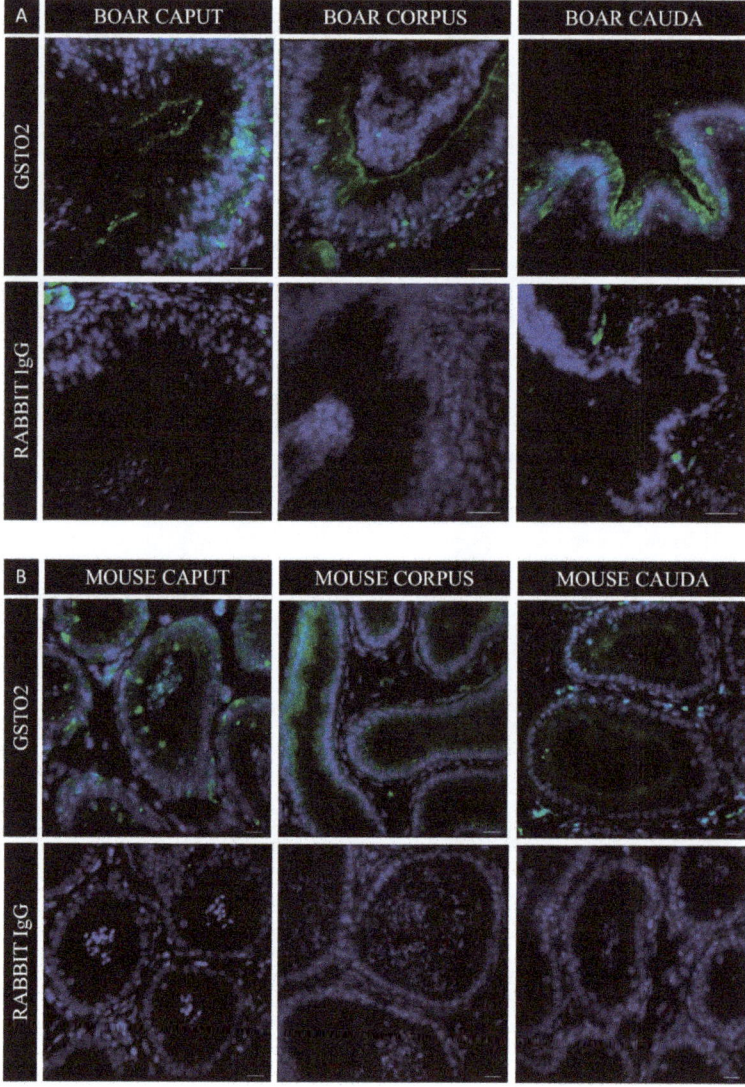

Figure 3. GSTO2 reactivity in histological sections of boar (**A**) and mouse (**B**) epididymis. In both species, caput, corpus and cauda epididymal sections show heightened GSTO2 reactivity at the luminal aspect of the epididymal epithelium with some fissures of reactivity towards the basal membrane. In mouse caput epididymal sections, some isolated pockets of reactivity can also be seen, spanning the length of the epididymal epithelium. Nuclear material was stained with DAPI. The bars in both (**A**) and (**B**) is representative of 20 μm.

3.2. The Functional Significance of GSTO2 During Capacitation

Many surface proteins secreted by the epididymis have been shown to function in the regulation of capacitation. Therefore, through functional inhibition, using a specific inhibitor that binds to the active site of GSTO enzymes (Figure S1), we sought to determine if GSTO2 has a role modulating some facet of the capacitation process. Inhibitor specificity was confirmed in both mouse and boar whole sperm through fluorescence gel electrophoresis (Figure S2).

Inhibition of GSTO2's catalytic site during in vitro capacitation in mice resulted in a dampening of the hallmark increase in tyrosine phosphorylation that occurs at the late stages of capacitation (Figure 4) without impairing sperm viability (Figure S3). These findings were also observed in boar spermatozoa (Figure S4). Most likely as a consequence of the diminished tyrosine phosphorylation events, mouse spermatozoa also demonstrated a significant decrease in their ability to successfully undergo acrosome exocytosis when GSTO enzymes were functionally inhibited prior to in vitro capacitation (Figure 5).

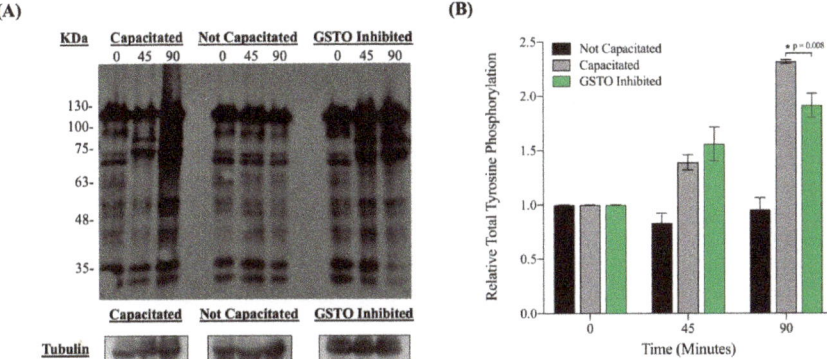

Figure 4. The level of protein tyrosine phosphorylation during in vitro capacitation in mouse spermatozoa. (**A**) Total protein tyrosine phosphorylation patterns at 0, 45 and 90 min after in vitro capacitation in DMSO-treated and capacitated (Capacitated), DMSO-treated and not capacitated (Not Capacitated) and GSTO-inhibited and capacitated (GSTO Inhibited) mouse spermatozoa samples. (**B**) The intensity of the total protein tyrosine phosphorylation levels, shown in the Western blot at each time point for each treatment, were quantified using Image J and normalized to the respective intensity of alpha tubulin. Total tyrosine phosphorylation relative intensities were further normalized so that all time 0 values were 1. All measurements are the averages of three trials, all performed with different mice. Error bars represent standard error and * signifies statistical significance of $p = 0.008$ determined by a one-way ANOVA.

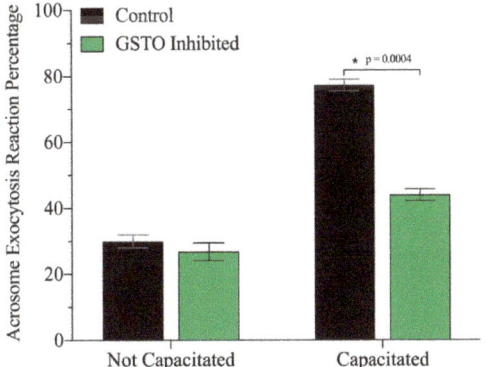

Figure 5. Sperm capacitation is reduced by the presence of the GSTO inhibitor when incubated in a capacitation medium. Sperm treatment was done through incubation for 25 min before in vitro capacitation was induced. The acrosome exocytosis reaction (a hallmark of capacitation) was induced using progesterone after 60 min of capacitation. Sperm samples were fixed and stained with PNA-647 and 4',6-diamidino-2-phenylindole (DAPI) and scored based on acrosome labelling. At least 200 sperm per treatment were assessed per trial and grouped into acrosome intact or acrosome reacting/reacted. Statistical significance was determined by multiple t-tests and is denoted by *, $p = 0.0004$.

When GSTO inhibited spermatozoa were analyzed with computer-aided sperm analysis significant decreases in total and progressive motility were observed (Figure 6, Panel A and B). A significant decrease in the overall curvilinear velocity was also observed in capacitated samples (Figure 6, panel C). Additionally, while a trend of higher linearity, a measured ratio of straight line and curvilinear velocity, was observed in GSTO inhibited samples they were not found to be significant (Figure 6, Panel D). These findings suggest that the spermatozoa treated with the GSTO inhibitor were not as successful in reaching the ideal state of hyperactive activity when compared to the controls.

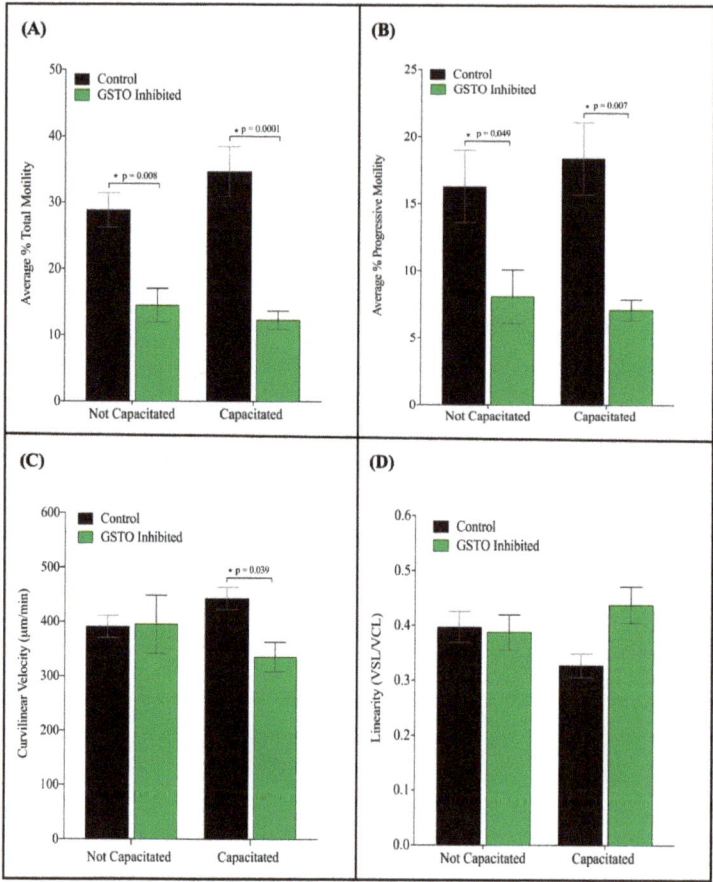

Figure 6. Computer-aided sperm analysis (CASA) of GSTO-inhibited and DMSO treated (Control) mouse spermatozoa after in vitro capacitation. Sperm motility parameters were analyzed on capacitated mouse spermatozoa that were either treated with DMSO (Control) or GSTO inhibitor prior to a 60-min incubation in capacitating medium at 37 degrees Celsius. The average total motility (**A**) was the combination of the progressive and non-progressive motility scores for each sample, whereas (**B**) shows solely the comparison of progressive motility. Both the total and progressive motility differences between the two treatment groups were statistically significant. The curvilinear velocity of GSTO inhibited sperm were also significantly decreased compared to controls when both treatment groups were capacitated (**C**). A higher linearity was also observed when spermatozoa were treated with the GSTO inhibitor (**D**), but the differences between the control and inhibited treatments were not found to be statistically significant. Error bars represent standard error. Statistical significance was determined using multiple t-tests and is denoted by *.

Further investigations looked into the peroxidation of lipids within the plasma membrane of mouse spermatozoa after in vitro capacitation using the BODIPY 581/591 C11 probe that fluoresces red in a neutral state but is modified to a green fluorescence when lipids undergo peroxidation. The ratio of green fluorescence intensity over total fluorescence intensity was used as a measure lipid peroxidation and revealed a significant increase in lipid peroxidation when GSTO activity was inhibited (Figure 7). These findings were reinforced by a significant increase in the overall cellular reactive oxygen species levels of spermatozoa treated with the GSTO inhibitor compared to controls (Figure S5). Lastly, when in vitro fertilization was performed in both mouse and swine, there was a significant decrease in the sperm's ability to successfully penetrate the oocyte (Figure 8).

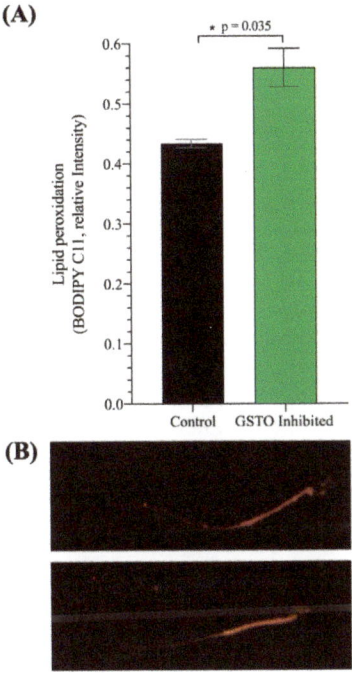

Figure 7. The assessment of sperm membrane lipid peroxidation after in vitro capacitation in mouse spermatozoa. (**A**) Lipid peroxidation levels were assessed based on the red and green fluorescence intensities of cells treated with the BODIPY 581/591 C11 probe. The relative intensity was calculated as the intensity of the green fluorescence over total fluorescence intensity. Spermatozoa were treated with the BODIPY 581/591 C11 probe and then incubated with either the GSTO Inhibitor or DMSO (Control) for 25 min before in vitro capacitation. After 60 min of in vitro capacitation, samples were fixed and imaged to determine the red and green fluorescence intensities of each cell. At least 200 cells were individually imaged in each treatment group for each trial, and a representative image from the control and GSTO inhibited groups is shown in (**B**). Three trials were performed, each with different mice, and error bars represent standard error. Statistical significance was determined using a t-test with Welch's correction and is denoted by *.

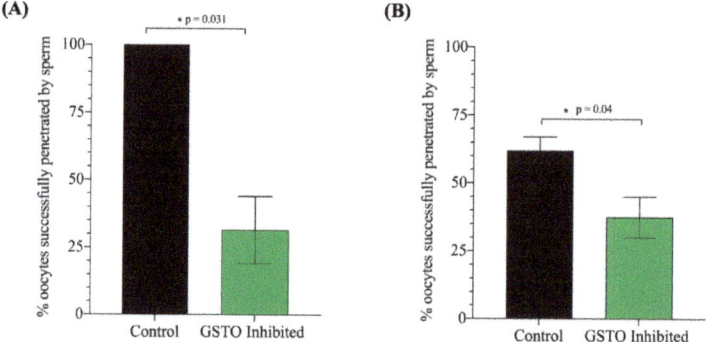

Figure 8. The assessment of sperm penetration during mouse and swine in vitro fertilization (IVF) using a low sperm concentration. Successful sperm penetration was assessed in both mouse (**A**) and swine (**B**) IVF models after 5–6 h of co-incubation in the fertilization droplet. Spermatozoa were pre-treated with either DMSO (Control) or GSTO inhibitor for 25 min before being placed in capacitation-inducing medium. Mouse sperm were incubated with cumulus–oophorus complexes at a concentration of 1×10^5/mL and boar sperm was incubated with in vitro matured swine oocytes at a concentration of 1×10^4/mL. Oocytes were washed and culture for 8 h (mouse) or 16 h (swine). Oocytes were fixed and stained with DAPI to assess pronuclear formation and sperm penetration. Statistical significance was assessed through a t-test with Welch's correction. The data represent the adjusted average of three replicates and error bars represent standard error. Statistical significance is denoted by *.

4. Discussion

The present study demonstrates, for the first time, the presence of GSTO2 on the surface of spermatozoa and its participation in the regulation of ROS levels during capacitation. Surface protein biotinylation, indirect immunofluorescence with anti-GSTO2 antibodies, and a GSTO-specific fluorescent inhibitor revealed the surface localization of two GSTO2 isoforms covering the plasmalemma of both mouse and boar spermatozoa. The absence of GSTO2 reactivity on the surface of eutherian spermatids at the time of spermiation, and its presence on epididymal spermatozoa, suggests that the origin of the surface GSTO2 isoforms differs from the isoforms characterized within the perinuclear theca region of the sperm head [34].

After spermatogenesis, the spermatozoa released from the testis are fully formed but not adequately primed to interact with the oocyte. Transport in the epididymis allows sperm cells to mature and gain surface enzymes and molecules required to function optimally. Sperm maturation is driven exclusively by external factors within the luminal microenvironment of the epididymis and occurs in the absence of transcriptional and translational activity by the sperm [28,29,42–46]. Therefore, many of the proteins, enzymes, chaperones and cytokines that collectively contribute to sperm protection and function are acquired through the release of secretory products from the principal cells into the epididymal lumen [28,29,45,47,48]. Our immunohistochemical analysis of both mouse and boar epididymal segments suggests that GSTO2 is likely a constituent of these secretions. Since GSTO2 is considered a cytosolic protein that is non-glycosylated, it most likely finds its way into the epididymal lumen through the apocrine secretory pathway.

The caput epididymis is of critical importance in promoting sperm maturation and has been shown to have greater apocrine secretory activity than the epididymal regions it precedes [49]. Thus, the predominance of GSTO2 within the apical poles of the principal cells in the caput of both the mouse and boar epididymis is supportive of its role in promoting sperm maturation and enhancing fertilization capacity. The presence of two isoforms of GSTO2 on the sperm surface indicates a difference in their functional roles and/or temporal association with the sperm surface. The absence of the lower molecular weight isoform from the surface of boar spermatozoa after in vitro capacitation suggests that it may be removed as the membrane reorganizes in preparation for sperm–oocyte interactions. Without the ability to differentiate between the isoforms, their specific organizations on the sperm surface cannot be fully realized. Moreover, while this is the first report of GSTO2 on the surface of eutherian spermatozoa, we cannot exclude the possibility that our GSTO fluorescent inhibitor is also interacting with a GSTO1 isoform, since the inhibitor does not differentiate between members of the GST Omega family. However, neither we nor others have been successful in identifying GSTO1 in eutherian sperm.

The regulation of oxidative stress levels within capacitation is a true balancing act. Spermatozoa optimally function on the knife's edge of beneficial versus detrimental oxidative stress levels and, therefore, must be equipped with the necessary antioxidant enzymes and substrates to effectively regulate their environment. When protective and antioxidant rich environmental barriers such as the seminal plasma are diluted within the female reproductive tract, spermatozoa must rely solely on membrane-bound, cytosolic, and mitochondrial antioxidant enzymes to maintain an effective oxidative equilibrium.

At low concentrations, reactive oxygen species (ROS) have been shown to positively affect capacitation and the acrosome exocytosis reaction and play central roles within most of the transduction pathways associated with the sperm acquiring its fertilizing ability [2–8,18]. Members of many of the oxidative-reductive regulating superfamilies' have been characterized as sperm-resident proteins, such as thioredoxins, glutaredoxins, glutathione peroxidases (GPXs) glutathione-S-transferases (GSTs) and peroxiredoxins (PRDXs) [23,31,50–55]. Located internally or on the surface of the sperm plasma membrane, these enzymes function in diverse roles regulating transduction cascades, cell signaling, sperm–oocyte interactions and oxidative stress, facilitated by antioxidant molecules such as glutathione (GSH) and ascorbic acid (AA) [17,25,53,54,56].

Ascorbic acid (AA) or vitamin C is a potent single electron donor that acts as a scavenger of ROS in most organ systems [53,57]. AA is used in energetically favorable oxidation reactions to neutralize reactive and damaging compounds that contain an unpaired electron, such as hydroxide [53]. In both the germinal epithelium and the epididymis, AA levels are significantly higher than that of blood plasma, and a high concentration of AA within the seminal plasma has been shown to be positively correlated with sperm count, sperm motility and normal sperm morphology [26,27,53,58]. Previous reports have also shown that AA is required to protect sperm from endogenous ROS production at all stages of development and maturation in both the germinal layer and the epididymis [59,60]. Therefore, its replenishment within epididymal transport and capacitation processes may prove vital in maintaining the oxidative-reductive homeostasis of sperm as they prepare for fertilization. The recycling of dehydroascorbate (DHA) into its reduced ascorbate state can be accomplished through various mechanisms, including the direct reduction by glutathione and the enzymatic reduction by various thiol transferases and NADPH-dependent reductases [61]. One such enzyme that can facilitate this reaction is glutathione-s-transferase omega 2 (GSTO2). Past reports have found that GSTO2 has the highest dehydroascorbate reductase functionality within mammalian systems, with a higher affinity for DHA than glutathione itself [30,37,62]. Therefore, in sperm maturation and capacitation, two highly oxidative processes that utilize AA as a major source of reducing power, the presence of a high affinity DHA recycling enzyme may be critical in maintaining the required concentration of AA to effectively defend against damaging levels of oxidative stress.

The sperm plasma membrane is a rich lipid bilayer that is highly susceptible to lipid peroxidation by oxidative stress [10–12,63]. Damage to the membrane structure has previously been shown to have wide reaching functional implications in sperm, as membrane fluidity and surface proteins have large roles in both capacitation and sperm–oocyte interactions [9,13–17]. Our results show that the absence of GSTO2's catalytic activity negatively impacts the spermatozoon's ability to prevent lipid peroxidation within the plasmalemma during capacitation, resulting in the impairment or dysregulation of membrane-associated processes such as the acrosome exocytosis reaction, sperm motility and, ultimately, the sperm's ability to penetrate the oocyte. Therefore, while GSTO2 may not have a direct role within each of these processes, it is possible that its functional involvement in the regulation of the membranes oxidative state may have indirect implications on the sperm's overall fitness.

Limitations of this study were that our functional inhibitor was permeable and did not discriminate between different GSTO enzymes and isoforms. However, to date, only GSTO2 has been shown to reside on or within spermatozoa. We recently reported that GSTO2 isoforms within the post-acrosomal and perforatorial regions of the perinuclear theca (PT) facilitate the post-fertilization nuclear decondensation and male nuclear transition [50]. For this reason, we did not investigate any events beyond sperm penetration where the PT isoforms would functionally participate. That these PT-resident GSTO2s could be involved in the capacitation process is unlikely, as they are firmly anchored to the PT and require harsh solubilization agents such as 1M NaOH to release them from the PT [34]. Due to this insolubility, their catalytic sites could be masked until solubilization in the oocyte cytoplasm during fertilization. Even if they were involved, their coverage compared to the surface GSTO2 is limited, and thus their functional contribution would be expected to be proportionally small.

5. Conclusions

Our investigations identify GSTO2 isoforms as functionally active surface-borne enzymes on mouse and boar spermatozoa. We propose that GSTO2 may facilitate the regulation of the oxidative environment of the plasmalemma indirectly, through the replenishment of antioxidant molecules such as ascorbic acid and glutathione. Overall, this study highlights the importance of oxidative-reductive regulation and demonstrates the wide-reaching negative implications its dysregulation can have on overall sperm fitness.

Supplementary Materials: The following are available online at http://www.mdpi.com/2076-3921/8/12/601/s1, Figure S1: Mouse and boar GSTO sequence comparison, Figure S2: GSTO inhibitor binding specificity in whole mouse and boar spermatozoa, Figure S3: The Assessment of sperm viability between treatment groups, Figure S4: The level of protein tyrosine phosphorylation during in vitro capacitation in boar spermatozoa, Figure S5: The difference in cellular reactive oxygen species (ROS) levels of GSTO Inhibited and Control capacitated spermatozoa.

Author Contributions: Conceptualization, L.E.H., P.S., C.O. and R.O.; Data curation, L.E.H., M.Z., J.M. and W.X.; Formal analysis, L.E.H.; Funding acquisition, R.O.; Investigation, L.E.H., M.Z., J.M., W.X. and C.O.; Methodology, L.E.H., M.Z., W.X., C.O. and R.O.; Supervision, P.S. and R.O.; Writing—original draft, L.E.H. and R.O.; Writing—review and editing, M.Z., J.M., P.S., C.O. and R.O.

Funding: This research was funded by the Canadian Institute of Health Research (84440) and the Natural Science and Engineering Research Council of Canada (RGPIN/05305) to RO and the Agriculture and Food Research Initiative Competitive Grant no. 2015-67015-23231 from the USDA National Institute of Food and Agriculture and seed funding from the Food for the 21st Century Program of the University of Missouri to PS.

Acknowledgments: We would like to express our thanks to the National Swine Research and Resource Center staff for their help with semen collection and process.

Conflicts of Interest: The authors declare no conflict of interest.

References

1. De Lamirande, E.; Leclerc, P.; Gagnon, C. Capacitation as a regulatory event that primes spermatozoa for the acrosome reaction and fertilization. *Mol. Hum. Reprod.* **1997**, *3*, 175–194. [CrossRef]
2. O'Flaherty, C.; de Lamirande, E.; Gagnon, C. Reactive oxygen species modulate independent protein phosphorylation pathways during human sperm capacitation. *Free Radic. Biol. Med.* **2006**, *40*, 1045–1055. [CrossRef]
3. Aitken, R.J. Free radicals, lipid peroxidation and sperm function. *Reprod. Fertil. Dev.* **1995**, *7*, 659–668. [CrossRef]
4. Aitken, R.J.; Paterson, M.; Fisher, H.; Buckingham, D.W.; van Duin, M. Redox regulation of tyrosine phosphorylation in human spermatozoa and its role in the control of human sperm function. *J. Cell Sci.* **1995**, *108 Pt 5*, 2017–2025.
5. Baumber, J.; Sabeur, K.; Vo, A.; Ball, B.A. Reactive oxygen species promote tyrosine phosphorylation and capacitation in equine spermatozoa. *Theriogenology* **2003**, *60*, 1239–1247. [CrossRef]

6. Roy, S.C.; Atreja, S.K. Production of superoxide anion and hydrogen peroxide by capacitating buffalo (Bubalus bubalis) spermatozoa. *Anim. Reprod. Sci.* **2008**, *103*, 260–270. [CrossRef] [PubMed]
7. de Lamirande, E.; O'Flaherty, C. Sperm Capacitation as an Oxidative Event BT—Studies on Men's Health and Fertility. In *Studies on Men's Health and Fertility*; Agarwal, A., Aitken, R.J., Alvarez, J.G., Eds.; Humana Press: Totowa, NJ, USA, 2012; pp. 57–94. ISBN 978-1-61779-776-7.
8. O'Flaherty, C.; Beorlegui, N.; Beconi, M.T. Participation of superoxide anion in the capacitation of cryopreserved bovine sperm. *Int. J. Androl.* **2003**, *26*, 109–114. [CrossRef] [PubMed]
9. Jones, R.; Mann, T.; Sherins, R.J. Adverse effects of peroxidized lipid on human spermatozoa. *Proc. R. Soc. Lond. Biol. Sci.* **1978**, *201*, 413–417.
10. Alvarez, C.; Storey, T.; Touchstone, C. Spontaneous Peroxidation and Production of hydrogen peroxide and superoxide in human spermatozoa. *J. Androl.* **1987**, *8*, 338–348. [CrossRef] [PubMed]
11. Aitken, R.J.; Fisher, H. Reactive Oxygen Species Generation and Human Spermatozoa: The Balance of Benefit and Risk. *BioEssays* **1994**, *16*, 259–267. [CrossRef]
12. Aitken, J.R.; Clarkson, J.S.; Fishel, S. Generation of Reactive Oxygen Species, Lipid Peroxidation, and Human Sperm Function. *Biol. Reprod.* **1989**, *41*, 183–197. [CrossRef] [PubMed]
13. Aitken, R.J.; Baker, M.A. Oxidative stress, sperm survival and fertility control. *Mol. Cell. Endocrinol.* **2006**, *250*, 66–69. [CrossRef] [PubMed]
14. Aitken, R.J. The Capacitation-Apoptosis Highway: Oxysterols and Mammalian Sperm Function. *Biol. Reprod.* **2011**, *85*, 9–12. [CrossRef]
15. Aitken, R.J. Reactive oxygen species as mediators of sperm capacitation and pathological damage. *Mol. Reprod. Dev.* **2017**, *84*, 1039–1052. [CrossRef] [PubMed]
16. De Lamirande, E.; Gagnon, C. Impact of reactive oxygen species on spermatozoa: A balancing act between beneficial and detrimental effects. *Hum. Reprod.* **1994**, *10*, 15–21. [CrossRef] [PubMed]
17. Griveau, J.F.; Dumont, E.; Renard, P.; Callegari, J.P.; Le Lannou, D. Reactive oxygen species, lipid peroxidation and enzymatic defence systems in human spermatozoa. *Reproduction* **1995**, *103*, 17–26. [CrossRef]
18. O'Flaherty, C.; de Lamirande, E.; Gagnon, C. Positive role of reactive oxygen species in mammalian sperm capacitation: Triggering and modulation of phosphorylation events. *Free Radic. Biol. Med.* **2006**, *41*, 528–540. [CrossRef]
19. De Lamirande, E.; Gagnon, C. Capacitation-associated production of superoxide anion by human spermatozoa. *Free Radic. Biol. Med.* **1995**, *18*, 487–495. [CrossRef]
20. Aitken, R.J.; Harkiss, D.; Knox, W.; Paterson, M.; Irvine, D.S. A novel signal transduction cascade in capacitating human spermatozoa characterised by a redox-regulated, cAMP-mediated induction of tyrosine phosphorylation. *J. Cell Sci.* **1998**, *111 Pt 5*, 645–656.
21. Griveau, J.F.; Renard, P.; Le Lannou, D. An in vitro promoting role for hydrogen peroxide in human sperm capacitation. *Int. J. Androl.* **1994**, *17*, 300–307. [CrossRef]
22. O'Flaherty, C.M.; Beorlegui, N.B.; Beconi, M.T. Reactive oxygen species requirements for bovine sperm capacitation and acrosome reaction. *Theriogenology* **1999**, *52*, 289–301. [CrossRef]
23. Aitken, R.J. Gpx5 protects the family jewels. *J. Clin. Investig.* **2009**, *119*, 1849–1851. [CrossRef] [PubMed]
24. Rhemrev, J.P.; van Overveld, F.W.; Haenen, G.R.; Teerlink, T.; Bast, A.; Vermeiden, J.P. Quantification of the nonenzymatic fast and slow TRAP in a postaddition assay in human seminal plasma and the antioxidant contributions of various seminal compounds. *J. Androl.* **2000**, *21*, 913–920. [PubMed]
25. Vernet, P.; Aitken, R.J.; Drevet, J.R. Antioxidant strategies in the epididymis. *Mol. Cell. Endocrinol.* **2004**, *216*, 31–39. [CrossRef]
26. Mancini, A.; Meucci, E.; Milardi, D.; Parroni, R.; Mordente, A.; Martorana, G.E.; De Marinis, L.; Littarru, G.P. Total antioxidant capacity of human seminal plasma. *Hum. Reprod.* **1996**, *11*, 1655–1660.
27. Lewis, S.E.M.; Boyle, P.M.; McKinney, K.A.; Young, I.S.; Thompson, W. Total antioxidant capacity of seminal plasma is different in fertile and infertile men. *Fertil. Steril.* **1995**, *64*, 868–870. [CrossRef]
28. Dacheux, J.L.; Druart, X.; Fouchecourt, S.; Syntin, P.; Gatti, J.L.; Okamura, N.; Dacheux, F. Role of epididymal secretory proteins in sperm maturation with particular reference to the boar. *J. Reprod. Fertil.* **1998**, *53*, 99–107.
29. Dacheux, J.L.; Belleannée, C.; Guyonnet, B.; Labas, V.; Teixeira-Gomes, A.P.; Ecroyd, H.; Druart, X.; Gatti, J.L.; Dacheux, F. The contribution of proteomics to understanding epididymal maturation of mammalian spermatozoa. *Syst. Biol. Reprod. Med.* **2012**, *58*, 197–210. [CrossRef]

30. Whitbread, A.K.; Masoumi, A.; Tetlow, N.; Schmuck, E.; Coggan, M.; Board, P.G. Characterization of the omega class of glutathione transferases. *Methods Enzymol.* **2005**, *401*, 78–99.
31. Hemachand, T.; Gopalakrishnan, B.; Salunke, D.M.; Totey, S.M.; Shaha, C. Sperm plasma-membrane-associated glutathione S-transferases as gamete recognition molecules. *J. Cell Sci.* **2002**, *115*, 2053–2065.
32. Shaha, C.; Gopalakrishnan, B. Biological Role of Glutathione S-Transferases on Sperm. In *Reproductive Immunology*; Springer: Dordrecht, The Netherlands, 2011; pp. 11–19.
33. Olshan, A.F.; Luben, T.J.; Hanley, N.M.; Perreault, S.D.; Chan, R.L.; Herring, A.H.; Basta, P.V.; DeMarini, D.M. Preliminary examination of polymorphisms of GSTM1, GSTT1, and GSTZ1 in relation to semen quality. *Mutat. Res./Fundam. Mol. Mech. Mutagenesis* **2010**, *688*, 41–46. [CrossRef] [PubMed]
34. Hamilton, L.E.; Acteau, G.; Xu, W.; Sutovsky, P.; Oko, R. The developmental origin and compartmentalization of glutathione-s-transferase omega 2 isoforms in the perinuclear theca of eutherian spermatozoa. *Biol. Reprod.* **2017**, *97*, 612–621. [CrossRef] [PubMed]
35. Board, P.G.; Coggan, M.; Chelvanayagam, G.; Easteal, S.; Jermiin, L.S.; Schulte, G.K.; Danley, D.E.; Hoth, L.R.; Griffor, M.C.; Kamath, A. V Identification, characterization, and crystal structure of the Omega class glutathione transferases. *J. Biol. Chem.* **2000**, *275*, 24798–24806. [CrossRef] [PubMed]
36. Nebert, D.W.; Vasiliou, V. Analysis of the glutathione S-transferase (GST) gene family. *Hum. Genom.* **2004**, *1*, 460. [CrossRef]
37. Schmuck, E.M.; Board, P.G.; Whitbread, A.K.; Tetlow, N.; Cavanaugh, J.A.; Blackburn, A.C.; Masoumi, A. Characterization of the monomethylarsonate reductase and dehydroascorbate reductase activities of Omega class glutathione transferase variants: Implications for arsenic metabolism and the age-at-onset of Alzheimer's and Parkinson's diseases. *Pharmacogenet. Genom.* **2005**, *15*, 493–501. [CrossRef]
38. Protopapas, N.; Hamilton, L.E.; Warkentin, R.; Xu, W.; Sutovsky, P.; Oko, R. The perforatorium and postacrosomal sheath of rat spermatozoa share common developmental origins and protein constituents. *Biol. Reprod.* **2019**, *100*, 1461–1472. [CrossRef]
39. Son, J.; Lee, J.J.; Lee, J.S.; Schller, A.; Chang, Y.T. Isozyme-specific fluorescent inhibitor of glutathione S-Transferase omega 1. *ACS Chem. Biol.* **2010**, *5*, 449–453. [CrossRef]
40. Laemmli, U.K. Cleavage of structural proteins during the assembly of the head of bacteriophage T4. *Nature* **1970**, *227*, 680–685. [CrossRef]
41. Zigo, M.; Jonáková, V.; Šulc, M.; Maňásková-Postlerová, P. Characterization of sperm surface protein patterns of ejaculated and capacitated boar sperm, with the detection of ZP binding candidates. *Int. J. Biol. Macromol.* **2013**, *61*, 322–328. [CrossRef]
42. Aitken, R.J.; Nixon, B.; Lin, M.; Koppers, A.J.; Lee, Y.H.; Baker, M.A. Proteomic changes in mammalian spermatozoa during epididymal maturation. *Asian J. Androl.* **2007**, *9*, 554–564. [CrossRef]
43. Cornwall, G.A. New insights into epididymal biology and function. *Hum. Reprod. Update* **2009**, *15*, 213–227. [CrossRef] [PubMed]
44. Cornwall, G.A. *Posttranslational Protein Modifications in the Reproductive System*; Springer: Berlin, Germany, 2014; Volume 759, pp. 159–180.
45. Guyonnet, B.; Dacheux, F.; Dacheux, J.L.; Gatti, J.L. The epididymal transcriptome and proteome provide some insights into new epididymal regulations. *J. Androl.* **2011**, *32*, 651–664. [CrossRef] [PubMed]
46. Dacheux, J.L.; Dacheux, F. New insights into epididymal function in relation to sperm maturation. *Reproduction* **2014**, *147*, R27–R42. [CrossRef] [PubMed]
47. Farkaš, R. Apocrine secretion: New insights into an old phenomenon. *Biochim. Biophys. Acta Gen. Subj.* **2015**, *1850*, 1740–1750. [CrossRef] [PubMed]
48. Hermo, L.; Jacks, D. Nature's ingenuity: Bypassing the classical secretory route via apocrine secretion. *Mol. Reprod. Dev.* **2002**, *63*, 394–410. [CrossRef] [PubMed]
49. Zhou, W.; De Iuliis, G.N.; Dun, M.D.; Nixon, B. Characteristics of the epididymal luminal environment responsible for sperm maturation and storage. *Front. Endocrinol.* **2018**, *9*, 59. [CrossRef] [PubMed]
50. Hamilton, L.E.; Suzuki, J.; Aguila, L.; Meinsohn, M.C.; Smith, O.E.; Protopapas, N.; Xu, W.; Sutovsky, P.; Oko, R. Sperm-borne glutathione-s-transferase omega 2 accelerates the nuclear decondensation of spermatozoa during fertilization in mice. *Biol. Reprod.* **2019**, *101*, 368–376. [CrossRef]
51. Miranda-Vizuete, A.; Sadek, C.M.; Jiménez, A.; Krause, W.J.; Sutovsky, P.; Oko, R. The mammalian testis-specific thioredoxin system. *Antioxid. Redox Signal.* **2004**, *6*, 25–40. [CrossRef]

52. Su, D.; Novoselov, S.V.; Sun, Q.A.; Moustafa, M.E.; Zhou, Y.; Oko, R.; Hatfield, D.L.; Gladyshev, V.N. Mammalian Selenoprotein Thioredoxin-glutathione Reductase. *J. Biol. Chem.* **2005**, *280*, 26491–26498. [CrossRef]
53. Krishnamoorthy, G.; Venkataraman, P.; Arunkumar, A.; Vignesh, R.C.; Aruldhas, M.M.; Arunakaran, J. Ameliorative effect of vitamins (α-tocopherol and ascorbic acid) on PCB (Aroclor 1254) induced oxidative stress in rat epididymal sperm. *Reprod. Toxicol.* **2007**, *23*, 239–245. [CrossRef]
54. O'Flaherty, C. Peroxiredoxins: Hidden players in the antioxidant defence of human spermatozoa. *Basic Clin. Androl.* **2014**, *24*, 4. [CrossRef] [PubMed]
55. Fanaei, H.; Khayat, S.; Halvaei, I.; Ramezani, V.; Azizi, Y.; Kasaeian, A.; Mardaneh, J.; Parvizi, M.R.; Akrami, M. Effects of ascorbic acid on sperm motility, viability, acrosome reaction and DNA integrity in teratozoospermic samples. *Iran. J. Reprod. Med.* **2014**, *12*, 103–110. [PubMed]
56. Song, G.J.; Norkus, E.P.; Lewis, V. Relationship between seminal ascorbic acid and sperm DNA integrity in infertile men. *Int. J. Androl.* **2006**, *29*, 569–575. [CrossRef] [PubMed]
57. Murugesan, P.; Muthusamy, T.; Balasubramanian, K.; Arunakaran, J. Studies on the protective role of vitamin C and E against polychlorinated biphenyl (Aroclor 1254)—Induced oxidative damage in Leydig cells. *Free Radic. Res.* **2005**, *39*, 1259–1272. [CrossRef] [PubMed]
58. Agarwal, A.; Said, T.M. Oxidative stress, DNA damage and apoptosis in male infertility: A clinical approach. *BJU Int.* **2005**, *95*, 503–507. [CrossRef] [PubMed]
59. Fraga, C.G.; Motchnik, P.A.; Shigenaga, M.K.; Helbock, H.J.; Jacob, R.A.; Ames, B.N. Ascorbic acid protects against endogenous oxidative DNA damage in human sperm. *Proc. Natl. Acad. Sci. USA* **1991**, *88*, 11003–11006. [CrossRef]
60. Dawson, E.B.; Harris, W.A.; Rankin, W.E.; Charpentier, L.A.; McGanity, W.J. Effect of Ascorbic Acid on Male Fertility. *Ann. N. Y. Acad. Sci.* **1987**, *498*, 312–323. [CrossRef]
61. May, J.; Asard, H. Ascorbate Recycling. In *Vitamin C: Its Functions and Biochemistry in Animals and Plants*; Asard, H., May, J., Smirnoff, N., Eds.; BIOS Scientific Publishers: Oxon, UK, 2004; pp. 153–175.
62. Board, P.G. The omega-class glutathione transferases: Structure, function, and genetics. *Drug Metab. Rev.* **2011**, *43*, 226–235. [CrossRef]
63. Jones, R.; Mann, T.; Sherins, R. Peroxidative breakdown of phospholipids in human spermatozoa, spermicidal properties of fatty acid peroxides, and protective action of seminal plasma. *Fertil. Steril.* **1979**, *31*, 531–537. [CrossRef]

© 2019 by the authors. Licensee MDPI, Basel, Switzerland. This article is an open access article distributed under the terms and conditions of the Creative Commons Attribution (CC BY) license (http://creativecommons.org/licenses/by/4.0/).

Review

From Past to Present: The Link Between Reactive Oxygen Species in Sperm and Male Infertility

Ana Izabel Silva Balbin Villaverde [1], Jacob Netherton [2] and Mark A. Baker [2,*]

1. Independent researcher, São Paulo 13000-000, Brazil; aivillaverde@hotmail.com
2. Department of Biological Science, University of Newcastle, Callaghan, NSW 2308, Australia; jacob.netherton@newcastle.edu.au
* Correspondence: Mark.Baker@newcastle.edu.au; Tel.: +61-2-4921-6143; Fax: +61-2-4921-6308

Received: 1 November 2019; Accepted: 26 November 2019; Published: 3 December 2019

Abstract: Reactive oxygen species (ROS) can be generated in mammalian cells via both enzymatic and non-enzymatic mechanisms. In sperm cells, while ROS may function as signalling molecules for some physiological pathways, the oxidative stress arising from the ubiquitous production of these compounds has been implicated in the pathogenesis of male infertility. In vitro studies have undoubtedly shown that spermatozoa are indeed susceptible to free radicals. However, many reports correlating ROS with sperm function impairment are based on an oxidative stress scenario created in vitro, lacking a more concrete observation of the real capacity of sperm in the production of ROS. Furthermore, sample contamination by leukocytes and the drawbacks of many dyes and techniques used to measure ROS also greatly impact the reliability of most studies in this field. Therefore, in addition to a careful scrutiny of the data already available, many aspects of the relationship between ROS and sperm physiopathology are still in need of further controlled and solid experiments before any definitive conclusions are drawn.

Keywords: dihydroethidium; lucigenin; luminol; tetrazolium salts; NADPH oxidase; cytochrome reductases

1. Introduction

The history of the relationship between reactive oxygen species (ROS) and spermatozoa starts with some fundamental experiments conducted by John MacLeod in 1943 [1]. Considering the knowledge accumulated till the present day, it is assumed that ROS, in high enough concentrations, can trigger peroxidative damage by the generation of reactive aldehydes, which are detrimental to cell function. This perception was demonstrated in different ways. However, in most studies, the negative effect of ROS on sperm quality was observed following external addition of ROS or exposure to ROS-generating in vitro systems, thus diverging from a physiological scenario. In this review, we discuss the many challenges in this field, including the various pitfalls associated with the techniques used for measuring ROS, which make it difficult to ascertain whether these compounds are a major factor contributing to male infertility or just metabolites playing a passive role. Nonetheless, novel and better methods for measuring ROS together with the current understanding of the pathways associated with peroxidative damage will certainly allow new insights into the involvement of oxidative stress in sperm function and male infertility.

2. The Foundation of the Link between ROS and Human Sperm

The earliest citation of the presence of ROS in spermatozoa comes from the laboratory of John MacLeod [1]. In 1943, MacLeod [1] decided to test the prevailing knowledge that the metabolism of human spermatozoa was exclusively dependent on glycolysis and that oxygen consumption was *"being of such small magnitude that it could not properly be interpreted as true respiration"* (cited from MacLeod [1]). Therefore, to investigate the existence of mitochondrial activity, MacLeod [1] used methylene blue as a redox sensor and observed that human sperm can reduce either glucose or succinate. In the case of succinate, the reduction of methylene blue is likely a consequence of the production of $FADH_2$ in the presence of succinic dehydrogenase (or electron transport chain Complex II), an enzyme of the mitochondrial respiratory complex that oxidizes succinate into fumarate. In addition, the oxidation of *p*-phenylenediamine by sperm cells was also observed in MacLeod's experiments, indicating the presence of cytochrome b, cytochrome c and cytochrome c oxidase. As such, this was the first evidence that sperm cells have indeed mitochondrial activity, or as better phrased by MacLeod, that they present an *"active cytochrome"* system.

Following these first observations, MacLeod [1] further examined the impact of high oxygen levels on sperm cells. For this purpose, he incubated human sperm in a 95% oxygen environment at 38 °C. Under these conditions, a drastic reduction in sperm motility occurred over time, which was completely prevented when the experiment was repeated in the presence of catalase, an enzyme that converts hydrogen peroxide (H_2O_2) into water and oxygen [1]. The notion here is that when forced to use oxidative phosphorylation, a toxic by-product is created in the form of H_2O_2. In fact, as revealed later by others, up to 0.2% of the oxygen used during mitochondrial respiration undergoes incomplete reduction, forming superoxide anion ($O_2\bullet^-$), which quickly reacts (dismutation) producing H_2O_2 [2] (for details see Figure 1). The latter can be fully reduced to water or may form oxygen radicals, such as the hydroxyl radical ($\bullet OH$), that are subsequently detrimental to sperm. Thus, the fundamental concept that ROS can negatively affect spermatozoa function was laid [3–6].

MacLeod [1] reasoned that spermatozoa were the major source of ROS, but later reports showed that leukocytes within sperm samples, a common feature among human ejaculates, were also involved in ROS production [7–9]. Leukocytes contain an NADPH-oxidase (NOX) that catalyses the production of $O_2\bullet^-$ by the oxidation of NAD(P)H [10]. The $O_2\bullet^-$ is then used to generate a wide range of reactive oxidants, with the main purpose of killing invading microorganisms [10]. However, this enzyme is so active that spermatozoa can be immobilised by as little as 6×10^5 stimulated leukocytes [8].

Motivated by the observations on the NOX activity of leukocytes, Whittington and Ford decided to reinvestigate the impact of high oxygen levels (i.e., 95% O_2 and 5% CO_2 versus 95% N_2 and 5% CO_2) using MacLeod's methodology. However, this time, sperm samples were freed of leukocytes following purification by Dynabeads [11]. Of interest, the leukocyte-free sperm populations were less affected by the high oxygen tensions and remained motile for over 6 h, showing only a reduction in curvilinear velocity. This finding clearly raises the question of whether sperm produce enough ROS to cause any significant cell damage.

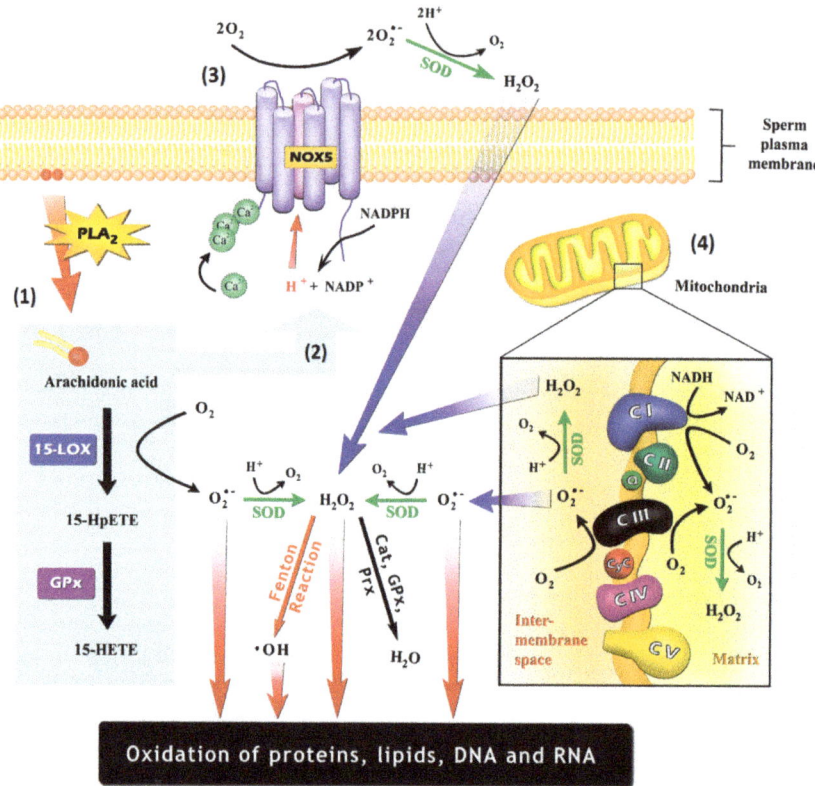

Figure 1. Possible mechanisms by which sperm cells may generate reactive oxygen species (ROS): (1) As a by-product of the oxidation of arachidonic acid (AA), which may be promoted by cyclooxygenases and lipoxygenases, such as arachidonate 15-lipoxygenase (LOX-15); (2) Through the stimulation of NADPH-oxidase (NOX) activity by AA itself or by their oxidation-generated metabolites, being 15-hydroperoxyeicosatetraenoic acid (15-HpETE) and 15-hydroxyeicosatetraenoic acid (15-HETE) potential inducers; (3) Generation by an NOX system, such as NADPH-oxidase isoform 5 (NOX5), which is embedded in the plasma membrane and is activated through an EF-hand Ca^{2+} binding domains; (4) Generation by the mitochondrial electron-transport chain, with the electron leakage within the ubiquinone binding sites in complex I (CI) and in complex III (CIII) being the most important mechanisms. Cat: catalase; CyC: cytochrome C; GPx: glutathione peroxidase; PLA2: phospholipase A2; Prx: peroxiredoxins; Q: ubiquinone; SOD: superoxide dismutase.

2.1. Spermatozoa and Their Susceptibility toward ROS

Regardless of the ROS source, the work developed by John MacLeod inspired a generation of andrologists to look at the susceptibility of spermatozoa towards these metabolites. Arguably, Thaddeus Mann was one of the first to realize the clinical significance of this association. In a landmark paper with Roy Jones and Dick Sherins, published in 1978, sperm cells showed motility loss when exposed to either exogenously introduced fatty acid, previously treated with UV light, or peroxidation of endogenous sperm phospholipids, induced by ascorbate and ferrous sulphate [12]. Both treatments are known to catalyse the oxidation of unsaturated fatty acids, forming unsaturated aldehydes such as acrolein, malonaldehyde (MDA) or 4-hydroxy-2-nonenal (4-HNE) [13]. Indeed, MDA production was confirmed by Jones et al. [12] using the thiobarbituric acid reacting substances test (TBARS). The active aldehydes formed by ROS can react with proteins through a Michael-type addition, particularly with the sulfur atom of cysteine, the imidizole nitrogen of histidine and the amine nitrogen of lysine [14]. Importantly, the toxic effect exerted by the covalently bounded aldehydes depends on the role of the adducted residue (e.g., protein structure and/or function) [15].

Currently, evidence shows that 4-HNE, a by-product of lipid peroxidation, can impair the activity of enzymes from the glycolytic and oxidative phosphorylation pathways, such as glyceraldehyde-3-phosphate-dehydrogenase (GAPDH) [16] and cytochrome-c oxidase [17]. In addition, 4-HNE has the potential to form adducts with A-kinase anchor protein 4 (AKAP4) and dynein heavy chain [18], two proteins involved in sperm motility [19,20]. These findings could explain the two main points observed by Jonas et al. [12], i.e., that sperm motility is impaired by reactive aldehydes and that necrozoospermic samples have higher MDA levels. Regardless of the adducts formed, the work developed by Jones et al. [12] appears to have sparked a major interest in the field of "ROS and defective spermatozoa" and, for the first time, offered a mechanism into why men may become infertile.

Following these initial observations, other reports have clearly confirmed that both ROS and aldehydes are detrimental to sperm function. However, something easily overlooked is the fact that most of the approaches within this theme use an exogenous source of ROS/aldehyde or force spermatozoa to generate ROS. For instance, much of the early work, reporting sperm motility loss, involved the use of exogenously added ascorbate plus ferrous ion [12,21–24], which induces the production of $O_2\bullet^-$, H_2O_2 and $\bullet OH$. Other examples include the addition of exogenous xanthine–xanthine oxidase [25–29], H_2O_2 [24,26,30,31], glucose and glucose oxidase [30], nitric oxide radical [32], menadione [24,33] and unsaturated aldehydes (acrolein, 4-HNE and MDA) [18,34–36]. All of these methods were shown to be detrimental to spermatozoa by the same pathway investigated by Jones et al. [12], in which the aldehydes formed by oxidation of unsaturated fatty acids lead to inhibition of sperm motility.

Although the external addition of the aforementioned compounds clearly affects sperm function, what remains a challenge to the field is the significance of this finding when we consider only the level of lipid peroxidation that occurs spontaneously in vitro and, most importantly, in vivo. Early studies with rabbit, mouse and human sperm have shown that spontaneous lipid peroxidation, based on MDA measurements, occurs at a slow rate, and factors such as temperature, oxygen tension and medium composition may greatly interfere [37–40]. In a work performed with stallion, the percentage of sperm naturally expressing 4-HNE increased from 53% to 86% over a 24 h incubation period under aerobic conditions [41]. In accordance, after 24 h, a slight increase in lipid peroxidation was detected in human sperm using the probe C11-BODIPY(581/591) [34]. These increments in lipid peroxidation were accompanied by a loss in sperm motility, which may limit sperm lifespan within the female tract [37–39,41]. Of interest, the lifetime of human sperm (i.e., time for complete loss of motility) was shown to be highly correlated with their level of superoxide dismutase (SOD) activity ($r = 0.97$) [40], strongly suggesting that peroxidation involving $O_2\bullet^-$ may play a major role in motility loss over time. Nevertheless, Aitken et al. [42] observed that SOD levels on both low- and high-density sperm populations, following Percoll separation, were negatively correlated with total motility after 24 h of incubation ($r = -0.303$ and $r = -0.338$, respectively). Although SOD activity was measured by different methods, one with acetylated ferricytochrome [40] and the other with lucigenin [42], this might not account for the contrasting data.

2.2. Polyunsaturated Fatty Acids Quantity and Sperm Susceptibility

It is quite clear that sperm motility is affected by ROS despite their source, and most likely this is related to lipid peroxidation. The question that arises from this observation is: Why are spermatozoa vulnerable in this regard? One argument put forward is that *"mammalian spermatozoa membranes are very sensitive to free radical-induced damage"* (cited from [43]) due to their high level of polyunsaturated fatty acids (PUFA). In whole ejaculates, the overall PUFA content in spermatozoa is between 36% and 39%, while in Percoll-purified sperm, this level reaches 48–52% of the total fatty acids [44]. This clearly demonstrates that spermatozoa are within the range of other PUFA-enriched tissues such as brain, retina and placenta (around 35%, 37% and 44% PUFA, respectively) [45–47]. In general, the hydrogen of the bisallylic methylene group (i.e., between two double bonds) has a weak bond energy (around 75 Kcal/mol) when compared to the ones present in allylic methylene groups and methylene groups that show bond dissociation energy of approximately 88 and 101 Kcal/mol, respectively [48,49]. Considering that bisallylic carbons are only present in PUFA, this intrinsic characteristic makes them more prone to peroxidation than monounsaturated and saturated fatty acids, therefore increasing the susceptible of PUFA-rich membranes such as those of sperm cells.

The most abundant PUFA in Percoll-purified human sperm was shown to be docosahexaenoic acid (DHA, 22:6n-3), followed by arachidonic acid (AA, 20:4n-6) and linoleic acid (LA, 18:2n-6), with around 34.5%, 10.5% and 6.5% of the total fatty acids, respectively [44]. Likewise, DHA is also the predominant PUFA in ruminant sperm cells [50,51]. High concentrations of DHA are also found in rod photoreceptors [52] and synaptosomes [53], where they likely modulate membrane properties, including "fluidity", flip-flop, membrane fusion and vesicle formation (reviewed by [54]). These properties are also known to be important for sperm function, thus making DHA a crucial membrane component for this cell type. To demonstrate its importance, Roqueta-Rivera and colleagues [55] showed that male mice depleted of the delta-6 desaturase enzyme, which participates in the synthesis of AA and DHA, have impaired fertility that can only be restored upon oral supplementation of DHA. Of interest, DHA presents higher oxidisability when compared to LA and AA due to their greater amount of bisallylic methylene groups [56]. Therefore, the benefits of having a singular high amount of DHA come at the expense of making sperm even more susceptible to lipid peroxidation.

The main α, β-unsaturated aldehyde formed by non-enzymatic oxidation of DHA is 4-hydroxy-2-hexenal (4-HHE), whereas the n-6 PUFA (e.g., LA acid and AA) generate 4-HNE [57]. These 4-hydroxyalkenals are very reactive and may serve as second toxic messengers, thus mediating the detrimental effects of oxidative stress upon sperm cells. For instance, even at femtomolar concentrations, 4-HHE is capable of inducing transition pore opening in mitochondria [58], which could be responsible for sperm motility loss and apoptotic changes [59,60]. However, despite its likely importance, the level of induced or spontaneous in vitro production of 4-HHE has never been examined in human sperm cells, yet it would theoretically be a more sensitive marker of oxidative stress. In contrast, 4-HNE has already been assessed and associated with a concomitant motility loss in stallion and human sperm [18,34,61,62].

Another by-product of the non-enzymatic oxidation of both n-3 and n-6 PUFA is the 3-carbon aldehyde MDA [63]. Although less toxic than 4-HNE, MDA is often used as a biomarker of lipid peroxidation due to its facile reaction with thiobarbituric acid. Nevertheless, the reliability of the TBARS test has been questioned by many, with one article stating that the "*MDA assay is not able to provide valid analytical data for biological samples due to its high reactivity and possibility of various cross-reactions with co-existing biochemicals*" [64]. Certainly, MDA levels have been found to be higher within infertile sperm [23,65–67], but the TBARS test has been used in all cases and, hence, further work is necessary to confirm these findings.

2.3. Leukocytes and Their Contribution to ROS Generation

Throughout the history of the relationship between ROS and sperm function, many have argued in favour of the hypothesis that the presence of seminal leukocytes is a confounding factor [68–71]. In this regard, reports correlating the number of white blood cells (WBC) within ejaculates and sperm dysfunction have shown both positive [68,69] and negative [72,73] correlations. In an intriguing study run by Harrison et al. [74], fertile men (i.e., fathered within 12 months) showed great variation in WBC counts, ranging from 0.5 to 16×10^6/mL of semen. It is worth mentioning that many of these fertile men were within the 95th percentile range of the WHO criteria that define leukocytospermia (i.e., more than 1×10^6 WBC/mL) [75]. On the other hand, within infertile men, the reported prevalence of leukocytospermia based on this cut-off value varies from 10% to around 20% [69,76]. These results show that leukocytospermia is not a strictly limiting factor for male fertility. In fact, Kaleli et al. [70] stated that "*leukocytospermia may have a favorable effect on some sperm functions at seminal leukocyte concentrations between 1 and 3×10^6/mL*".

Data regarding the impact of leukocyte on semen quality are always difficult to interpret because it is hard to predict: (1) when these cells had entered the seminal compartment; (2) whether they were activated; and (3) when and how they were activated. Normally, spermatozoa only encounter a large number of WBC upon ejaculation, and significant numbers of leukocytes are rarely seen in the lumina of the seminiferous or epididymal tubules [77]. In addition, upon ejaculation, when WBC generally contact sperm cells, seminal plasma is also present, thus protecting sperm with its antioxidant compounds [78,79]. Nevertheless, as soon as seminal plasma is removed, leukocytes may damage the spermatozoa, a tendency easily verified by the strong association between the presence of leukocytes in washed sperm preparations and in vitro fertilization (IVF) rates [80].

3. The Free Radical-Generating Systems in Sperm

One question still pending concerns the exact nature of the enzymatic systems responsible for free radical production in sperm cells (Figure 1). In a major review by Agarwal et al. [81], the authors indicate two ways spermatozoa may generate ROS, being: (1) an NADPH-oxidase system embedded in the plasma membrane [82]; and (2) an NADH-dependent oxidoreductase (diaphorase) at the level of mitochondria [83].

3.1. The Potential for an NADPH–Oxidase System in Sperm

The NOX hypothesis for sperm was conceived on the basis of two main observations. Firstly, ionophore A23187 was shown to increase the ROS-dependent chemiluminescent signal of either oligozoospermic samples [84] or capacitated sperm [85], indicating the action of a Ca^{2+}-dependent NOX. Secondly, the addition of NAD(P)H to sperm suspensions can generate a dose-dependent increase in luminol–peroxidase signal and in nitro blue tetrazolium (NBT) reduction, indirectly suggesting $O_2\bullet^-$ production [86,87]. In line with this theory, the NAD(P)H-dependent lucigenin signal was effectively inhibited by the addition of copper, zinc, diphenyleneiodonium (DPI) and SOD [86–89].

Following these previous observations, studies performed on equine and human sperm presented the NADPH-oxidase isoform 5 (NOX5) as one potential candidate for the ROS-generating system (Figure 1) [90–92]. This NOX isoform contains EF-hand Ca^{2+} binding domains, being activated by Ca^{2+} [90,93]. The mRNA expression of NOX5 is first detected in pachytene spermatocytes (human [90]), whereas the protein can be visualized in the developing spermatid (equine [91]). In human spermatozoa, a NOX5 antibody demonstrated cross-reactivity in the flagellum, neck and acrosome regions [92], with higher reactivity in asthenozoospermic men [94]. Additionally, Armstrong et al. [95] demonstrated that sperm NOX5 has a lower ROS-producing capacity when compared to WBC, and its activation is probably independent of protein kinase C. Despite these results, some points have not yet been satisfactorily addressed by previous reports, such as the possibility of leukocyte contamination in sperm samples, discrepancies in molecular weight and a lack of mass spectrometry evidence on the abundance of NOX5 in sperm [96–98]. Additionally, NOX5 is not found in rodents, which limits deeper pathophysiological studies.

Furthermore and in contrast to the results presented so far, the addition of NAD(P)H was also reported not to stimulate $O_2\bullet^-$ production when the superoxide-dependent probe 2-methyl-6-(*p*-methoxyphenyl)-3,7-dihydroimidazo [1,2-a] pyrazine-3-one (MCLA) [99] and the electron spin method [100] were used, therefore questioning the existence of a NOX activity in sperm.

This contradiction was later elucidated when our laboratory successfully identified the enzymes responsible for the NAD(P)H-dependent lucigenin signal as cytochrome p450 reductase (CP450R) [101] and cytochrome b5-reductase (Cb5R) [102]. CP450R (acting preferably on NAD(P)H) and Cb5R (with higher affinity for NADH) are both capable of a direct one-electron reduction of either lucigenin (Figure 2) or tetrazolium salts (e.g., NBT and WST-1) (Figure 3), thus easily explaining why these probes can evoke a signal with NAD(P)H, whilst other methods had failed (i.e., MCLA and electron spin resonance). In addition, the reduction of lucigenin and tetrazolium salts by these enzymes forms unstable radicals that may also produce $O_2\bullet^-$, which are essential for signal generation (Figures 2 and 3; for more detail see [101] and [102]). Due to the latter, SOD has the ability to inhibit the NAD(P)H-dependent lucigenin chemiluminescence and the tetrazolium salt formation generated by CP450R and Cb5R (Figures 2 and 3). Unfortunately, this inhibition by SOD is similar to the one expected when ROS is generated by NOX activity. Furthermore, like NOX, CP450R and Cb5R are also flavoproteins and, therefore, susceptible to DPI inhibition (Figures 2 and 3). Taken together, these inhibition tests are not suitable to differentiate whether the lucigenin and the tetrazolium salt signals were generated by cytochrome and/or NOX activity.

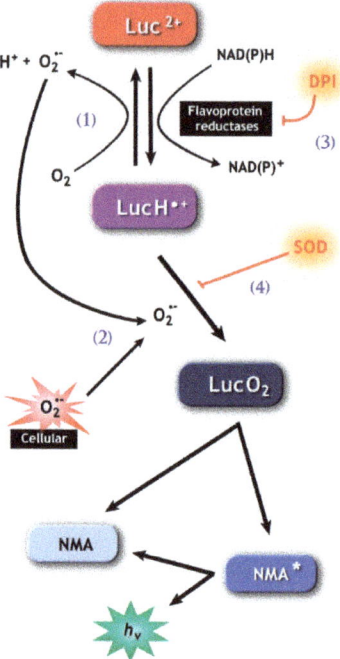

Figure 2. The pathways involved in lucigenin (bis-N-methylacridinium nitrate) chemiluminescence. (1), Lucigenin (Luc^{2+}) can be reduced to the lucigenin cation radical ($LucH\bullet^-$) by flavoprotein reductases, including cytochrome b5-reductase (Cb5R) and cytochrome p450 reductase (CP450R). $LucH\bullet^-$ can be autoxidize back to lucigenin resulting in the production of (superoxide anion) $O_2\bullet^-$, or (2) can react with $O_2\bullet^-$, forming lucigenin dioxetane ($LucO_2$). The latter spontaneously decomposes into N-methylacridone (NMA) that generates the chemiluminescence signal. Note that the signal can be abolished by diphenyleneiodonium (DPI) (3), which inhibits flavoprotein reductases, and by superoxide dismutase (SOD) (4), which consumes the $O_2\bullet^-$ necessary for the formation of $LucO_2$

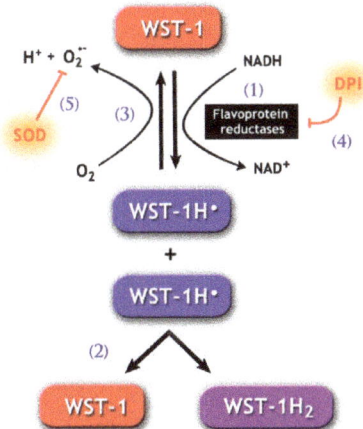

Figure 3. Chemical pathways for the 2-[4-iodophenyl]-3-[4-nitrophenyl]-5-[2,4-disulfophenyl]-2H tetrazolium monosodium salt (WST-1) assay. WST-1 can be reduced by electron transport from NADH via flavoprotein reductases, such as Cb5R and CP450R (1), forming the WST-1 radical (WST-1H•). The latter goes through disproportionation (2), which generates the reduced soluble purple formazan product (WST-1H$_2$) detected by spectrophotometry methods. Notably, WST-1H• may also react with molecular oxygen, forming $O_2\bullet^-$ (3). The generation of WST-1H• can be prevented by the addition of DPI (4), a flavoprotein inhibitor. Likewise, SOD (5) may inhibit the formation of WST-1H$_2$, because the reduction of $O_2\bullet^-$ concentration by SOD increases the autoxidation of WST-1H•, therefore reducing the latter's availability for the formation of WST-1H$_2$.

3.2. Other Enzymatic Sources of ROS in Sperm

Given the compelling pieces of evidence supporting the importance of ROS in sperm physiopathology, studies are still needed in order to determine the involvement of other sperm enzymes in the production of $O_2\bullet^-$. Besides NOX, the oxidative metabolism of AA, the second most abundant PUFA in human sperm cells [44], by cyclooxygenases (COX) and lipoxygenases (LOX) is also an important ROS-generating source. In this case, ROS can be generated as a by-product of AA oxidation [103] and/or as a result of NOX activation by either AA itself [104,105] or its LOX- and COX-generated metabolites [106,107] (Figure 1). The connection between LOX metabolites and ROS generation by NOX may be true for sperm cells. For instance, in mice germ cells, the inhibition of the isoform 15-LOX by PD146176 resulted in the reduction of ROS production within these cells [108]. Although not further investigated by Bromfield et al. [108], a study developed with Jurkat cells reported that 15-LOX metabolites may be involved in NOX stimulation [109].

3.3. Sperm Mitochondria and ROS Generation

In mammalian cells, another potential enzymatic source of ROS is the mitochondrial electron-transport chain. Under normal conditions, around 0.1–0.2% of the electrons passing the respiratory chain may leak and react with oxygen molecules, mainly forming $O_2\bullet^-$ [110]. Electron leakage may occur in several sites within the respiratory chain, being the ubiquinone binding sites in complex I (Q-binding site; $O_2\bullet^-$ is produced on the matrix side) and in complex III (Q$_o$ site; $O_2\bullet^-$ is produced in the intermembrane space) the most important ones [111,112] (Figure 1). In sperm, the specific inhibition of electron transport in complex I (by rotenone) and complex III (by antimycin-A) showed that these cells are also capable of producing ROS in these mitochondrial sites [113]. However, it is still unclear whether the ROS produced by mitochondria exerts specific physiological and/or pathological roles in sperm. For equine sperm, mitochondrial ROS were reported to positively correlate with sperm motility and velocity, probably due to an intense oxidative phosphorylation activity [114].

Nevertheless, contrasting with the equine species, human spermatozoa greatly rely on glycolysis for ATP production, with little contribution of oxidative phosphorylation [115]. For this reason, the interference of mitochondrial ROS in human sperm function may be less obvious. Of interest, defective human sperm have been shown to spontaneously generate mitochondrial ROS to a point that sperm motility may be affected [113].

4. ROS Measurement Techniques and Their Reliability

Throughout the literature on sperm and ROS, many statements and theories are still controversial and in need of re-examination and rectification. In part, this is due to some limitations and drawbacks which may be seen with the techniques and probes commonly used to evaluate ROS in living cells. Currently, no probe offers an unbiased measurement of ROS, with an ideal high reactivity and specificity for one ROS species. Importantly, this leads to a scenario in which unrealistic conclusions about the relationship between ROS and sperm function can be made. An extensive discussion on this theme can be found elsewhere (see [116]). In this review, we will limit the discussion to the probes that are more commonly used in the field of spermatology (Table 1).

Table 1. Characteristics and limiting factors of the probes commonly used to detect ROS in sperm cells.

Probe	Method	Characteristics and Limiting Factors
Tetrazolium salts	Colorimetric	Nitro blue tetrazolium (NBT) is the most commonly used one Low sensitivity to detect ROS Low specificity for $O_2\bullet^-$ detection, with various intracellular reductases being able to generate the same response Autoxidation can generate $O_2\bullet^-$
Lucigenin	Chemiluminescence	More specific for extracellular $O_2\bullet^-$ Inability to detect $O_2\bullet^-$ at low level Low specificity for $O_2\bullet^-$ detection. Signal can be triggered by various nucleophiles and reducing agents, being sensitive to changes in the reductase activity within the tested systems. Reduced radical can generate $O_2\bullet^-$
Luminol/HRP	Chemiluminescence	Allows the detection of both intra- and extracellular ROS Reacts with several electron-donor compounds, showing indiscriminate recognition of numerous free radicals The luminol radical formed by various univalent oxidants can form $O_2\bullet^-$ through autoxidation Susceptible to various interferences in biological systems, such as poor ROS detection at neutral pH and absorption of the emitted light (400 nm) by some biomolecules
DHE	Fluorescence HPLC and LC–MS	Used to detect intracellular $O_2\bullet^-$ Highly specific for $O_2\bullet^-$ detection, producing 2-hydroxyethidium (2-OH-E+); however, the majority of DHE reacts with other oxidants, resulting in the production of ethidium (E+) Both by-products of non-specific (E+) and specific (2-OH-E+) oxidation have overlapping fluorescence properties, thus not allowing distinction by fluorescence methods. For specific $O_2\bullet^-$ quantification, 2-OH-E+ must be measured by techniques such as HPLC and LC-MS

Dihydroethidium (DHE); high-performance liquid chromatography (HPLC); horseradish peroxidase (HRP); liquid chromatography–mass spectrometry (LC-MS); superoxide anion ($O_2\bullet^-$).

4.1. Lucigenin and Tetrazolium Salts

The use of NAD(P)H in conjunction with either lucigenin or tetrazolium salt techniques has been previously discussed here in Section 3.1. In this case, the main concern is that several tissue reductases, including sperm cytochromes (CP450R and Cb5R) [101,102], can reduce both probes and, therefore, lead to artefactual NAD(P)H-dependent reduction and the generation of $O_2\bullet^-$ by autoxidation [117,118] (Figures 2 and 3). However, despite these consistent factors, many studies have used this approach to indirectly report the presence of $O_2\bullet^-$ in sperm and further correlate it with semen quality [119–121], capacitation [122], hyperactivation [123], DNA integrity [120,124,125], apoptosis [120], IVF outcomes [121], among others. Yet, caution and a deep understanding of the limitations of both detection methods must guide the interpretation of these data.

4.2. Luminol/HRP

Luminol (5-amino-2,3-dihydro-1,4-phthalazine-dione) present the advantage of having a high sensitivity and the capacity to detect both intra- and extracellular ROS [82,118]. To react with $O_2\bullet^-$, luminol is first converted into an intermediate radical by a one-electron oxidation normally mediated by H_2O_2 [126,127] and enhanced by the addition of horseradish peroxidase (HRP) [82,128] (Figure 4). One major limitation is the fact that the luminol radical reacts not only with $O_2\bullet^-$ but also with various compounds capable of donating an electron [126,127], thus showing indiscriminate recognition of several free radicals. In addition, other complex and difficult-to-control factors, such as the formation of $O_2\bullet^-$ by the luminol radical, may influence the chemiluminescence of this probe [118,127,129]. Therefore, according to Zhang and colleagues [118], it is "*unwise to monitor the dynamics of free radical generation in cells or systems with this probe alone*".

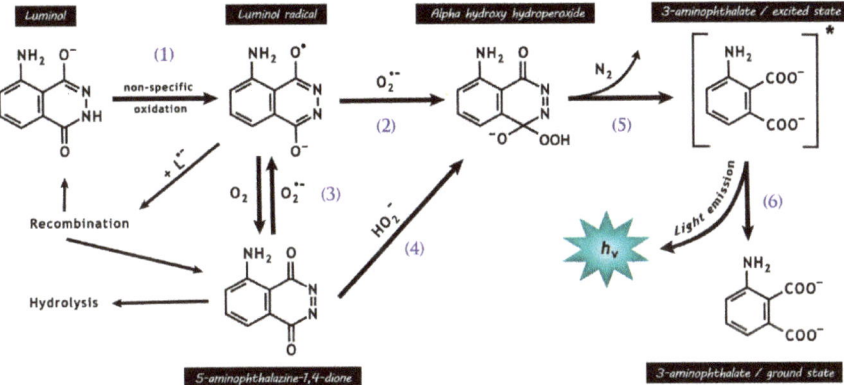

Figure 4. Chemical reactions responsible for luminol chemiluminescence. Luminol is first oxidized by many radicals (e.g., •OH and $CO_3\bullet^-$, except $O_2\bullet^-$) and peroxidases, forming the luminol radical (L•$^-$) (1). L•$^-$ then reacts with $O_2\bullet^-$, forming the short-lived intermediate hydroperoxide (2). Molecular oxygen may be reduced to $O_2\bullet^-$ by L•$^-$ (3), with a rate around seven orders of magnitude lower than that for reaction (2), resulting in the production of 5-aminophthalazine-1,4-dione. The latter may also form the intermediate hydroperoxide by the addiction of hydrogen peroxide anions (4). The intermediate hydroperoxide is quickly decomposed to 3-aminophyhalane in an excited state (5), which emits light on relaxation to the ground state (6).

Previous studies have used the luminol-based technique to suggest that pathological spermatozoa (e.g., amorphous heads, damaged acrosomes and retained cytoplasmic droplets) generate higher amounts of ROS than their normal counterparts [130,131]. Nevertheless, one possible interpretation for these data is that luminol–HRP reacts with sperm containing luminol-reactive metabolites not yet specified. This is reinforced by the fact that the retention of an excess of residual cytoplasm, a common feature of abnormal sperm, is associated with higher ROS measurements [132]. It is important to note that the excess of residual cytoplasm may contain higher amounts of the metabolites responsible for luminol signal, therefore not directly related to ROS production.

4.3. Dihydroethidium

Dihydroethidium (DHE), also called hydroethidine (HE), has been branded as a superoxide indicator and, when combined with the hexyl triphenylphosphonium cation (MitoSOX™ Red), it can specifically detect mitochondrial ROS. Oxidation of DHE by intracellular $O_2^{\bullet-}$ forms 2-hydroxyethidium (2-OH-E$^+$), that emits a red fluorescence with excitation at 510 nm [133] (Figure 5). However, DHE is also susceptible to non-specific oxidation by other oxidants (e.g., H_2O_2, •OH), generating ethidium (E$^+$), a compound with fluorescence characteristics similar to those of 2-OH-E$^+$ [134]. For this reason, because both by-products of specific (2-OH-E$^+$) and non-specific (E$^+$) oxidation of DHE have overlapping fluorescence, quantification of $O_2^{\bullet-}$ by this means is not possible when only fluorescence-based techniques are used. Of concern, many reports have used these methods to assess $O_2^{\bullet-}$ in sperm cells, thus not considering the potential contribution of the non-specific oxidation of DHE via alternative pathways [135–138].

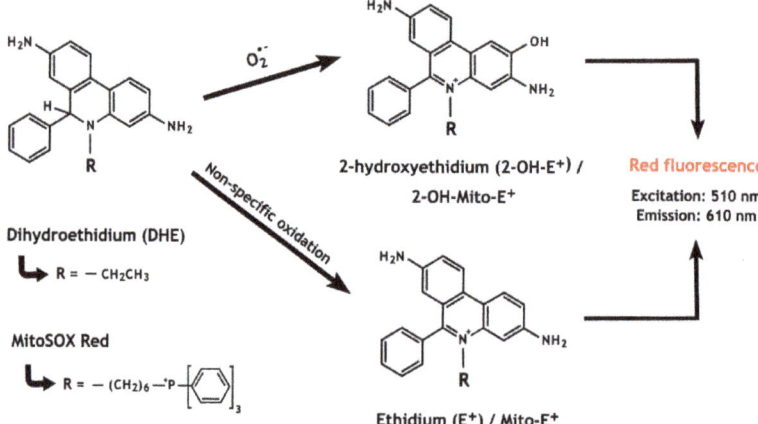

Figure 5. The chemistry behind the ROS detection methods based on dihydroethidium and MitoSOX Red oxidation. The non-specific oxidation, which forms ethidium, is predominant over the superoxide anion-induced reaction that results in the formation of 2-hydroxyethidium. Notably, both oxidized by-products present overlapping fluorescence properties.

An alternative to unambiguously confirm the presence of intracellular $O_2^{\bullet-}$ is to separately identify both 2-OH-E$^+$ and E$^+$ with techniques such as high-performance liquid chromatography (HPLC) and liquid chromatography–mass spectrometry (LC–MS) [139,140]. Using HPLC and a reversed-phase column, the 2-OH-E$^+$ and E$^+$ peaks can be separated and resolved, allowing $O_2^{\bullet-}$ quantification [140]. To the best of our knowledge, this methodology has only been used to analyse menadione-treated spermatozoa [33] and has never been used to compare the level of ROS spontaneously generated by normal and pathological sperm cells. Recently, we have used the LC–MS/MS approach to investigate sperm $O_2^{\bullet-}$ generation during in vitro incubation [141]. As previously reported, we also observed an increase in DHE over time. However, this was not accompanied by an increment in 2-OH-E$^+$ levels but was rather a consequence of an increase in the level of E$^+$ (i.e., not related to $O_2^{\bullet-}$ generation). Our finding clearly shows the importance of distinguishing 2-OH-E$^+$ from E$^+$ when assessing $O_2^{\bullet-}$ production.

5. Conclusions

From past to present, the knowledge gathered over the many years of study in this field offers us a few lessons that need to be taken into account. Firstly, to avoid any interference when assessing the production of ROS by sperm cells, an efficient removal of leukocytes from the samples is mandatory.

Their presence will always cast doubt and potentially lead to data misinterpretation, as clearly evidenced by the work of Whittington and Ford [11]. Secondly, many of the methods used to assess ROS in sperm cells present drawbacks and limitations during application, possibly obfuscating the true nature of the involvement of free radicals in sperm physiology and male infertility. The rational use of probes and sometimes the adoption of more than one method are recommended for a better assessment of ROS in cells. An indirect assessment of oxidative stress may also be done by the analysis of the products originated from lipid (MDA, 4-HNE, HHE) [23,61,62,67] and DNA oxidation (DNA base adduct 8-hydroxy-2'-deoxyguanosine) [142–144].

Finally, although sperm are susceptible to in vitro induced and exogenous sources of ROS and its by-products, the in vivo relevance of these compounds needs further clarity. Of interest, considering only the ROS produced by sperm, our laboratory has recently found that neither $O_2\bullet^-$ nor other free radicals, which lead to 4-HNE production, are responsible for motility loss during incubation [141]. In addition, a clear distinction must be made between the physiological versus the pathological roles of ROS in sperm. While a subtle increase in ROS may be necessary for sperm function such as in capacitation, the relationship between sperm abnormality and ROS may arise from a redox imbalance within the different environments to which sperm are subjected, especially in testis [145]. However, before any definitive conclusions are made, more studies using refined methodologies to look at the level of spontaneous ROS generation or lipid peroxidation in fertile and infertile males are required. In addition, while measurements of both 4-HNE and MDA have been performed in spermatozoa, the levels of 4-HHE, perhaps a more important aldehyde, still need to be evaluated.

Funding: This research was funded by National Health and Medical Research Council grant number 1182948.

Conflicts of Interest: The authors declare no conflict of interest.

Abbreviations

2-[4-iodophenyl]-3-[4-nitrophenyl]-5-[2,4-disulfophenyl]-2H tetrazolium monosodium salt (WST-1); 2-hydroxyethidium (2-OH-E$^+$); 2-methyl-6-(*p*-methoxyphenyl)-3,7-dihydroimidazo [1,2-a] pyrazine-3-one (MCLA); 4-hydroxy-2-hexenal (4-HHE); 4-hydroxy-2-nonenal (4-HNE); A-kinase anchor protein 4 (AKAP4); arachidonic acid (AA); cyclooxygenases (COX); cytochrome b5-reductase (Cb5R); cytochrome p450 reductase (CP450R); dihydroethidium (DHE); diphenyleneiodonium (DPI); docosahexaenoic acid (DHA); ethidium (E$^+$); in vitro fertilization (IVF); glyceraldehyde-3-phosphate-dehydrogenase (GAPDH); high-performance liquid chromatography (HPLC); horseradish peroxidase (HRP); hydroethidine (HE); hydrogen peroxide (H_2O_2); hydroxyl radical ($\bullet$OH); linoleic acid (LA); lipoxygenases (LOX); liquid chromatography mass spectrometry (LC-MS); malonaldehyde (MDA); NADPH-oxidase (NOX); NADPH-oxidase isoform 5 (NOX5); nitro blue tetrazolium (NBT); polyunsaturated fatty acids (PUFA); reactive oxygen species (ROS); superoxide anion ($O_2\bullet^-$); superoxide dismutase (SOD); thiobarbituric acid reacting substances test (TBARS); white blood cells (WBC).

References

1. MacLeod, J. The role of oxygen in the metabolism and motility of human spermatozoa. *Am. J. Physiol.* **1943**, *138*, 512–518. [CrossRef]
2. Cadenas, E.; Boveris, A.; Ragan, C.I.; Stoppani, A.O. Production of superoxide radicals and hydrogen peroxide by NADH-ubiquinone reductase and ubiquinol-cytochrome c reductase from beef-heart mitochondria. *Arch. Biochem. Biophys.* **1977**, *80*, 248–257. [CrossRef]
3. Aitken, J.; Fisher, H. Reactive oxygen species generation and human spermatozoa: The balance of benefit and risk. *Bioessays* **1994**, *16*, 259–267. [CrossRef]
4. Aitken, R.J. Free radicals, lipid peroxidation and sperm function. *Reprod. Fertil. Dev.* **1995**, *7*, 659–668. [CrossRef]
5. Tremellen, K. Oxidative stress and male infertility—A clinical perspective. *Hum. Reprod. Update* **2008**, *14*, 243–258. [CrossRef]
6. Ko, E.Y.; Sabanegh, E.S.; Agarwal, A. Male infertility testing: Reactive oxygen species and antioxidant capacity. *Fertil. Steril.* **2014**, *102*, 1518–1527. [CrossRef]
7. Kessopoulou, E.; Tomlinson, M.J.; Barratt, C.L.; Bolton, A.E.; Cooke, I.D. Origin of reactive oxygen species in human semen: Spermatozoa or leucocytes? *J. Reprod. Fertil.* **1992**, *94*, 463–470. [CrossRef]

8. Kovalski, N.N.; de Lamirande, E.; Gagnon, C. Reactive oxygen species generated by human neutrophils inhibit sperm motility: Protective effect of seminal plasma and scavengers. *Fertil. Steril.* **1992**, *58*, 809–816. [CrossRef]
9. Aitken, R.J.; Buckingham, D.W.; Brindle, J.; Gomez, E.; Baker, H.W.; Irvine, D.S. Analysis of sperm movement in relation to the oxidative stress created by leukocytes in washed sperm preparations and seminal plasma. *Hum. Reprod.* **1995**, *10*, 2061–2071. [CrossRef]
10. Baehner, R.L.; Nathan, D.G. Leukocyte oxidase: Defective activity in chronic granulomatous disease. *Science* **1967**, *155*, 835–836. [CrossRef]
11. Whittington, K.; Ford, W. The effect of incubation periods under 95% oxygen on the stimulated acrosome reaction and motility of human spermatozoa. *Mol. Hum. Reprod.* **1998**, *4*, 1053–1057. [CrossRef]
12. Jones, R.; Mann, T.; Sherins, R. Adverse effects of peroxidized lipid on human spermatozoa. *Proc. R. Soc. Lond. B Biol. Sci.* **1978**, *201*, 413–417.
13. Kenaston, C.B.; Wilbur, K.M.; Ottolenghi, A.; Bernheim, F. Comparison of methods for determining fatty acid oxidation produced by ultraviolet irradiation. *J. Am. Oil Chem. Soc.* **1955**, *32*, 33–35. [CrossRef]
14. Esterbauer, H.; Schaur, R.J.; Zollner, H. Chemistry and biochemistry of 4-hydroxynonenal, malonaldehyde and related aldehydes. *Free Radic. Biol. Med.* **1991**, *11*, 81–128. [CrossRef]
15. Pizzimenti, S.; Ciamporcero, E.; Daga, M.; Pettazzoni, P.; Arcaro, A.; Cetrangolo, G.; Minelli, R.; Dianzani, C.; Lepore, A.; Gentile, F.; et al. Interaction of aldehyde derived from lipid peroxidation and membrane proteins. *Front. Physiol.* **2013**, *4*, 242. [CrossRef]
16. Uchida, K.; Stadtman, E.R. Covalent attachment of 4-hydroxynonenal to glyceraldehyde-3-phosphate dehydrogenase. A possible involvement of intra- and intermolecular cross-linking reaction. *J. Biol. Chem.* **1993**, *268*, 6388–6393.
17. Musatov, A.; Carroll, C.A.; Liu, Y.C.; Henderson, G.I.; Weintraub, S.T.; Robinson, N.C. Identification of bovine heart cytochrome c oxidase subunits modified by the lipid peroxidation product 4-hydroxy-2-nonenal. *Biochemistry* **2002**, *41*, 8212–8220. [CrossRef]
18. Baker, M.A.; Weinberg, A.; Hetherington, L.; Villaverde, A.I.; Velkov, T.; Baell, J.; Gordon, C.P. Defining the mechanism by which the reactive oxygen species by-product, 4-hydroxynonenal, affects human sperm cell function. *Biol. Reprod.* **2015**, *92*, 108–112. [CrossRef]
19. Neesen, J.; Kirschner, R.; Ochs, M.; Schmiedl, A.; Habermann, B.; Mueller, C.; Holstein, A.F.; Nuesslein, T.; Adham, I.; Engel, W. Disruption of an inner arm dynein heavy chain gene results in asthenozoospermia and reduced ciliary beat frequency. *Hum. Mol. Genet.* **2001**, *10*, 1117–1128. [CrossRef]
20. Miki, K.; Willis, W.D.; Brown, P.R.; Goulding, E.H.; Fulcher, K.D.; Eddy, E.M. Targeted disruption of the Akap4 gene causes defects in sperm flagellum and motility. *Dev. Biol.* **2002**, *248*, 331–342. [CrossRef]
21. Jones, R.; Mann, T. Lipid peroxidation in spermatozoa. *Proc. R. Soc. Lond. B Biol. Sci.* **1973**, *184*, 103–107.
22. Aitken, R.J.; Clarkson, J.S.; Fishel, S. Generation of reactive oxygen species, lipid peroxidation, and human sperm function. *Biol. Reprod.* **1989**, *41*, 183–197. [CrossRef]
23. Rao, B.; Soufir, J.; Martin, M.; David, G. Lipid peroxidation in human spermatozoa as relate to midpiece abnormalities and motility. *Mol. Reprod. Dev.* **1989**, *24*, 127–134.
24. Guthrie, H.D.; Welch, G.R. Using fluorescence-activated flow cytometry to determine reactive oxygen species formation and membrane lipid peroxidation in viable boar spermatozoa. *Methods Mol. Biol.* **2010**, *594*, 163–171.
25. Nissen, H.; Kreysel, H. Superoxide dismutase in human semen. *Klin. Wochenschr.* **1983**, *61*, 63–65. [CrossRef]
26. De Lamirande, E.; Gagnon, C. Reactive oxygen species and human spermatozoa: I. Effects on the motility of intact spermatozoa and on sperm axonemes. *J. Androl.* **1992**, *13*, 368.
27. Aitken, R.; Buckingham, D.; Harkiss, D. Use of a xanthine oxidase free radical generating system to investigate the cytotoxic effects of reactive oxygen species on human spermatozoa. *J. Reprod. Fertil.* **1993**, *97*, 441–450. [CrossRef]
28. De Lamirande, E.; Cagnon, C. Human sperm hyperactivation and capacitation as parts of an oxidative process. *Free Radic. Biol. Med.* **1993**, *14*, 157–166. [CrossRef]
29. Baumber, J.; Ball, B.A.; Gravance, C.G.; Medina, V.; Davies-Morel, M.C. The effect of reactive oxygen species on equine sperm motility, viability, acrosomal integrity, mitochondrial membrane potential, and membrane lipid peroxidation. *J. Androl.* **2000**, *21*, 895–902.

30. Bize, I.; Santander, G.; Cabello, P.; Driscoll, D.; Sharpe, C. Hydrogen peroxide is involved in hamster sperm capacitation in vitro. *Biol. Reprod.* **1991**, *44*, 398–403. [CrossRef]
31. Duru, N.K.; Morshedi, M.; Oehninger, S. Effects of hydrogen peroxide on DNA and plasma membrane integrity of human spermatozoa. *Fertil. Steril.* **2000**, *74*, 1200–1207. [CrossRef]
32. Zini, A.; de Lamirande, E.; Gagnon, C. Low levels of nitric oxide promote human sperm capacitation in vitro. *J. Androl.* **1995**, *16*, 424–431.
33. De Iuliis, G.N.; Wingate, J.K.; Koppers, A.J.; McLaughlin, E.A.; Aitken, R.J. Definitive evidence for the nonmitochondrial production of superoxide anion by human spermatozoa. *J. Clin. Endocrinol. Metab.* **2006**, *91*, 1968–1975. [CrossRef]
34. Aitken, R.J.; Whiting, S.; de Iuliis, G.N.; McClymont, S.; Mitchell, L.A.; Baker, M.A. Electrophilic aldehydes generated by sperm metabolism activate mitochondrial reactive oxygen species generation and apoptosis by targeting succinate dehydrogenase. *J. Biol. Chem.* **2012**, *287*, 33048–33060. [CrossRef]
35. Moazamian, R.; Polhemus, A.; Connaughton, H.; Fraser, B.; Whiting, S.; Gharagozloo, P.; Aitken, R.J. Oxidative stress and human spermatozoa: Diagnostic and functional significance of aldehydes generated as a result of lipid peroxidation. *Mol. Hum. Reprod.* **2015**, *21*, 502–515. [CrossRef]
36. Hall, S.E.; Aitken, R.J.; Nixon, B.; Smith, N.D.; Gibb, Z. Electrophilic aldehyde products of lipid peroxidation selectively adduct to heat shock protein 90 and arylsulfatase A in stallion spermatozoa. *Biol. Reprod.* **2017**, *96*, 107–121.
37. Alvarez, J.G.; Storey, B.T. Spontaneous lipid peroxidation in rabbit epididymal spermatozoa: Its effect on sperm motility. *Biol. Reprod.* **1982**, *27*, 1102–1108. [CrossRef]
38. Alvarez, J.G.; Storey, B.T. Assessment of cell damage caused by spontaneous lipid peroxidation in rabbit spermatozoa. *Biol. Reprod.* **1984**, *30*, 323–331. [CrossRef]
39. Alvarez, J.G.; Storey, B.T. Spontaneous lipid peroxidation in rabbit and mouse epididymal spermatozoa: Dependence of rate on temperature and oxygen concentration. *Biol. Reprod.* **1985**, *32*, 342–351. [CrossRef]
40. Alvarez, J.G.; Touchstone, J.C.; Blasco, L.; Storey, B.T. Spontaneous lipid peroxidation and production of hydrogen peroxide and superoxide in human spermatozoa Superoxide dismutase as major enzyme protectant against oxygen toxicity. *J. Androl.* **1987**, *8*, 338–348. [CrossRef]
41. Gibb, Z.; Lambourne, S.R.; Curry, B.J.; Hall, S.E.; Aitken, R.J. Aldehyde dehydrogenase plays a pivotal role in the maintenance of stallion sperm motility. *Biol. Reprod.* **2016**, *94*, 133. [CrossRef]
42. Aitken, R.J.; Buckingham, D.W.; Carreras, A.; Irvine, D.S. Superoxide dismutase in human sperm suspensions: Relationship with cellular composition, oxidative stress, and sperm function. *Free Radic. Biol. Med.* **1996**, *21*, 495–504. [CrossRef]
43. Maneesh, M.; Jayalekshmi, H. Role of reactive oxygen species and antioxidants on pathophysiology of male reproduction. *Indian J. Clin. Biochem.* **2006**, *21*, 80–89. [CrossRef]
44. Lenzi, A.; Picardo, M.; Gandini, L.; Dondero, F. Lipids of the sperm plasma membrane: From polyunsaturated fatty acids considered as markers of sperm function to possible scavenger therapy. *Hum. Reprod. Update* **1996**, *2*, 246–256. [CrossRef]
45. Martinez, M.; Ballabriga, A.; Gil-Gibernau, J.J. Lipids of the developing human retina: I. Total fatty acids, plasmalogens, and fatty acid composition of ethanolamine and choline phosphoglycerides. *J. Neurosci. Res.* **1988**, *20*, 484–490. [CrossRef]
46. Al, M.D.; Hornstra, G.; van der Schouw, Y.T.; Bulstra-Ramakers, M.T.; Huisjes, H.J. Biochemical EFA status of mothers and their neonates after normal pregnancy. *Early Hum. Dev.* **1990**, *24*, 239–248. [CrossRef]
47. Yehuda, S.; Rabinovitz, S.; Mostofsky, D.I. Essential fatty acids are mediators of brain biochemistry and cognitive functions. *J. Neurosci. Res.* **1999**, *56*, 565–570. [CrossRef]
48. Gardner, H.W. Oxygen radical chemistry of polyunsaturated fatty acids. *Free Radic. Biol. Med.* **1989**, *7*, 65–86. [CrossRef]
49. Koppenol, W.H. Oxyradical reactions: From bond-dissociation energies to reduction potentials. *FEBS Lett.* **1990**, *264*, 165–167. [CrossRef]
50. Neill, A.R.; Masters, C.J. Metabolism of fatty acids by bovine spermatozoa. *Biochem. J.* **1972**, *127*, 375–385. [CrossRef]
51. Neill, A.R.; Masters, C.J. Metabolism of fatty acids by ovine spermatozoa. *J. Reprod. Fertil.* **1973**, *34*, 279–287. [CrossRef]

52. Anderson, R.E. Lipids of ocular tissues. IV. A comparison of the phospholipids from the retina of six mammalian species. *Exp. Eye Res.* **1970**, *10*, 339–344. [CrossRef]
53. Breckenridge, W.C.; Gombos, G.; Morgan, I.G. The lipid composition of adult rat brain synaptosomal plasma membranes. *Biochim. Biophys. Acta* **1972**, *266*, 695–707. [CrossRef]
54. Stillwell, W.; Wassall, S.R. Docosahexaenoic acid: Membrane properties of a unique fatty acid. *Chem. Phys. Lipids* **2003**, *126*, 1–27. [CrossRef]
55. Roqueta-Rivera, M.; Stroud, C.K.; Haschek, W.M.; Akare, S.J.; Segre, M.; Brush, R.S.; Agbaga, M.P.; Anderson, R.E.; Hess, R.A.; Nakamura, M.T. Docosahexaenoic acid supplementation fully restores fertility and spermatogenesis in male delta-6 desaturase-null mice. *J. Lipid Res.* **2010**, *51*, 360–367. [CrossRef]
56. Cosgrove, J.P.; Church, D.F.; Pryor, W.A. The kinetics of the autoxidation of polyunsaturated fatty acids. *Lipids* **1987**, *22*, 299–304. [CrossRef]
57. Catalá, A. Five decades with polyunsaturated fatty acids: Chemical synthesis, enzymatic formation, lipid peroxidation and its biological effects. *J. Lipids* **2013**, *2013*, 710290. [CrossRef]
58. Kristal, B.S.; Park, B.K.; Yu, B.P. 4-Hydroxyhexenal is a potent inducer of the mitochondrial permeability transition. *J. Biol. Chem.* **1996**, *271*, 6033–6038. [CrossRef]
59. Ortega Ferrusola, C.; González Fernández, L.; Salazar Sandoval, C.; Macías García, B.; Rodríguez Martínez, H.; Tapia, J.A.; Peña, F.J. Inhibition of the mitochondrial permeability transition pore reduces "apoptosis like" changes during cryopreservation of stallion spermatozoa. *Theriogenology* **2010**, *74*, 458–465. [CrossRef]
60. Uribe, P.; Cabrillana, M.E.; Fornés, M.W.; Treulen, F.; Boguen, R.; Isachenko, V.; Isachenko, E.; Sánchez, R.; Villegas, J.V. Nitrosative stress in human spermatozoa causes cell death characterized by induction of mitochondrial permeability transition-driven necrosis. *Asian J. Androl.* **2018**, *20*, 600–607.
61. Aitken, R.J.; Gibb, Z.; Mitchell, L.A.; Lambourne, S.R.; Connaughton, H.S.; de Iuliis, G.N. Sperm motility is lost in vitro as a consequence of mitochondrial free radical production and the generation of electrophilic aldehydes but can be significantly rescued by the presence of nucleophilic thiols. *Biol. Reprod.* **2012**, *87*, 110. [CrossRef]
62. Martin Muñoz, P.; Ortega Ferrusola, C.; Vizuete, G.; Plaza Dávila, M.; Rodriguez Martinez, H.; Peña, F.J. Depletion of intracellular thiols and increased production of 4-hydroxynonenal that occur during cryopreservation of stallion spermatozoa lead to caspase activation, loss of motility, and cell death. *Biol. Reprod.* **2015**, *93*, 143.
63. Esterbauer, H.; Cheeseman, K.H. Determination of aldehydic lipid peroxidation products: Malonaldehyde and 4-hydroxynonenal. *Methods Enzymol.* **1990**, *186*, 407–421.
64. Khoubnasabjafari, M.; Ansarin, K.; Jouyban, A. Reliability of malondialdehyde as a biomarker of oxidative stress in psychological disorders. *BioImpacts* **2015**, *5*, 123–127.
65. Suleiman, S.A.; Ali, M.E.; Zaki, Z.; El-Malik, E.; Nasr, M. Lipid peroxidation and human sperm motility: Protective role of vitamin E. *J. Androl.* **1996**, *17*, 530–537.
66. Keskes-Ammar, L.; Feki-Chakroun, N.; Rebai, T.; Sahnoun, Z.; Ghozzi, H.; Hammami, S.; Zghal, K.; Fki, H.; Damak, J.; Bahloul, A. Sperm oxidative stress and the effect of an oral vitamin E and selenium supplement on semen quality in infertile men. *Arch. Androl.* **2003**, *49*, 83–94. [CrossRef]
67. Tavilani, H.; Doosti, M.; Saeidi, H. Malondialdehyde levels in sperm and seminal plasma of asthenozoospermic and its relationship with semen parameters. *Clin. Chim. Acta* **2005**, *356*, 199–203. [CrossRef]
68. Wolff, H.; Anderson, D.J. Immunohistologic characterization and quantitation of leukocyte subpopulations in human semen. *Fertil. Steril.* **1988**, *49*, 497–504. [CrossRef]
69. Wolff, H.; Politch, J.A.; Martinez, A.; Haimovici, F.; Hill, J.A.; Anderson, D.J. Leukocytospermia is associated with poor semen quality. *Fertil. Steril.* **1990**, *53*, 528–536. [CrossRef]
70. Kaleli, S.; Öçer, F.; Irez, T.; Budak, E.; Aksu, M.F. Does leukocytospermia associate with poor semen parameters and sperm functions in male infertility? The role of different seminal leukocyte concentrations. *Eur. J. Obstet. Gynecol. Reprod. Biol.* **2000**, *89*, 185–191. [CrossRef]
71. Saleh, R.A.; Agarwal, A. Oxidative stress and male infertility: From research bench to clinical practice. *J. Androl.* **2002**, *23*, 737–752.
72. El-Demiry, M.I.; Young, H.; Elton, R.A.; Hargreave, T.B.; James, K.; Chisholm, G.D. Leucocytes in the ejaculate from fertile and infertile men. *Br. J. Urol.* **1986**, *58*, 715–720. [CrossRef]
73. Kung, A.; Ho, P.; Wang, C. Seminal leucocyte subpopulations and sperm function in fertile and infertile Chinese men. *Int. J. Androl.* **1993**, *16*, 189–194. [CrossRef]

74. Harrison, P.; Barratt, C.; Robinson, A.; Kessopoulou, E.; Cooke, I. Detection of white blood cell populations in the ejaculates of fertile men. *Am. J. Reprod. Immunol.* **1991**, *19*, 95–98. [CrossRef]
75. WHO, *Laboratory Manual for the Examination of Human Semen and Sperm-Cervical Mucus Interation*; Cambridge University Press: Cambridge, UK, 2010.
76. Wolff, H. The biologic significance of white blood cells in semen. *Fertil. Steril.* **1995**, *63*, 1143–1157.
77. Flickinger, C.J.; Bush, L.A.; Howards, S.S.; Herr, J.C. Distribution of leukocytes in the epithelium and interstitium of four regions of the Lewis rat epididymis. *Anat. Rec.* **1997**, *248*, 380–390. [CrossRef]
78. Mohammad Eid Hammadeh, M.E.; Filippos, A.A.; Hamad, M.F. Reactive oxygen species and antioxidant in seminal plasma and their impact on male fertility. *Int. J. Fertil. Steril.* **2009**, *3*, 87–110.
79. Aitken, R.J.; Baker, M.A. Oxidative stress, spermatozoa and leukocytic infiltration: Relationships forged by the opposing forces of microbial invasion and the search for perfection. *Am. J. Reprod. Immunol.* **2013**, *100*, 11–19. [CrossRef]
80. Krausz, C.; Mills, C.; Rogers, S.; Tan, S.; Aitken, R.J. Stimulation of oxidant generation by human sperm suspensions using phorbol esters and formyl peptides: Relationships with motility and fertilization in vitro. *Fertil. Steril.* **1994**, *62*, 599–605. [CrossRef]
81. Agarwal, A.; Saleh, R.A.; Bedaiwy, M.A. Role of reactive oxygen species in the pathophysiology of human reproduction. *Fertil. Steril.* **2003**, *79*, 829–843. [CrossRef]
82. Aitken, R.J.; Buckingham, D.W.; West, K.M. Reactive oxygen species and human spermatozoa: Analysis of the cellular mechanisms involved in luminol- and lucigenin-dependent chemiluminescence. *J. Cell. Physiol.* **1992**, *151*, 466–477. [CrossRef]
83. Gavella, M.; Lipovac, V. NADH-dependent oxido-reductase (diaphorase) activity and isozyme pattern of sperm in infertile men. *Arch. Androl.* **1992**, *28*, 135–141. [CrossRef]
84. Aitken, R.J.; Clarkson, J.S.; Hargreave, T.B.; Irvine, D.S.; Wu, F.C. Analysis of the relationship between defective sperm function and the generation of reactive oxygen species in cases of oligozoospermia. *J. Androl.* **1989**, *10*, 214–220. [CrossRef]
85. De Lamirande, E.; Tsai, C.; Harakat, A.; Gagnon, C. Involvement of reactive oxygen species in human sperm acrosome reaction induced by A23187, lysophosphatidylcholine, and biological fluid ultrafiltrates. *J. Androl.* **1998**, *19*, 585–594.
86. Aitken, R.J.; Fisher, H.M.; Fulton, N.; Gomez, E.; Knox, W.; Lewis, B.; Irvine, S. Reactive oxygen species generation by human spermatozoa is induced by exogenous NADPH and inhibited by the flavoprotein inhibitors diphenylene iodonium and quinacrine. *Mol. Reprod. Dev.* **1997**, *47*, 468–482. [CrossRef]
87. Vernet, P.; Fulton, N.; Wallace, C.; Aitken, R.J. Analysis of reactive oxygen species generating systems in rat epididymal spermatozoa. *Biol. Reprod.* **2001**, *65*, 1102–1113. [CrossRef]
88. Aitken, R.J.; Vernet, P. Maturation of redox regulatory mechanisms in the epididymis. *J. Reprod. Fertil. Suppl.* **1998**, *53*, 109–118.
89. Aitken, R.J.; Ryan, A.L.; Baker, M.A.; McLaughlin, E.A. Redox activity associated with the maturation and capacitation of mammalian spermatozoa. *Free Radic. Biol. Med.* **2004**, *36*, 994–1010. [CrossRef]
90. Bánfi, B.; Molnár, G.; Maturana, A.; Steger, K.; Hegedûs, B.; Demaurex, N.; Krause, K.H. A Ca(2+)-activated NADPH oxidase in testis, spleen, and lymph nodes. *J. Biol. Chem.* **2001**, *276*, 37594–37601. [CrossRef]
91. Sabeur, K.; Ball, B.A. Characterization of NADPH oxidase 5 in equine testis and spermatozoa. *Reproduction* **2007**, *134*, 263–270. [CrossRef]
92. Musset, B.; Clark, R.A.; DeCoursey, T.E.; Petheo, G.L.; Geiszt, M.; Chen, Y.; Cornell, J.E.; Eddy, C.A.; Brzyski, R.G.; El Jamali, A. NOX5 in human spermatozoa expression, function, and regulation. *J. Biol. Chem.* **2012**, *287*, 9376–9388. [CrossRef]
93. Bánfi, B.; Tirone, F.; Durussel, I.; Knisz, J.; Moskwa, P.; Molnár, G.Z.; Krause, K.H.; Cox, J.A. Mechanism of Ca2+ activation of the NADPH oxidase 5 (NOX5). *J. Biol. Chem.* **2004**, *279*, 18583–18591. [CrossRef]
94. Vatannejad, A.; Tavilani, H.; Sadeghi, M.R.; Karimi, M.; Lakpour, N.; Amanpour, S.; Shabani Nashtaei, M.; Doosti, M. Evaluation of the NOX5 protein expression and oxidative stress in sperm from asthenozoospermic men compared to normozoospermic men. *J. Endocrinol. Invest.* **2019**, *42*, 1181–1189. [CrossRef]
95. Armstrong, J.S.; Bivalcqua, T.J.; Chamulitrat, W.; Sikka, S.; Hellstrom, W.J. A comparison of the NADPH oxidase in human sperm and white blood cells. *Int. J. Androl.* **2002**, *25*, 223–229. [CrossRef]

96. Baker, M.A.; Reeves, G.; Hetherington, L.; Müller, J.; Baur, I.; Aitken, R.J. Identification of gene products present in Triton X-100 soluble and insoluble fractions of human spermatozoa lysates using LC-MS/MS analysis. *Proteom. Clin. Appl.* **2007**, *1*, 524–532. [CrossRef]
97. Baker, M.A.; Naumovski, N.; Hetherington, L.; Weinberg, A.; Velkov, T.; Aitken, R.J. Head and flagella subcompartmental proteomic analysis of human spermatozoa. *Proteomics* **2013**, *13*, 61–74. [CrossRef]
98. Wang, G.; Guo, Y.; Zhou, T.; Shi, X.; Yu, J.; Yang, Y.; Wu, Y.; Wang, J.; Liu, M.; Chen, X.; et al. In-depth proteomic analysis of the human sperm reveals complex protein compositions. *J. Proteom.* **2013**, *79*, 114–122. [CrossRef]
99. De Lamirande, E.; Harakat, A.; Gagnon, C. Human sperm capacitation induced by biological fluids and progesterone, but not by NADH or NADPH, is associated with the production of superoxide anion. *J. Androl.* **1998**, *19*, 215–225.
100. Richer, S.C.; Ford, W.C. A critical investigation of NADPH oxidase activity in human spermatozoa. *Mol. Hum. Reprod.* **2001**, *7*, 237–244. [CrossRef]
101. Baker, M.A.; Krutskikh, A.; Curry, B.J.; McLaughlin, E.A.; Aitken, R.J. Identification of cytochrome P450-reductase as the enzyme responsible for NADPH-dependent lucigenin and tetrazolium salt reduction in rat epididymal sperm preparations. *Biol. Reprod.* **2004**, *71*, 307–318. [CrossRef]
102. Baker, M.A.; Krutskikh, A.; Curry, B.J.; Hetherington, L.; Aitken, R.J. Identification of cytochrome-b5 reductase as the enzyme responsible for NADH-dependent lucigenin chemiluminescence in human spermatozoa. *Biol. Reprod.* **2005**, *73*, 334–342. [CrossRef]
103. Katsuki, H.; Okuda, S. Arachidonic acid as a neurotoxic and neurotrophic substance. *Prog. Neurobiol.* **1995**, *46*, 607–636. [CrossRef]
104. Shiose, A.; Sumimoto, H. Arachidonic acid and phosphorylation synergistically induce a conformational change of p47phox to activate the phagocyte NADPH oxidase. *J. Biol. Chem.* **2000**, *275*, 13793–13801. [CrossRef]
105. Kim, C.; Dinauer, M.C. Impaired NADPH oxidase activity in Rac2-deficient murine neutrophils does not result from defective translocation of p47phox and p67phox and can be rescued by exogenous arachidonic acid. *J. Biol. Chem.* **2006**, *79*, 223–234.
106. De Carvalho, D.D.; Sadok, A.; Bourgarel-Rey, V.; Gattacceca, F.; Penel, C.; Lehmann, M.; Kovacic, H. Nox1 downstream of 12-lipoxygenase controls cell proliferation but not cell spreading of colon cancer cells. *Int. J. Cancer* **2008**, *122*, 1757–1764. [CrossRef]
107. Cho, K.J.; Seo, J.M.; Kim, J.H. Bioactive lipoxygenase metabolites stimulation of NADPH oxidases and reactive oxygen species. *Mol. Cells* **2011**, *32*, 1–5. [CrossRef]
108. Bromfield, E.G.; Mihalas, B.P.; Dun, M.D.; Aitken, R.J.; McLaughlin, E.A.; Walters, J.L.; Nixon, B. Inhibition of arachidonate 15-lipoxygenase prevents 4-hydroxynonenal-induced protein damage in male germ cells. *Biol. Reprod.* **2017**, *96*, 598–609. [CrossRef]
109. Kumar, K.A.; Arunasree, K.M.; Roy, K.R.; Reddy, N.P.; Aparna, A.; Reddy, G.V.; Reddanna, P. Effects of (15S)-hydroperoxyeicosatetraenoic acid and (15S)-hydroxyeicosatetraenoic acid on the acute- lymphoblastic-leukaemia cell line Jurkat: Activation of the Fas-mediated death pathway. *Biotechnol. Appl. Biochem.* **2009**, *52*, 121–133. [CrossRef]
110. Tahara, E.B.; Navarete, F.D.; Kowaltowski, A.J. Tissue-, substrate-, and site-specific characteristics of mitochondrial reactive oxygen species generation. *Free Radic. Biol. Med.* **2009**, *46*, 1283–1297. [CrossRef]
111. Brand, M.D. The sites and topology of mitochondrial superoxide production. *Exp. Gerontol.* **2010**, *45*, 466–472. [CrossRef]
112. Dröse, S.; Brandt, U. Molecular mechanisms of superoxide production by the mitochondrial respiratory chain. *Adv. Exp. Med. Biol.* **2012**, *748*, 145–169.
113. Koppers, A.J.; de Iuliis, G.N.; Finnie, J.M.; McLaughlin, E.A.; Aitken, R.J. Significance of mitochondrial reactive oxygen species in the generation of oxidative stress in spermatozoa. *J. Clin. Endocrinol. Metab.* **2008**, *93*, 3199–3207. [CrossRef]
114. Gibb, Z.; Lambourne, S.R.; Aitken, R.J. The paradoxical relationship between stallion fertility and oxidative stress. *Biol. Reprod.* **2014**, *91*, 77. [CrossRef]

115. Nascimento, J.M.; Shi, L.Z.; Tam, J.; Chandsawangbhuwana, C.; Durrant, B.; Botvinick, E.L.; Berns, M.W. Comparison of glycolysis and oxidative phosphorylation as energy sources for mammalian sperm motility, using the combination of fluorescence imaging, laser tweezers, and real-time automated tracking and trapping. *J. Cell. Physiol.* **2008**, *217*, 745–751. [CrossRef]
116. Wardman, P. Fluorescent and luminescent probes for measurement of oxidative and nitrosative species in cells and tissues: Progress, pitfalls, and prospects. *Free Radic. Biol. Med.* **2007**, *43*, 995–1022. [CrossRef]
117. Aitken, R.J. Nitroblue tetrazolium (NBT) assay. *Reprod. Biomed. Online* **2018**, *36*, 90–91. [CrossRef]
118. Zhang, Y.; Daia, M.; Yuan, Z. Methods for the detection of reactive oxygen species. *Anal. Methods* **2018**, *38*, 1–17. [CrossRef]
119. Said, T.M.; Agarwal, A.; Sharma, R.K.; Mascha, E.; Sikka, S.C.; Thomas, A.J., Jr. Human sperm superoxide anion generation and correlation with semen quality in patients with male infertility. *Fertil. Steril.* **2004**, *82*, 871–877. [CrossRef]
120. Tunc, O.; Thompson, J.; Tremellen, K. Development of the NBT assay as a marker of sperm oxidative stress. *Int. J. Androl.* **2010**, *33*, 13–21. [CrossRef]
121. Pujol, A.; Obradors, A.; Esteo, E.; Costilla, B.; García, D.; Vernaeve, V.; Vassena, R. Oxidative stress level in fresh ejaculate is not related to semen parameters or to pregnancy rates in cycles with donor oocytes. *J. Assist. Reprod. Genet.* **2016**, *33*, 529–534. [CrossRef]
122. Donà, G.; Fiore, C.; Andrisani, A.; Ambrosini, G.; Brunati, A.; Ragazzi, E.; Armanini, D.; Bordin, L.; Clari, G. Evaluation of correct endogenous reactive oxygen species content for human sperm capacitation and involvement of the NADPH oxidase system. *Hum. Reprod.* **2011**, *26*, 3264–3273. [CrossRef]
123. McKinney, K.A.; Lewis, S.E.; Thompson, W. Reactive oxygen species generation in human sperm: Luminol and lucigenin chemiluminescence probes. *Arch. Androl.* **1996**, *36*, 119–125. [CrossRef]
124. Said, T.M.; Agarwal, A.; Sharma, R.K.; Thomas, A.J.; Sikka, S.C. Impact of sperm morphology on DNA damage caused by oxidative stress induced by β-nicotinamide adenine dinucleotide phosphate. *Fertil. Steril.* **2005**, *83*, 95–103. [CrossRef]
125. Gosálvez, J.; Coppola, L.; Fernández, J.L.; López-Fernández, C.; Góngora, A.; Faundez, R.; Kim, J.; Sayme, N.; de la Casa, M.; Santiso, R.; et al. Multi-centre assessment of nitroblue tetrazolium reactivity in human semen as a potential marker of oxidative stress. *Reprod. Biomed. Online* **2017**, *34*, 513–521. [CrossRef]
126. Merényi, G.; Lind, J.; Eriksen, T.E. Luminol chemiluminescence: Chemistry, excitation, emitter. *J. Biolumin. Chemilumin.* **1990**, *5*, 53–56. [CrossRef]
127. Faulkner, K.; Fridovich, I. Luminol and lucigenin as detectors for O2.-. *Free Radic. Biol. Med.* **1993**, *15*, 447–451. [CrossRef]
128. Prichard, P.M.; Cormier, M.J. Studies on the mechanism of the horseradish peroxidase catalyzed luminescent peroxidation of luminol. *Biochem. Biophys. Res. Commun.* **1968**, *3*, 131–136. [CrossRef]
129. Vilim, V.; Wilhelm, J. What do we measure by a luminol-dependent chemiluminescence of phagocytes? *Free Radic. Biol. Med.* **1989**, *6*, 623–629. [CrossRef]
130. Gil-Guzman, E.; Ollero, M.; Lopez, M.C.; Sharma, R.K.; Alvarez, J.G.; Thomas, A.J., Jr.; Agarwal, A. Differential production of reactive oxygen species by subsets of human spermatozoa at different stages of maturation. *Hum. Reprod.* **2001**, *16*, 1922–1930. [CrossRef]
131. Aziz, N.; Saleh, R.A.; Sharma, R.K.; Lewis-Jones, I.; Esfandiari, N.; Thomas, A.J., Jr.; Agarwal, A. Novel association between sperm reactive oxygen species production, sperm morphological defects, and the sperm deformity index. *Fertil. Steril.* **2004**, *81*, 349–354. [CrossRef]
132. Aitken, J.; Krausz, C.; Buckingham, D. Relationships between biochemical markers for residual sperm cytoplasm, reactive oxygen species generation, and the presence of leukocytes and precursor germ cells in human sperm suspensions. *Mol. Reprod. Dev.* **1994**, *39*, 268–279. [CrossRef]
133. Zhao, H.; Kalivendi, S.; Zhang, H.; Joseph, J.; Nithipatikom, K.; Vásquez-Vivar, J.; Kalyanaraman, B. Superoxide reacts with hydroethidine but forms a fluorescent product that is distinctly different from ethidium: Potential implications in intracellular fluorescence detection of superoxide. *Free Radic. Biol. Med.* **2003**, *34*, 1359–1368. [CrossRef]
134. Zielonka, J.; Kalyanaraman, B. Hydroethidine- and MitoSOX-derived red fluorescence is not a reliable indicator of intracellular superoxide formation: Another inconvenient truth. *Free Radic. Biol. Med.* **2010**, *48*, 983–1001. [CrossRef]

135. Burnaugh, L.; Sabeur, K.; Ball, B. Generation of superoxide anion by equine spermatozoa as detected by dihydroethidium. *Theriogenology* **2007**, *67*, 580–589. [CrossRef]
136. Espinoza, J.; Schulz, M.; Sánchez, R.; Villegas, J. Integrity of mitochondrial membrane potential reflects human sperm quality. *Andrologia* **2009**, *41*, 51–54. [CrossRef]
137. Aitken, R.J.; Hanson, A.R.; Kuczera, L. Electrophoretic sperm isolation: Optimization of electrophoresis conditions and impact on oxidative stress. *Hum. Reprod.* **2011**, *26*, 1955–1964. [CrossRef]
138. Mahfouz, R.Z.; du Plessis, S.S.; Aziz, N.; Sharma, R.; Sabanegh, E.; Agarwal, A. Sperm viability, apoptosis, and intracellular reactive oxygen species levels in human spermatozoa before and after induction of oxidative stress. *Fertil. Steril.* **2010**, *93*, 814–821. [CrossRef]
139. Zhao, H.; Joseph, J.; Fales, H.M.; Sokoloski, E.A.; Levine, R.L.; Vasquez-Vivar, J.; Kalyanaraman, B. Detection and characterization of the product of hydroethidine and intracellular superoxide by HPLC and limitations of fluorescence. *Proc. Natl. Acad. Sci. USA* **2005**, *102*, 5727–5732. [CrossRef]
140. Kalyanaraman, B.; Dranka, B.P.; Hardy, M.; Michalski, R.; Zielonka, J. HPLC-based monitoring of products formed from hydroethidine-based fluorogenic probes—The ultimate approach for intra- and extracellular superoxide detection. *Biochim. Biophys. Acta* **2014**, *1840*, 739–744. [CrossRef]
141. Netherton, J.K.; Hetherington, L.; Ogle, R.A.; Mazloumi, M.; Velkov, T.; Villaverde, A.I.S.B.; Tanphaichitr, N.; Baker, M.A. Mass Spectrometry reveals new insights into the production of superoxide anions and 4-hydroxynonenal adducted proteins in human sperm. *Proteomics* **2019**, accepted.
142. Kodama, H.; Yamaguchi, R.; Fukuda, J.; Kasai, H.; Tanaka, T. Increased oxidative deoxyribonucleic acid damage in the spermatozoa of infertile male patients. *Fertil. Steril.* **1997**, *68*, 519–524. [CrossRef]
143. Aitken, R.J.; de Iuliis, G.N.; Finnie, J.M.; Hedges, A.; McLachlan, R.I. Analysis of the relationships between oxidative stress, DNA damage and sperm vitality in a patient population: Development of diagnostic criteria. *Hum. Reprod.* **2010**, *25*, 2415–2426. [CrossRef]
144. Guz, J.; Gackowski, D.; Foksinski, M.; Rozalski, R.; Zarakowska, E.; Siomek, A.; Szpila, A.; Kotzbach, M.; Kotzbach, R.; Olinski, R. Comparison of oxidative stress/DNA damage in semen and blood of fertile and infertile men. *PLoS ONE* **2013**, *8*, e68490. [CrossRef]
145. Asadi, N.; Bahmani, M.; Kheradmand, A.; Rafieian-Kopaei, M. The impact of oxidative stress on testicular function and the role of antioxidants in improving it: A review. *J. Clin. Diagn. Res.* **2017**, *11*, IE01–IE05. [CrossRef]

 © 2019 by the authors. Licensee MDPI, Basel, Switzerland. This article is an open access article distributed under the terms and conditions of the Creative Commons Attribution (CC BY) license (http://creativecommons.org/licenses/by/4.0/).

Review

The Importance of Oxidative Stress in Determining the Functionality of Mammalian Spermatozoa: A Two-Edged Sword

Robert J. Aitken [1,2,*] and Joel R. Drevet [3]

1. Priority Research Centre for Reproductive Sciences, Faculty of Science and Faculty of Health and Medicine, The University of Newcastle, Callaghan, NSW 2308, Australia
2. Hunter Medical Research Institute, New Lambton Heights, NSW, 2305, Australia
3. GReD Institute, INSERM U1103—CNRS UMR6293—Université Clermont Auvergne, Faculty of Medicine, CRBC building, 28 place Henri Dunant, 63001 Clermont-Ferrand, France; joel.drevet@uca.fr
* Correspondence: john.aitken@newcastle.edu.au

Received: 3 January 2020; Accepted: 21 January 2020; Published: 27 January 2020

Abstract: This article addresses the importance of oxidative processes in both the generation of functional gametes and the aetiology of defective sperm function. Functionally, sperm capacitation is recognized as a redox-regulated process, wherein a low level of reactive oxygen species (ROS) generation is intimately involved in driving such events as the stimulation of tyrosine phosphorylation, the facilitation of cholesterol efflux and the promotion of cAMP generation. However, the continuous generation of ROS ultimately creates problems for spermatozoa because their unique physical architecture and unusual biochemical composition means that they are vulnerable to oxidative stress. As a consequence, they are heavily dependent on the antioxidant protection afforded by the fluids in the male and female reproductive tracts and, during the precarious process of insemination, seminal plasma. If this antioxidant protection should be compromised for any reason, then the spermatozoa experience pathological oxidative damage. In addition, situations may prevail that cause the spermatozoa to become exposed to high levels of ROS emanating either from other cells in the immediate vicinity (particularly neutrophils) or from the spermatozoa themselves. The environmental and lifestyle factors that promote ROS generation by the spermatozoa are reviewed in this article, as are the techniques that might be used in a diagnostic context to identify patients whose reproductive capacity is under oxidative threat. Understanding the strengths and weaknesses of ROS-monitoring methodologies is critical if we are to effectively identify those patients for whom treatment with antioxidants might be considered a rational management strategy.

Keywords: male infertility; oxidative stress; lipid peroxidation; sperm biology

1. Introduction

In a landmark paper published in 1943, McLeod [1] demonstrated that the incubation of human spermatozoa under conditions of high oxygen tension precipitated a loss of motility that could be reversed by the presence of catalase. The clear implication of these findings, that human spermatozoa are vulnerable to hydrogen peroxide attack was confirmed many years later when human spermatozoa were exposed to the mixture of reactive oxygen species (ROS) generated by the xanthine oxidase system, a known mediator of sperm oxidative stress in vivo and in vitro [2,3]. Exposure in vitro was found to induce a loss of fertilizing potential and, ultimately, motility, via mechanisms that could be completely reversed by concomitant exposure to catalase, which specifically catalyzes the conversion of hydrogen peroxide to oxygen and water, but was exacerbated by superoxide dismutase, which catalyzes the dismutation of superoxide anion to hydrogen peroxide [3]. Excessively high levels of spontaneous ROS generation were subsequently shown to be associated with the defective sperm function encountered in cases of human infertility [4,5]. The mechanism by which such oxidative stress induced defective sperm function was further shown to be linked to the capacity of these reactive oxygen metabolites to stimulate lipid peroxidation in spermatozoa [6–8]. These cells are particularly vulnerable to this process because they are richly endowed with polyunsaturated fatty acids, the double bonds of which facilitate the hydrogen abstraction process that initiates the peroxidation cascade [8]. The causative links between oxidative stress, lipid peroxidation and sperm function have subsequently been confirmed many times by independent groups for human spermatozoa [9–13] and the spermatozoa of every other mammalian species examined including the bull [14], stallion [15,16], pig [17,18], dog [19], cat [20], etc. This fundamental concept, pioneered by such luminaries as Jones and Mann [21], Gagnon [22] and Storey [6] is fundamental to our understanding of the aetiology of defective sperm function. Of course, it is not the only factor responsible for compromising the fertilizing potential of mammalian spermatozoa and it is important not to overstate the case. Many other factors are potentially involved in suppressing a function such as sperm motility, including excess intracellular calcium [23], dephosphorylation of phosphatidyl-inositol-3-kinase [24,25], exposure to motility-inhibiting proteins in semen, including seminal amyloid [26] and seminogelin [27], and exposure to sperm immobilizing antibodies [28]. However, the relative significance of oxidative stress as a major factor in the aetiology of defective sperm function has been clearly demonstrated in the few antioxidant studies that have included measures of lipid peroxidation (malondialdehyde [MDA]) in their evaluation schedule as well as a placebo control group [29]. In one such study, patients exhibiting asthenozoospermia (motility ≤ 40%) were treated with antioxidants (100 mg vitamin E) or placebo, 3 times a day for either 6 months or until the patient's partner was diagnosed as pregnant, whichever came first. This study demonstrated that treatment with vitamin E significantly improved sperm motility while significantly reducing MDA levels in the spermatozoa in a manner that could not be replicated by administering a placebo formulation. A second, independent, placebo-controlled study drew the same conclusions; treatment with a combination of vitamin E and selenium for 3 months successfully decreasing MDA concentrations and concomitantly improving semen quality [30]. A further study which did not include a placebo control group but did measure MDA levels in seminal plasma, also confirmed the importance of lipid peroxidation in the pathophysiology of defective sperm function. Treatment for 90 days with an antioxidant preparation decreased seminal MDA and statistically improved all elements of the conventional semen profile including motility, concentration and morphology [31]. Another similar study, also concluded that antioxidant treatment (N-acetylcysteine in this case) for 3 months could simultaneously decrease seminal MDA levels while significantly improving sperm concentration and motility in the ejaculate [32]. Thus overall, there is a wide-ranging consensus based on thousands of patients that male infertility is associated with significantly elevated levels of MDA in semen which is, in turn, negatively correlated with levels of seminal antioxidant protection and key aspects of semen quality including sperm count, motility and morphology [33,34].

The apparent effectiveness of antioxidant therapy in suppressing oxidative stress while improving semen quality demonstrates the causative nature of these associations. It should be emphasized in passing that antioxidant trials that do not involve measurement of oxidative stress (sadly the majority) are essentially worthless for the very reason that not all defective sperm function is oxidatively induced. Clearly this is an area for further research involving properly controlled clinical trials with antioxidant formulations that are based upon a knowledge of sperm biochemistry and the bioavailability of the administered antioxidants within the male reproductive tract. This has been achieved in animal models where the cause of the infertility is unequivocally oxidative in nature. Thus, using the GPx5$^{-/-}$ knockout model, mice can be generated in which the reproductive pathology observed is entirely dependent on the creation of a localized stress within the epididymis. Treatment of these animals with a carefully formulated antioxidant preparation resulted in a return of oxidative DNA-damaged spermatozoa to control levels. In the same study, the loss of fertility observed as a consequence of scrotal heat stress could also be completely abrogated by pre-treating the animals with the same antioxidant formulation [35].

So, there is an overwhelming volume of evidence indicating a cascade of causal interactions between oxidative stress, peroxidative damage to sperm lipids and DNA and impaired sperm function. A number of questions arise in the wake of such findings including: what are the situations in which sperm oxidative stress occurs? Is the oxidative stress systemic or localized? Where does the oxidative stress come from in terms of cell type and biochemical pathway? What species of ROS are involved? How can we best measure them? In the remainder of this review, we shall attempt to address these key issues.

2. What Are the Situations in Which Sperm Oxidative Stress Occurs?

2.1. Capacitation and Hyperactivation

There are multiple lines of evidence demonstrating that spermatozoa are professional generators of ROS because of the fundamental role these molecules play in the induction of sperm capacitation. Biochemically, one of the major pathways through which ROS promote capacitation is via the redox regulation of tyrosine phosphorylation. The significance of ROS in this context appears to apply to all mammalian species examined including man [36–38], rat [39], mouse [40], buffalo [41], bull [42] and stallion [43]. The mechanisms underpinning this redox effect on protein tyrosine phosphorylation are multifaceted and involve the stimulation of cAMP generation, the inhibition of tyrosine phosphatase activity and, the modulation of additional signal transduction cascades, including SRC (Rous sarcoma oncogene)—and ERK (Extracellular Receptor Kinase)—mediated pathways (Figure 1) [38,43–45]. It has also been demonstrated that the formation of oxysterols during sperm capacitation facilitates one of the hallmarks of sperm capacitation, the removal of cholesterol from the sperm plasma membrane [46]. Such a change is thought to enhance the fluidity of the plasma membrane, promoting critical intermolecular interactions that promote the development of a capacitated state.

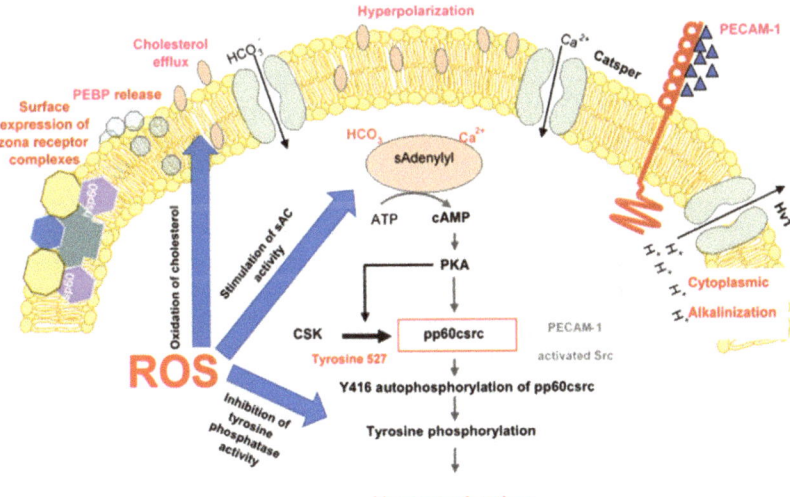

Figure 1. The role of reactive oxygen species in the induction of sperm capacitation. The latter is a complex process involving hyperpolarization of the sperm plasma membrane, cytoplasmic alkalinisation as a consequence of proton extrusion via the Hv1 proton channel, calcium entry via Catsper and a global increase in tyrosine phosphorylation mediated by cAMP. ROS (reactive oxygen species) are involved in several aspects of capacitation including cholesterol oxidation and extrusion, stimulation of soluble adenylyl cyclase (sAC) and suppression of tyrosine phosphatase activity.

One specific aspect of sperm biology driven by redox activity during sperm capacitation is the onset of hyperactivated motility. This work was pioneered by Gagnon and de Lamirande [47] who demonstrated that hyperactivation is a redox-mediated event, possibly reflecting the ability of ROS to induce tyrosine phosphorylation in the fibrous sheath of the sperm tail [48].

2.2. Capacitation and Sperm-Egg Interaction

Another redox-regulated aspect of sperm function that has received little attention is sperm-zona interaction. When human spermatozoa are incubated with ferrous ion promoters to enhance lipid peroxidation, there is, at levels of peroxidation that are still compatible with full viability and motility, a dramatic increase in the ability of these cells to bind to the zona pellucida [8]. This phenomenon has been used to enhance levels of fertilization in a mouse in vitro fertilization system in which the induction of sublethal levels of lipid peroxidation was shown to significantly increase the number of spermatozoa binding to the zona pellucida [49]. How the induction of peroxidative damage to the sperm plasma membrane enhances sperm-zona binding is unknown. It does not appear to involve a generalized increase in sperm adhesiveness because no amount of lipid peroxidation will enhance zona binding if the latter has been precoated with an anti-zona antibody (unpublished observations). During fertilization, the oocyte is also thought to generate ROS and may play an active role in the process of zona hardening, a proposed component of the block-to-polyspermy. Zona hardening is thought to be induced by peroxidases released during the cortical granule reaction enhancing the creation of molecular cross-links within zona pellucida with the aid of a hydrogen peroxide burst associated with fertilization [50]. This mechanism has been elegantly demonstrated in the case of the sea urchin fertilization envelope where the dual oxidase, Udx1, has been shown to generate the hydrogen peroxide associated with the hardening of this membrane [51]. The existence of an analogous process in mammalian oocytes seems likely; however, while these cells are known to contain a variety of oxidases capable of generating ROS [52], their role during fertilization remains largely unexplored. Whatever mechanisms are involved, it is clear that many aspects of sperm biology including tyrosine phosphorylation, cholesterol exclusion, hyperactivation and sperm-egg interaction are redox regulated.

As a result, capacitating spermatozoa in either the female reproductive tract or in vitro can be thought of as under physiological oxidative stress. In order to protect the spermatozoa during this critical time in their life history, sophisticated antioxidant defense mechanisms have developed, involving such key players as glutathione-S-transferase omega 2 and peroxiredoxin 6 [53,54]. As a consequence of such strategies, spermatozoa can engage in a redox-regulated capacitation cascade without fear of succumbing to the oxidative stresses involved. However, if a spermatozoon should fail to find an egg and the capacitation period is prolonged, even these defensive mechanisms are ultimately overwhelmed and the cell, now in a state of "over-capacitation", enters a senescence pathway culminating in the enhanced release of ROS from the mitochondria and cell death [24]. Thus, capacitation and senescence can be regarded as components of redox-regulated continuum [55].

2.3. Inadequate Antioxidant Protection from Seminal Plasma

As a biological response to oxidative stress, seminal plasma has evolved one of the most powerful antioxidant fluids known to man, replete with a range of antioxidant enzymes and small molecular mass free radical scavengers that, combined, generate a level of total antioxidant power that is estimated to be 10× higher than blood [56]. This antioxidant cocktail includes catalase, superoxide dismutase (SOD), glutathione peroxidase, glutathione-S-transferase and peroxiredoxins as well as water-soluble (uric acid, hypotaurine, tyrosine, polyphenols, vitamin C, ergothioneine and glutathione) and fat-soluble (all-*trans*-retinoic acid, all-*trans*-retinol, α-tocopherol, carotenoids and coenzyme Q10) scavengers [56–58]. There are now many studies indicating that there is a consistent negative relationship between the levels of antioxidant protection provided by seminal plasma and the appearance of male infertility as well as the incidence of miscarriage [33,59]. A recent development in this field has been the suggestion that the measurement of oxidation-reduction potential (ORP) in human semen samples is predictive of oxidative stress [60]. It will be of interest to determine whether such ORP measurements are providing diagnostic information over and above the measurement of total antioxidant potential. It will also be fascinating to determine whether the decreased seminal antioxidant protection observed in cases of male infertility is a result of local [61] or systemic [62] pro-oxidant factors.

2.4. Leucocyte Infiltration

The antioxidant properties of seminal plasma are particularly important in protecting spermatozoa from the oxidative stress created by infiltrating leukocytes, particularly neutrophils. Up to the point of ejaculation, spermatozoa in the seminiferous and epididymal tubules would have had little, if any, direct contact with activated phagocytic leukocytes. However following ejaculation, they become exposed to phagocytes, originating from the urethra and secondary sexual organs, which will be significantly elevated in cases of genital tract infection. These seminal leukocytes are in an activated state, generating free radicals and influencing seminal redox balance as reflected in several oxido-sensitive indices [63]. However, as long as seminal plasma is present, the spermatozoa are protected by the antioxidants contained therein [64,65]. However, when seminal plasma is removed, the leukocytes have free reign to attack the spermatozoa and limit their capacity for movement and fertilization. The presence of contaminating leukocytes, even in low numbers, in the washed sperm suspensions used for IVF therapy has been shown to have a profound impact on the success of this form of therapy [66]. Possible solutions to this problem include the incorporation of selected antioxidants into the sperm culture media used for IVF (e.g., glutathione, N-acetylcysteine, hypotaurine, etc.) or the targeted removal of the leukocyte population using magnetic beads or ferrrofluids coated with antibodies against the common leukocyte antigen, CD45 [67,68].

2.5. Cryostorage

Another situation in which oxidative stress is known to play a key role in the disruption of normal sperm function is cryopreservation. This relationship was first pointed out in the 1940s by the pioneering work of Tosic and Walton [69] who showed that the addition of an egg-yolk based extender to bovine spermatozoa resulted in a loss of motility that was dependent on the presence of a sperm amino acid oxidase generating high levels of hydrogen peroxide in response to the aromatic amino acids in egg yolk (particularly phenylalanine). In this scenario, the loss of membrane integrity on the part of non-viable spermatozoa enabled the aromatic amino acids in egg yolk to gain access to the intracellular oxidase and thereby generate quantities of hydrogen peroxide that could suppress the motility of viable cells in the immediate vicinity. A similar situation has been observed in ram [70] and stallion [71] spermatozoa. While human spermatozoa also contain an aromatic amino acid oxidase, it is not bound and becomes rapidly lost from cells if their plasma membrane integrity has been compromised [72]. One of the obvious solutions to this amino acid oxidase problem in domestic animals has been to incorporate catalase into the sperm cryopreservation medium [73].

While the damage induced in mammalian spermatozoa by cryopreservation is clearly multifactorial (involving, to various degrees, cold shock, osmotic disruption and intracellular ice crystal formation) oxidative stress is clearly a significant factor in the mediation of cryostorage injury. For this reason, there is intense interest in the use of antioxidants to promote cryosurvival. Recent studies have, for example, indicated that the specific activity of SOD in seminal plasma is related to the freezability of stallion, jackass and dog spermatozoa [74,75]. Loss of SOD from human semen samples is also thought to be a key factor in determining their ability to survive cryostorage [76]. Consistent with these observations, supplementation of cryopreservation media with both SOD and catalase has been found to enhance the post-thaw motility of human spermatozoa [77]. Exactly the same has been found for porcine and rooster spermatozoa cryopreserved in the presence of supplementary SOD and catalase [78,79]. SOD mimetics have also been shown to improve the cryopreservation of alpaca and ram spermatozoa and to improve subsequent blastocyst formation rates in the goat [80–82]. The combination of SOD mimetic and catalase was, predictably, more effective than mimetic alone [83].

In addition to the use of antioxidant enzymes to curtail oxidative stress during cryopreservation a large number of small molecular mass free radical scavengers have also been deployed for this purpose; entry of the terms "cryopreservation" and "antioxidants" into PubMed yields over 3000 references and the area has recently been thoroughly reviewed [84]. The antioxidants assessed included ascorbic acid, α-tocopherol, reduced glutathione, zinc oxide nanoparticles, zinc sulphate, resveratrol, quercetin, melatonin, L-carnitine, coenzyme Q, hypotaurine/taurine and butylated hydroxytoluene. In general, such antioxidants have a beneficial effect on sperm survival and functionality following cryopreservation although there is still much to be done to create the optimal antioxidant blend for protecting the spermatozoa of individual species.

2.6. Lifestyle Exposures

Another reason for spermatozoa to become oxidatively stressed relates to environmental and lifestyle exposures that either directly promote ROS generation or suppress levels of intrinsic antioxidant protection. A classic example is smoking. If men smoke heavily, their entire physiology is oxidatively stressed, as reflected by lower levels of antioxidants, such as ascorbate, in seminal plasma as well as enhanced levels of ROS generation by the spermatozoa. A 48% increase in seminal leukocyte concentration in male smokers also contributes to the level of redox stress experienced by the spermatozoa [85]. A particular pathological feature of cigarette smoking is that it generates a significant increase in oxidative DNA damage in spermatozoa as reflected by elevated levels of 8-hydroxy-2'-deoxyguanosine (8-OHdG) [86]. This increase in oxidative DNA damage may be attributable to both the above-mentioned depletion of antioxidant protection as well as the suppressive impact of cadmium (a critical constituent of cigarette smoke) on OGG-1 (8-oxoguanine-DNA glycosylase-1; the first enzyme in the base excision repair pathway responsible for removing 8-OHdG adducts from the genome) during spermatogenesis [87]. The high levels of 8-OHdG present in the spermatozoa of male smokers has, in turn, been associated with the increased childhood cancer rates observed in the offspring of male smokers [88,89]. Interestingly, we have identified a region of chromosome 15 (15q13–15q14) as a particular hot spot for oxidative DNA damage in human spermatozoa [90] and this is the very region of the genome that has recently been identified as contributing to the aetiology of acute lymphoblastic leukemia [91], one of the childhood cancers that we know to be associated with paternal smoking [92].

Obesity is another lifestyle factor responsible for inducing a state of oxidative stress in spermatozoa. This condition is associated with a generalized pro-inflammatory state associated with systemic oxidative stress, antioxidant depletion and oxidative sperm DNA damage [93]. Fortunately, this situation can be reversed, at least in mice, by the concomitant administration of micronutrients (zinc, selenium, lycopene, vitamins E and C, folic acid, and green tea extract) to counter the oxidative stress [94].

Different frequencies of electromagnetic radiation have also been suggested to induce oxidative stress in the male germ line. The impact of radiofrequency electromagnetic radiation (RF-EMR) has recently been reviewed and supports the general consensus that, like obesity, this form of radiation can induce ROS generation, reduce antioxidant protection and increase sperm DNA damage [95]. Moreover, a 2-step mechanism has been proposed to explain this phenomenon; in the first step, RF-EMR induces displacement of electrons from the electron transport chain in the inner mitochondrial membrane. These leaked electrons are immediately taken up by the universal electron acceptor, oxygen, to generate superoxide anion. In the second step, the latter dismutates into the powerful oxidant, hydrogen peroxide, which then initiates a lipid peroxidation cascade culminating in the generation of cytotoxic electrophilic aldehydes such as 4-hydroxynonenal (4-HNE). This aldehyde then immediately seeks out the nucleophilic centers of proteins in the immediate vicinity, alkylating the latter at the vulnerable amino acids, cysteine, lysine and histidine. The alkylation of proteins in the electron transport chain, such as succinic acid dehydrogenase, further disrupts the flow of electrons within the inner mitochondrial membrane leading to yet more ROS generation and exacerbating the state of oxidative stress, ultimately compromising both the functional competence of the spermatozoa and the integrity of their DNA [96,97]. The clinical significance of these findings can be found in several epidemiological studies demonstrating significant links between mobile phone usage and semen quality [95].

Importantly, it has also been clearly demonstrated that these RF-EMR effects are not mediated by increased heat. This is critical because raised intratesticular temperature is another means by which EMR can influence sperm quality. Such thermo-sensitivity is reflected in the location of the testes in a scrotal sac resulting in an intratesticular temperature of 34 °C–35 °C, 2–3 °C below the core body temperature of 37 °C. If intratesticular temperatures are raised by, for example, experimental cryptorchidism, then there is a rapid cessation of germ cell differentiation triggered by a sudden wave of Fas-mediated apoptosis and autophagy in pachytene spermatocytes and spermatids via mechanisms that can be reversed by the administration of antioxidants [98,99]. Similarly, the loss of fertility induced by warming the scrotal mouse testes to 42 °C for 30 min can be completely reversed by the administration of antioxidants [35]. Another situation in which raised intratesticular temperature is evident is varicocele. This condition is also known to be associated with a loss of sperm function and DNA integrity as a consequence of oxidative stress [100,101]. Surgical ligation of the left internal spermatic vein to correct the varicosity has been shown to reduce oxidative stress parameters and to improve sperm quality and the concomitant administration of antioxidants has been shown to facilitate this process [102,103]. As we enter an era of elevated ambient temperatures associated with global climate change, we might anticipate an increased incidence of male infertility, as well as an elevated risk of a paternally-induced mutations in the offspring, as a consequence of oxidative DNA damage to the spermatozoa. It is already known that exposure to elevated summer temperatures suppresses human seminal quality and that oxidative stress is a major mediator of this change [104,105]. Such exposure will be of particular significance to livestock species where exposure to elevated ambient temperatures might not only suppress fertility but also disrupt the true breeding of selected genetic traits [106].

2.7. Environmental Pollutants

Another situation associated with the creation of oxidative stress in the male germ line involves exposure to a wide range of environmental chemicals that are capable of directly stressing spermatozoa and inducing ROS generation. One example that has been recently highlighted is the preservative, parabens, which is present in many commercial aqueous products including vaginal lubricants. This mixture of parabenzoic acid esters is capable of stimulating the generation of mitochondrial and cytosolic ROS, inhibiting sperm motility and viability in a dose-dependent manner. The ability of individual parabens to activate ROS generation and induce oxidative DNA damage was related to alkyl chain length. At the concentrations used clinically, methylparaben inhibited sperm motility and affected cell viability while augmenting ROS production and oxidative DNA damage [107]. Similarly, the commonly encountered environmental toxicants, phthalate esters and bisphenol A (BPA), are known to possess a capacity to induce oxidative stress in spermatozoa by virtue of their ability to activate ROS generation [108,109]. In the case of BPA, the induction of oxidative stress is associated with a premature acrosome reaction, loss of sperm motility, reduced viability, disturbed ionic balance, and alterations of the sperm proteome [109]. The causative involvement of ROS in the pathological changes induced by exposure to BPA has been demonstrated by virtue of the ability of antioxidants (reduced glutathione and α-tocopherol) to reduce its pathological impact [110]. The related toxicants bisphenol F and bisphenol S have also been shown to disrupt reproductive function via an oxidative mechanism [111,112]. The sperm mitochondria appear to play a key role in the genesis of ROS in this model, as part of a self-perpetuating redox cycle that culminates in DNA damage and the induction of apoptosis [113]. In the case of phthalate esters, we have employed an invertebrate model (*Galeolaria caespitosa*) to show that these compounds not only induce high levels of oxidative stress in the spermatozoa but also have an impact upon the developmental normality of the embryo via an epigenetic mechanism that has not previously been reported [114]. In these studies, addition of dibutyl phthalate (DBP) to *Galeolaria* spermatozoa resulted in a highly significant dose-dependent inhibitory effect on fertilization and embryogenesis. At low levels of DBP exposure, fertilization could occur but the resulting embryos exhibited a disrupted pattern of cleavage and chromosome segregation resulting in the genesis of abnormal embryos. Such abnormalities were associated with the induction of oxidative stress in the spermatozoa associated with the suppression of SOD activity and formation of electrophilic lipid aldehydes (4-HNE). The latter were subsequently found to bind to the acrosome and sperm centriole. Since the latter is responsible for orchestrating cell division in the embryo, we propose that 4-HNE adduction has a significant impact on the ability of the sperm centrioles to serve as microtubule organizing centers in the zygote, impairing both the normal segregation of chromosomes during mitosis and impeding the cytoskeletal changes that underpin the process of cell division [110]. Whether similar mechanisms underpin the observed association between oxidative stress in the male germ line and developmental abnormalities in human embryos that culminate in repeated early miscarriage [59,115] is currently an open question that has not yet been addressed. It is known that ROS generation and DNA fragmentation are significantly elevated in the spermatozoa of female partners experiencing recurrent early pregnancy loss [116] however the importance of 4HNE adduction of sperm centriolar proteins in the aetiology of this condition is unknown.

2.8. Iatrogenic Stress and Sperm Preparation

A final scenario for the creation of oxidative stress in spermatozoa involves iatrogenic damage associated with the techniques we are currently using to separate spermatozoa from seminal plasma for IVF purposes. As indicated above, seminal plasma has evolved to protect spermatozoa from oxidative stress generated during the ejaculatory process when the spermatozoa are suddenly shifted from a low- to a high- oxygen tension environment contaminated with activated neutrophils and macrophages that are actively generating ROS. The most effective sperm isolation strategies are therefore those where the spermatozoa are isolated directly from semen rather than from a washed pellet, since in the latter situation, leukocytes are able to attack the spermatozoa without any of the protection normally afforded by seminal antioxidants. Swim up from semen, discontinuous density gradient centrifugation and electrophoretic isolation all fulfil this condition and generally generate high quality spermatozoa for insemination [117,118]. Discontinuous density gradient centrifugation has been reported to increase DNA damage in certain cases possibly because of the presence of transition metals such as iron and copper in the commercial colloidal silicon preparations used to create such gradients [119]. While susceptibility to the presence of such metals appears to vary from sample to sample [120] such impacts can be readily addressed by the incorporation of a metal chelators such as EDTA into the gradients [119].

3. What Types of ROS are Involved?

Given the importance of oxidative stress in determining the functionality of mammalian spermatozoa, it is reasonable to ask which forms of ROS are involved and how the offending species might be sensitively assessed for diagnostic purposes. The first point to make is that ROS, as their name implies, are extremely reactive molecules that are generated in all complex cellular systems and react readily not just with vulnerable substrates including lipids, proteins and DNA, but also with each other. Classically, superoxide anion has a half-life at physiological pH of a few seconds and is rapidly removed from biological systems via the action of SOD to create hydrogen peroxide. This process is biologically important because it converts a relatively inert, non-membrane permeant free radical anion into a membrane permeant oxidant that will readily interact with appropriate substrates. Superoxide anion will also interact with another free radical species generated by spermatozoa, nitric oxide (NO), to generate a powerful oxidant, the peroxynitrite anion (ONOO−). It has been proposed that the combined action of these oxidants, peroxynitrite and hydrogen peroxide, drive the oxidative processes responsible for the regulation of sperm capacitation [55]. The fact that scavengers of both hydrogen peroxide (catalase) and peroxynitrite (uric acid) can suppress capacitation in different species adds weight to this argument [121–125]. The complexity of reactive oxygen metabolites involved in regulating sperm functionality increases still further in the presence of transition metals which can catalyze the breakdown of lipid peroxides. This process generates lipid peroxyl and alkoxyl radicals that actively participate in the hydrogen abstraction process that promotes the lipid peroxidation chain reaction. The latter inevitably leads to the generation of small-molecular mass lipid aldehydes, such as 4 HNE, that bind to the mitochondria and stimulate yet more ROS generation in a self-perpetuating cycle [96,126].

The fundamental point here is that we must be careful not to oversimplify the chemistry responsible for the physiological oxidative drive to capacitation or the creation of pathological oxidative stress. There are likely to be many different radical and non-radical species involved in these processes originating from a wide variety of different sources. Mitochondrial ROS is clearly an important contributor [127] and recent data supporting a role for lipoxygenase in this process are exciting [128] and supported by the finding that unesterified unsaturated fatty acids such as arachidonic acid are potent triggers for ROS generation by human spermatozoa [129]. Since spontaneous ROS generation by human spermatozoa is correlated with their free polyunsaturated fatty acid content [130] this pathway may well be particularly important in the aetiology of excessive ROS generation by defective human spermatozoa. Other potential pathways of ROS generation by human spermatozoa include reduced nicotinamide adenine dinucleotide phosphate (NADPH) oxidases, particularly Nox 5 [131,132] and other poorly characterized plasma membrane redox systems, identified using the redox active probe, WST-1 [133]. There can be no doubt that spermatozoa are vulnerable to oxidative stress and that this susceptibility is exacerbated by the ability of spermatozoa to generate ROS from multiple sites and to increase this activity under conditions of stress. Resolving the extent to which the oxidative damage observed in cases of male infertility is a reflection of active ROS generation by the spermatozoa themselves (as when there is local abundance of free unesterified polyunsaturated fatty acids, for example) and/or a passive consequence of oxidative stress generated systemically (in response to obesity or cigarette smoking) is a key question that will have to be addressed in future studies.

4. How Can We Best Measure Oxidative Stress in the Germ Line?

If oxidative stress is such an important contributor to male infertility, it is critical that we develop robust methods to diagnose this condition within the infertile population. Where the oxidative stress is systemic, a direct measurement of lipid peroxides in seminal plasma seems to be the current method of choice. The measurement of MDA in seminal plasma has been found to reflect a variety of parameters associated with oxidative stress including DNA damage in spermatozoa, their capacity to generate ROS and both protein carbonyl and nitrotyrosine expression in semen [134]. As indicated above, seminal MDA has also been used as a diagnostic criterion in preparation for antioxidant therapy [29]. To date, there are no reports of 4-HNE levels in seminal plasma being used to diagnose oxidative stress even though this aldehyde is likely to be a more sensitive marker of lipid peroxidation than MDA [135].

Measurement of ROS generation by human spermatozoa is made particularly difficult because of the various oxygen metabolites involved and the low levels of ROS generation compared with contaminating cell types, particularly neutrophils (Figure 2). Extensive use has been made of luminol- and lucigenin-dependent chemiluminescence for diagnostic purposes and we have written extensively on the underlying chemistry of these reagents and their shortcomings [136,137]. For example, we have demonstrated that the one electron reduction of lucigenin required to generate chemiluminescence can be achieved by reductases, such cytochrome b5 reductase and cytochrome P450 reductase using NADH or NADPH as electron donors respectively. The diagnostic value of this probe therefore probably lies more in its ability to reflect the volume of residual cytoplasm retained by defective spermatozoa than their capacity for ROS generation via an NAD(P)H oxidase [138]. Exactly the same argument applies to the nitroblue tetrazolium assay which cannot be used as a probe for ROS under any circumstances [139]. Such shortcomings do not mean that potential ROS-generating entities such as NADPH oxidase have no role to play in the pathophysiology of spermatozoa, only that we currently lack the diagnostic methods needed to demonstrate what that significance might be.

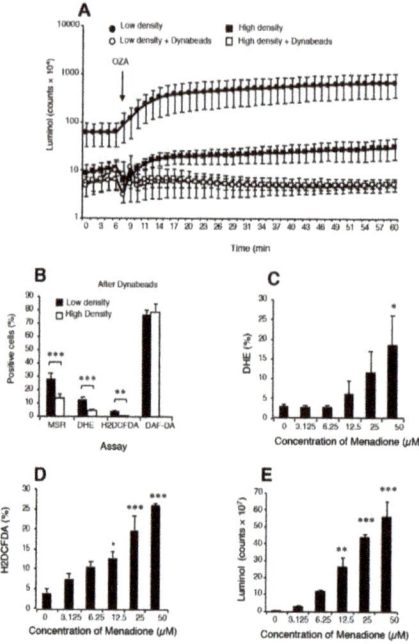

Figure 2. Analysis of spontaneous ROS generation by spermatozoa isolated from the high and low density regions of Percoll gradients. (**A**) Although luminol–horse radish peroxidase (HRP) dependent chemiluminescence could differentiate between high- (functional) and low- (dysfunctional) density sperm populations, such discrimination was completely lost when leukocytes were removed using CD45-coated Dynabeads; $n = 6$. (**B**) Following the removal of contaminating leukocytes, several ROS sensitive probes (MSR, DHE and H2DCFDA) but not the NO probe, DAF-DA, could discriminate the differences in sperm quality associated with high and low density Percoll populations. The most effective probe in this context was MSR, in keeping with the key role that mitochondria play in the aetiology of defective sperm function; $n = 12$. If leukocyte-free sperm suspensions were triggered to generate significant ROS using the redox cycling quinone, menadione (vitamin K), then several of the probes used in diagnostic andrology including MSR (not shown) (**C**) DHE, (**D**) H2DCFDA, and even (**E**) luminol/peroxidase could clearly detect a dose-dependent redox signal. Indeed, these dose-dependent analyses reveal that luminol and H2DCFDA were actually more sensitive than DHE in this regard. Overall, this analysis suggests that while probes such as luminol/peroxidase can clearly detect extracellular ROS generation, in practice their output is heavily influenced by the presence of contaminating leukocytes. Detecting differences in the spontaneous redox activity of the *spermatozoa* can, as indicated in panel B, only be achieved by flow cytometry using probes such as MSR, DHE and H2DCFDA (136). In panels (**A**,**E**), the chemiluminescence results are presented as counts per minute generated by the luminometer's photomultiplier while in panels (**B**–**D**), the results are presented as the percentage of the sperm population exhibiting a positive response by flow cytometry. Abbreviations: MitoSOX Red (MSR), dihydroethidium (DHE), dichlorodihydrofluorescein diacetate (H2DCFDA), 4,5-diaminofluorescein diacetate (DAF-DA). OZA = opsonized zymogen to activate any phagocytic leucocytes in the cell suspension. Significance values: * $p < 0.05$; ** $p < 0.01$ *** $p < 0.001$.

Luminol is a different story and one that is often poorly understood. Luminol requires a one electron oxidation to create the luminol radical which is the *sine qua non* for chemiluminescence [140]. In granulocytes, the primary oxidation event depends on the action of myeloperoxidase and is probably mediated by the powerful oxidant, hypochlorous acid. In the case of spermatozoa, an intracellular peroxidase again appears to be responsible for mediating luminol-dependent chemiluminescence

(LDCL) although the limited availability of peroxidase activity means that the spontaneous signal is low [141]. In order to improve the sensitivity of the assay, horse radish peroxidase has been used to sensitize the assay for the generation of extracellular hydrogen peroxide [141]. As the major source of extracellular hydrogen peroxide in the human ejaculate is contaminating leukocytes, the LDCL picture is characteristically dominated by these cells (Figure 2). If the leukocyte population is selectively removed using magnetic beads coated with antibodies against CD45 (the common leukocyte antigen) the luminol-peroxidase signal is reduced to background levels and no difference can be detected two high- and low-density Percoll sperm populations of differing quality [136]. However if, under these same leukocyte-free conditions, the spermatozoa are exposed to a reagent that will induce extracellular hydrogen peroxide release, such as the redox-cycling napthoquinone, menadione (Figure 2), then a very powerful luminol signal is generated [136]. Such results demonstrate that it is not so much the capacity of the luminol-peroxidase system to detect extracellular ROS that is open to question but rather the ability of this probe to detect the low levels of ROS released extracellularly by spermatozoa in the face of leukocyte contamination.

In order to generate the kind of sensitivity needed to detect differences in relative levels of spontaneous ROS generation associated with variations in sperm function, flow cytometry protocols have to be used. This methodology allows the operator to focus exclusively on the sperm population while any contaminating cells, such as precursor germ cells or leukocytes, can be carefully gated out. Under these conditions, leukocyte-free populations of spermatozoa from the high (functional sperm) and low (dysfunctional sperm) regions of discontinuous Percoll gradients can be readily distinguished on the basis of their reactivity with 3 probes (Figure 2), MitoSox red (mitochondrial ROS generation) dihydroethidium (total intracellular ROS generation) and H2DCFDA (dichlorodihydrofluorescein diacetate targeting intracellular oxidants such as hydrogen peroxide and, to a lesser extent, peroxynitrite). Although questions are occasionally asked about the specificity of these reagents for specific forms of ROS, the dynamic interchangeability of individual oxidants and free radicals means that such considerations are irrelevant in a diagnostic context. The fact is that these probes can detect differences in redox activity that are highly correlated with defective sperm function and therefore they have significant clinical value [136]. By contrast, use of more sophisticated techniques such as mass spectrometry that may be definitive but lack the sensitivity to detect the low levels intracellular ROS associated with spermatozoa are not helpful, no matter how impressive their powers of resolution.

5. Conclusions

In conclusion, the importance of oxidative stress as a major modulator of mammalian sperm function is incontrovertible. ROS have a positive role to play in driving the cascade of biochemical changes associated with sperm capacitation through their ability to control tyrosine phosphatase activity, stimulate cAMP generation and mediate the oxidation and ultimate release of cholesterol from the sperm plasma membrane. However, for obvious structural reasons these cells have very little defense against oxidative stress and are, thus, highly dependent on the powerful antioxidant properties of the fluids that surround them following their release from the germinal epithelium (seminiferous and epididymal tubule fluid and, for a short while at the moment of insemination, seminal plasma). If this extracellular antioxidant protection should be compromised for genetic, environmental or lifestyle reasons then the male germ line comes under oxidative attack. Oxidative stress may also be exacerbated by a variety of conditions that promote ROS generation by cells in the immediate vicinity (e.g., leukocyte infiltration secondary to infection) or by the spermatozoa themselves (exposure to RF-EMR, the stresses associated with cryopreservation, heat and exposure to a wide range of xenobiotics including bisphenol A, phthalate esters, parabens, etc). The complex combination of factors responsible for creating oxidative stress in the male germ line may well vary from patient to patient and is still not fully explored. Understanding the nature of these factors is significant because it will help guide the preconception care of such patients. Ensuring the functionality and genetic integrity of spermatozoa prior to conception is important, not just because it may promote the chances of conception but because it will minimize the genetic and epigenetic mutational load carried by the offspring and in so doing promote their long-term health trajectory.

Author Contributions: R.J.A. conceived the article and prepared an initial draft which was then reviewed, edited and revised by J.R.D. All authors have read and agreed to the published version of the manuscript.

Funding: This research received no external funding.

Conflicts of Interest: The authors declare no conflict of interest.

References

1. MacLeod, J. The role of oxygen in the metabolism and motility of human spermatozoa. *Am. J. Physiol* **1943**, *138*, 512–518. [CrossRef]
2. Sanocka, D.; Miesel, R.; Jedrzejczak, P.; Chełmonska-Soyta, A.C.; Kurpisz, M. Effect of reactive oxygen species and the activity of antioxidant systems on human semen; association with male infertility. *Int. J. Androl.* **1997**, *20*, 255–264. [CrossRef]
3. Aitken, R.J.; Buckingham, D.; Harkiss, D. Use of a xanthine oxidase free radical generating system to investigate the cytotoxic effects of reactive oxygen species on human spermatozoa. *J. Reprod. Fertil.* **1993**, *97*, 441–450. [CrossRef]
4. Aitken, R.J.; Clarkson, J.S. Cellular basis of defective sperm function and its association with the genesis of reactive oxygen species by human spermatozoa. *J. Reprod. Fertil.* **1987**, *81*, 459–469. [CrossRef]
5. Aitken, R.J.; Clarkson, J.S.; Hargreave, T.B.; Irvine, D.S.; Wu, F.C. Analysis of the relationship between defective sperm function and the generation of reactive oxygen species in cases of oligozoospermia. *J. Androl.* **1989**, *10*, 214–220. [CrossRef]
6. Jones, R.; Mann, T.; Sherins, R. Peroxidative breakdown of phospholipids in human spermatozoa, spermicidal properties of fatty acid peroxides, and protective action of seminal plasma. *Fertil. Steril.* **1979**, *31*, 531–537. [CrossRef]
7. Alvarez, J.G.; Touchstone, J.C.; Blasco, L.; Storey, B.T. Spontaneous lipid peroxidation and production of hydrogen peroxide and superoxide in human spermatozoa. Superoxide dismutase as major enzyme protectant against oxygen toxicity. *J. Androl.* **1987**, *8*, 338–348. [CrossRef] [PubMed]
8. Aitken, R.J.; Clarkson, J.S.; Fishel, S. Generation of reactive oxygen species, lipid peroxidation, and human sperm function. *Biol. Reprod.* **1989**, *41*, 183–197. [CrossRef] [PubMed]
9. Mayorga-Torres, B.J.M.; Camargo, M.; Cadavid, Á.P.; du Plessis, S.S.; Cardona Maya, W.D. Are oxidative stress markers associated with unexplained male infertility? *Andrologia* **2017**, *49*, e12659. [CrossRef] [PubMed]
10. Williams, A.C.; Ford, W.C. Relationship between reactive oxygen species production and lipid peroxidation in human sperm suspensions and their association with sperm function. *Fertil. Steril.* **2005**, *83*, 929–936. [CrossRef] [PubMed]
11. Verma, A.; Kanwar, K.C. Effect of vitamin E on human sperm motility and lipid peroxidation in vitro. *Asian J. Androl.* **1999**, *1*, 151–154. [PubMed]
12. Oborna, I.; Wojewodka, G.; De Sanctis, J.B.; Fingerova, H.; Svobodova, M.; Brezinova, J.; Hajduch, M.; Novotny, J.; Radova, L.; Radzioch, D. Increased lipid peroxidation and abnormal fatty acid profiles in seminal and blood plasma of normozoospermic males from infertile couples. *Hum. Reprod.* **2010**, *25*, 308–316. [CrossRef] [PubMed]
13. Benedetti, S.; Tagliamonte, M.C.; Catalani, S.; Primiterra, M.; Canestrari, F.; De Stefani, S.; Palini, S.; Bulletti, C. Differences in blood and semen oxidative status in fertile and infertile men, and their relationship with sperm quality. *Reprod. Biomed. Online* **2012**, *25*, 300–306. [CrossRef] [PubMed]
14. Brouwers, J.F.; Gadella, B.M. In situ detection and localization of lipid peroxidation in individual bovine sperm cells. *Free Radic. Biol. Med.* **2003**, *35*, 1382–1391. [CrossRef]
15. Peña, F.J.; O'Flaherty, C.; Ortiz Rodríguez, J.M.; Martín Cano, F.E.; Gaitskell-Phillips, G.L.; Gil, M.C.; Ortega Ferrusola, C. Redox regulation and oxidative stress: The particular case of the stallion spermatozoa. *Antioxidants* **2019**, *8*, 567. [CrossRef]
16. Aitken, R.J.; Gibb, Z.; Mitchell, L.A.; Lambourne, S.R.; Connaughton, H.S.; De Iuliis, G.N. Sperm motility is lost in vitro as a consequence of mitochondrial free radical production and the generation of electrophilic

aldehydes but can be significantly rescued by the presence of nucleophilic thiols. *Biol. Reprod.* **2012**, *87*, 110. [CrossRef]
17. Guthrie, H.D.; Welch, G.R. Using fluorescence-activated flow cytometry to determine reactive oxygen species formation and membrane lipid peroxidation in viable boar spermatozoa. *Methods Mol. Biol.* **2010**, *594*, 163–171.
18. Awda, B.J.; Mackenzie-Bell, M.; Buhr, M.M. Reactive oxygen species and boar sperm function. *Biol. Reprod.* **2009**, *81*, 553–561. [CrossRef]
19. Cassani, P.; Beconi, M.T.; O'Flaherty, C. Relationship between total superoxide dismutase activity with lipid peroxidation, dynamics and morphological parameters in canine semen. *Anim. Reprod. Sci.* **2005**, *86*, 163–173. [CrossRef]
20. Jara, B.; Merino, O.; Sánchez, R.; Risopatrón, J. Positive effect of butylated hydroxytoluene (BHT) on the quality of cryopreserved cat spermatozoa. *Cryobiology* **2019**, *89*, 76–81. [CrossRef]
21. Jones, R.; Mann, T. Lipid peroxides in spermatozoa; formation, rôle of plasmalogen, and physiological significance. *Proc. R. Soc. Lond. B Biol. Sci.* **1976**, *193*, 317–333.
22. Iwasaki, A.; Gagnon, C. Formation of reactive oxygen species in spermatozoa of infertile patients. *Fertil. Steril.* **1992**, *57*, 409–416. [CrossRef]
23. Hong, C.Y.; Chiang, B.N.; Turner, P. Calcium ion is the key regulator of human sperm function. *Lancet* **1984**, *2*, 1449–1451. [CrossRef]
24. Koppers, A.J.; Mitchell, L.A.; Wang, P.; Lin, M.; Aitken, R.J. Phosphoinositide 3-kinase signalling pathway involvement in a truncated apoptotic cascade associated with motility loss and oxidative DNA damage in human spermatozoa. *Biochem. J.* **2011**, *436*, 687–698. [CrossRef] [PubMed]
25. Fernandez, M.C.; Yu, A.; Moawad, A.R.; O'Flaherty, C. Peroxiredoxin 6 regulates the phosphoinositide 3-kinase/AKT pathway to maintain human sperm viability. *Mol. Hum. Reprod.* **2019**, *25*, 787–796. [CrossRef]
26. Kumar, V.; Kumar, P.G.; Yadav, J.K. Impact of semen-derived amyloid (SEVI) on sperm viability and motility: Its implication in male reproductive fitness. *Eur. Biophys. J.* **2019**, *48*, 659–671. [CrossRef]
27. Mitra, A.; Richardson, R.T.; O'Rand, M.G. Analysis of recombinant human semenogelin as an inhibitor of human sperm motility. *Biol. Reprod.* **2010**, *82*, 489–496. [CrossRef]
28. Nakagawa, K.; Yamano, S.; Kamada, M.; Maegawa, M.; Tokumura, A.; Irahara, M.; Saito, H. Sperm-immobilizing antibodies suppress an increase in the plasma membrane fluidity of human spermatozoa. *Fertil. Steril.* **2004**, *82*, 1054–1058. [CrossRef]
29. Suleiman, S.A.; Ali, M.E.; Zaki, Z.M.; el-Malik, E.M.; Nasr, M.A. Lipid peroxidation and human sperm motility: Protective role of vitamin E. *J. Androl.* **1996**, *17*, 530–537.
30. Keskes-Ammar, L.; Feki-Chakroun, N.; Rebai, T.; Sahnoun, Z.; Ghozzi, H.; Hammami, S.; Zghal, K.; Fki, H.; Damak, J.; Bahloul, A. Sperm oxidative stress and the effect of an oral vitamin E and selenium supplement on semen quality in infertile men. *Arch. Androl.* **2003**, *49*, 83–94. [CrossRef]
31. Barik, G.; Chaturvedula, L.; Bobby, Z. Role of oxidative stress and antioxidants in male infertility: An interventional study. *J. Hum. Reprod. Sci.* **2019**, *12*, 204–209. [CrossRef] [PubMed]
32. Jannatifar, R.; Parivar, K.; Roodbari, N.H.; Nasr-Esfahani, M.H. Effects of N-acetyl-cysteine supplementation on sperm quality, chromatin integrity and level of oxidative stress in infertile men. *Reprod. Biol. Endocrinol.* **2019**, *17*, 24. [CrossRef] [PubMed]
33. Subramanian, V.; Ravichandran, A.; Thiagarajan, N.; Govindarajan, M.; Dhandayuthapani, S.; Suresh, S. Seminal reactive oxygen species and total antioxidant capacity: Correlations with sperm parameters and impact on male infertility. *Clin. Exp. Reprod. Med.* **2018**, *45*, 88–93. [CrossRef] [PubMed]
34. Huang, C.; Cao, X.; Pang, D.; Li, C.; Luo, Q.; Zou, Y.; Feng, B.; Li, L.; Cheng, A.; Chen, Z. Is male infertility associated with increased oxidative stress in seminal plasma? A-meta analysis. *Oncotarget* **2018**, *9*, 24494–24513. [CrossRef] [PubMed]
35. Gharagozloo, P.; Gutiérrez-Adán, A.; Champroux, A.; Noblanc, A.; Kocer, A.; Calle, A.; Pérez-Cerezales, S.; Pericuesta, E.; Polhemus, A.; Moazamian, A.; et al. A novel antioxidant formulation designed to treat male infertility associated with oxidative stress: Promising preclinical evidence from animal models. *Hum. Reprod.* **2016**, *31*, 252–262. [CrossRef] [PubMed]
36. Aitken, R.J.; Paterson, M.; Fisher, H.; Buckingham, D.W.; van Duin, M. Redox regulation of tyrosine phosphorylation in human spermatozoa and its role in the control of human sperm function. *J. Cell Sci.* **1995**, *108*, 2017–2025.

37. Leclerc, P.; de Lamirande, E.; Gagnon, C. Regulation of protein-tyrosine phosphorylation and human sperm capacitation by reactive oxygen derivatives. *Free Radic. Biol. Med.* **1997**, *22*, 643–656. [CrossRef]
38. Aitken, R.J.; Harkiss, D.; Knox, W.; Paterson, M.; Irvine, D.S. A novel signal transduction cascade in capacitating human spermatozoa characterised by a redox-regulated, cAMP-mediated induction of tyrosine phosphorylation. *J. Cell Sci.* **1998**, *111*, 645–656.
39. Lewis, B.; Aitken, R.J. A redox-regulated tyrosine phosphorylation cascade in rat spermatozoa. *J. Androl.* **2001**, *22*, 611–622.
40. Ecroyd, H.W.; Jones, R.C.; Aitken, R.J. Endogenous redox activity in mouse spermatozoa and its role in regulating the tyrosine phosphorylation events associated with sperm capacitation. *Biol. Reprod.* **2003**, *69*, 347–354. [CrossRef]
41. Roy, S.C.; Atreja, S.K. Effect of reactive oxygen species on capacitation and associated protein tyrosine phosphorylation in buffalo (*Bubalus bubalis*) spermatozoa. *Anim. Reprod. Sci.* **2008**, *107*, 68–84. [CrossRef] [PubMed]
42. Rivlin, J.; Mendel, J.; Rubinstein, S.; Etkovitz, N.; Breitbart, H. Role of hydrogen peroxide in sperm capacitation and acrosome reaction. *Biol. Reprod.* **2004**, *70*, 518–522. [CrossRef]
43. Baumber, J.; Sabeur, K.; Vo, A.; Ball, B.A. Reactive oxygen species promote tyrosine phosphorylation and capacitation in equine spermatozoa. *Theriogenology* **2003**, *60*, 1239–1247. [CrossRef]
44. Baker, M.A.; Hetherington, L.; Aitken, R.J. Identification of SRC as a key PKA-stimulated tyrosine kinase involved in the capacitation-associated hyperactivation of murine spermatozoa. *J. Cell Sci.* **2006**, *119*, 3182–3192. [CrossRef] [PubMed]
45. O'Flaherty, C.; de Lamirande, E.; Gagnon, C. Reactive oxygen species modulate independent protein phosphorylation pathways during human sperm capacitation. *Free Radic. Biol. Med.* **2006**, *40*, 1045–1055. [CrossRef] [PubMed]
46. Brouwers, J.F.; Boerke, A.; Silva, P.F.; Garcia-Gil, N.; van Gestel, R.A.; Helms, J.B.; van de Lest, C.H.; Gadella, B.M. Mass spectrometric detection of cholesterol oxidation in bovine sperm. *Biol. Reprod.* **2011**, *85*, 128–136. [CrossRef] [PubMed]
47. de Lamirande, E.; Gagnon, C. Human sperm hyperactivation and capacitation as parts of an oxidative process. *Free Radic. Biol. Med.* **1993**, *14*, 157–166. [CrossRef]
48. Yunes, R.; Doncel, G.F.; Acosta, A.A. Incidence of sperm-tail tyrosine phosphorylation and hyperactivated motility in normozoospermic and asthenozoospermic human sperm samples. *Biocell* **2003**, *27*, 29–36.
49. Kodama, H.; Kuribayashi, Y.; Gagnon, C. Effect of sperm lipid peroxidation on fertilization. *J. Androl.* **1996**, *17*, 151–157.
50. Miesel, R.; Drzejczak, P.J.; Kurpisz, M. Oxidative stress during the interaction of gametes. *Biol. Reprod.* **1993**, *49*, 918–923. [CrossRef]
51. Wong, J.L.; Créton, R.; Wessel, G.M. The oxidative burst at fertilization is dependent upon activation of the dual oxidase Udx1. *Dev. Cell* **2004**, *7*, 801–814. [CrossRef] [PubMed]
52. Aitken, R.J. Impact of oxidative stress on male and female germ cells; implications for fertility. *Reproduction* **2019**. [CrossRef] [PubMed]
53. Hamilton, L.E.; Zigo, M.; Mao, J.; Xu, W.; Sutovsky, P.; O'Flaherty, C.; Oko, R. GSTO2 isoforms participate in the oxidative regulation of the plasmalemma in eutherian spermatozoa during capacitation. *Antioxidants* **2019**, *8*. [CrossRef] [PubMed]
54. O'Flaherty, C. Peroxiredoxin 6: The protector of male fertility. *Antioxidants* **2018**, *7*, 173. [CrossRef]
55. Aitken, R.J.; Baker, M.A.; Nixon, B. Are sperm capacitation and apoptosis the opposite ends of a continuum driven by oxidative stress. *Asian J. Androl.* **2015**, *17*, 633–639. [CrossRef]
56. Rhemrev, J.P.; van Overveld, F.W.; Haenen, G.R.; Teerlink, T.; Bast, A.; Vermeiden, J.P. Quantification of the nonenzymatic fast and slow TRAP in a post-addition assay in human seminal plasma and the antioxidant contributions of various seminal compounds. *J. Androl.* **2000**, *21*, 913–920.
57. Lazzarino, G.; Listorti, I.; Bilotta, G.; Capozzolo, T.; Amorini, A.M.; Longo, S.; Caruso, G.; Lazzarino, G.; Tavazzi, B.; Bilotta, P. Water- and fat-soluble antioxidants in human seminal plasma and serum of fertile males. *Antioxidants* **2019**, *8*, 96. [CrossRef]
58. O'Flaherty, C. Peroxiredoxins: Hidden players in the antioxidant defence of human spermatozoa. *Basic Clin. Androl.* **2014**, *24*, 4. [CrossRef]

59. Kamkar, N.; Ramezanali, F.; Sabbaghian, M. The relationship between sperm DNA fragmentation, free radicals and antioxidant capacity with idiopathic repeated pregnancy loss. *Reprod. Biol.* **2018**, *18*, 330–335. [CrossRef]
60. Agarwal, A.; Henkel, R.; Sharma, R.; Tadros, N.N.; Sabanegh, E. Determination of seminal oxidation-reduction potential (ORP) as an easy and cost-effective clinical marker of male infertility. *Andrologia* **2018**, *50*, e12914. [CrossRef]
61. Twigg, J.; Irvine, D.S.; Houston, P.; Fulton, N.; Michael, L.; Aitken, R.J. Iatrogenic DNA damage induced in human spermatozoa during sperm preparation: Protective significance of seminal plasma. *Mol. Hum. Reprod.* **1998**, *4*, 439–445. [CrossRef] [PubMed]
62. Fraga, C.G.; Motchnik, P.A.; Wyrobek, A.J.; Rempel, D.M.; Ames, B.N. Smoking and low antioxidant levels increase oxidative damage to sperm DNA. *Mutat. Res.* **1996**, *351*, 199–203. [CrossRef]
63. Sanocka, D.; Fraczek, M.; Jedrzejczak, P.; Szumała-Kakol, A.; Kurpisz, M. Male genital tract infection: An influence of leukocytes and bacteria on semen. *J. Reprod. Immunol.* **2004**, *62*, 111–124. [CrossRef] [PubMed]
64. Aitken, R.J.; West, K.; Buckingham, D. Leukocytic infiltration into the human ejaculate and its association with semen quality, oxidative stress, and sperm function. *J. Androl.* **1994**, *15*, 343–352.
65. Aitken, R.J.; Baker, M.A. Oxidative stress, spermatozoa and leukocytic infiltration: Relationships forged by the opposing forces of microbial invasion and the search for perfection. *J. Reprod. Immunol.* **2013**, *100*, 11–19. [CrossRef]
66. Krausz, C.; Mills, C.; Rogers, S.; Tan, S.L.; Aitken, R.J. Stimulation of oxidant generation by human sperm suspensions using phorbol esters and formyl peptides: Relationships with motility and fertilization in vitro. *Fertil. Steril.* **1994**, *62*, 599–605. [CrossRef]
67. Aitken, R.J.; Buckingham, D.W.; West, K.; Brindle, J. On the use of paramagnetic beads and ferrofluids to assess and eliminate the leukocytic contribution to oxygen radical generation by human sperm suspensions. *Am. J. Reprod. Immunol.* **1996**, *35*, 541–551. [CrossRef]
68. Baker, H.W.; Brindle, J.; Irvine, D.S.; Aitken, R.J. Protective effect of antioxidants on the impairment of sperm motility by activated polymorphonuclear leukocytes. *Fertil. Steril.* **1996**, *65*, 411–419. [CrossRef]
69. Tosic, J.; Walton, A. Formation of hydrogen peroxide by spermatozoa and its inhibitory effect of respiration. *Nature* **1946**, *158*, 485. [CrossRef]
70. Upreti, G.C.; Jensen, K.; Munday, R.; Duganzich, D.M.; Vishwanath, R.; Smith, J.F. Studies on aromatic amino acid oxidase activity in ram spermatozoa: Role of pyruvate as an antioxidant. *Anim. Reprod. Sci.* **1998**, *51*, 275–287. [CrossRef]
71. Aitken, J.B.; Naumovski, N.; Curry, B.; Grupen, C.G.; Gibb, Z.; Aitken, R.J. Characterization of an L-amino acid oxidase in equine spermatozoa. *Biol. Reprod.* **2015**, *92*, 125. [CrossRef]
72. Houston, B.; Curry, B.; Aitken, R.J. Human spermatozoa possess an IL4I1 l-amino acid oxidase with a potential role in sperm function. *Reproduction* **2015**, *149*, 587–596. [CrossRef]
73. Arslan, H.O.; Herrera, C.; Malama, E.; Siuda, M.; Leiding, C.; Bollwein, H. Effect of the addition of different catalase concentrations to a TRIS-egg yolk extender on quality and in vitro fertilization rate of frozen-thawed bull sperm. *Cryobiology* **2019**, *91*, 40–52. [CrossRef] [PubMed]
74. Papas, M.; Catalan, J.; Barranco, I.; Arroyo, L.; Bassols, A.; Yeste, M.; Miró, J. Total and specific activities of superoxide dismutase (SOD) in seminal plasma are related with the cryotolerance of jackass spermatozoa. *Cryobiology* **2019**. [CrossRef] [PubMed]
75. Papas, M.; Catalán, J.; Fernandez-Fuertes, B.; Arroyo, L.; Bassols, A.; Miró, J.; Yeste, M. Specific activity of superoxide dismutase in stallion seminal plasma is related to sperm cryotolerance. *Antioxidants* **2019**, *8*, 539. [CrossRef] [PubMed]
76. Alvarez, J.G.; Storey, B.T. Evidence for increased lipid peroxidative damage and loss of superoxide dismutase activity as a mode of sublethal cryodamage to human sperm during cryopreservation. *J. Androl.* **1992**, *13*, 232–241. [PubMed]
77. Rossi, T.; Mazzilli, F.; Delfino, M.; Dondero, F. Improved human sperm recovery using superoxide dismutase and catalase supplementation in semen cryopreservation procedure. *Cell Tissue Bank* **2001**, *2*, 9–13. [CrossRef]
78. Roca, J.; Rodríguez, M.J.; Gil, M.A.; Carvajal, G.; Garcia, E.M.; Cuello, C.; Vazquez, J.M.; Martinez, E.A. Survival and in vitro fertility of boar spermatozoa frozen in the presence of superoxide dismutase and/or catalase. *J. Androl.* **2005**, *26*, 15–24.

79. Amini, M.R.; Kohram, H.; Zare-Shahaneh, A.; Zhandi, M.; Sharideh, H.; Nabi, M.M. The effects of different levels of catalase and superoxide dismutase in modified Beltsville extender on rooster post-thawed sperm quality. *Cryobiology* **2015**, *70*, 226–232. [CrossRef]
80. Forouzanfar, M.; Abid, A.; Hosseini, S.M.; Hajian, M.; Nasr Esfahani, M.H. Supplementation of sperm cryopreservation media with cell permeable superoxide dismutase mimetic agent (MnTE) improves goat blastocyst formation. *Cryobiology* **2013**, *67*, 394–397. [CrossRef]
81. Santiani, A.; Evangelista, S.; Valdivia, M.; Risopatrón, J.; Sánchez, R. Effect of the addition of two superoxide dismutase analogues (Tempo and Tempol) to alpaca semen extender for cryopreservation. *Theriogenology* **2013**, *79*, 842–846. [CrossRef] [PubMed]
82. Forouzanfar, M.; Fekri Ershad, S.; Hosseini, S.M.; Hajian, M.; Ostad-Hosseini, S.; Abid, A.; Tavalaee, M.; Shahverdi, A.; Vosough Dizaji, A.; Nasr Esfahani, M.H. Can permeable super oxide dismutase mimetic agents improve the quality of frozen-thawed ram semen? *Cryobiology* **2013**, *66*, 126–130. [CrossRef] [PubMed]
83. Shafiei, M.; Forouzanfar, M.; Hosseini, S.M.; Esfahani, M.H. The effect of superoxide dismutase mimetic and catalase on the quality of postthawed goat semen. *Theriogenology* **2015**, *83*, 1321–1327. [CrossRef] [PubMed]
84. Len, J.S.; Koh, W.S.D.; Tan, S.X. The roles of reactive oxygen species and antioxidants in cryopreservation. *Biosci. Rep.* **2019**, *39*. [CrossRef]
85. Opuwari, C.S.; Henkel, R.R. An update on oxidative damage to spermatozoa and oocytes. *Biomed. Res. Int.* **2016**, *2016*, 9540142. [CrossRef]
86. Shen, H.-M.; Chia, S.E.; Ni, Z.Y.; New, A.L.; Lee, B.L.; Ong, C.N. Detection of oxidative DNA damage in human sperm and the association with cigarette smoking. *Reprod. Toxicol.* **1997**, *11*, 675–680. [CrossRef]
87. Youn, C.K.; Kim, S.H.; Lee, D.Y.; Song, S.H.; Chang, I.Y.; Hyun, J.W.; Chung, M.H.; You, H.J. Cadmium down-regulates human OGG1 through suppression of Sp1 activity. *J. Biol. Chem.* **2005**, *280*, 25185–25195. [CrossRef]
88. Aitken, R.J. Not every sperm is sacred; a perspective on male infertility. *Mol. Hum. Reprod.* **2018**, *24*, 287–298. [CrossRef]
89. Lee, K.M.; Ward, M.H.; Han, S.; Ahn, H.S.; Kang, H.J.; Choi, H.S.; Shin, H.Y.; Koo, H.H.; Seo, J.J.; Choi, J.E.; et al. Paternal smoking, genetic polymorphisms in CYP1A1 and childhood leukemia risk. *Leuk. Res.* **2009**, *33*, 250–258. [CrossRef]
90. Xavier, M.J.; Nixon, B.; Roman, S.D.; Scott, R.J.; Drevet, J.R.; Aitken, R.J. Paternal impacts on development: Identification of genomic regions vulnerable to oxidative DNA damage in human spermatozoa. *Hum. Reprod.* **2019**, *34*, 1876–1890. [CrossRef]
91. Heerema, N.A.; Sather, H.N.; Sensel, M.G.; La, M.K.; Hutchinson, R.J.; Nachman, J.B.; Reaman, G.H.; Lange, B.J.; Steinherz, P.G.; Bostrom, B.C.; et al. Abnormalities of chromosome bands 15q13-15 in childhood acute lymphoblastic leukemia. *Cancer* **2002**, *94*, 1102–1110. [CrossRef] [PubMed]
92. Cao, Y.; Lu, J.; Lu, J. Paternal smoking before conception and during pregnancy is associated with an increased risk of childhood acute lymphoblastic leukemia: A systematic review and meta-analysis of 17 case-control studies. *J. Pediatr. Hematol. Oncol.* **2020**, *42*, 32–40. [CrossRef] [PubMed]
93. Pearce, K.L.; Hill, A.; Tremellen, K.P. Obesity related metabolic endotoxemia is associated with oxidative stress and impaired sperm DNA integrity. *Basic Clin. Androl.* **2019**, *29*, 6. [CrossRef] [PubMed]
94. McPherson, N.O.; Shehadeh, H.; Fullston, T.; Zander-Fox, D.L.; Lane, M. Dietary micronutrient supplementation for 12 days in obese male mice restores sperm oxidative stress. *Nutrients* **2019**, *11*, 2196. [CrossRef]
95. Houston, B.J.; Nixon, B.; King, B.V.; De Iuliis, G.N.; Aitken, R.J. The effects of radiofrequency electromagnetic radiation on sperm function. *Reproduction* **2016**, *152*, R263–R276. [CrossRef]
96. Aitken, R.J.; Whiting, S.; De Iuliis, G.N.; McClymont, S.; Mitchell, L.A.; Baker, M.A. Electrophilic aldehydes generated by sperm metabolism activate mitochondrial reactive oxygen species generation and apoptosis by targeting succinate dehydrogenase. *J. Biol. Chem.* **2012**, *287*, 33048–33060. [CrossRef]
97. De Iuliis, G.N.; Newey, R.J.; King, B.V.; Aitken, R.J. Mobile phone radiation induces reactive oxygen species production and DNA damage in human spermatozoa in vitro. *PLoS ONE* **2009**, *4*, e6446. [CrossRef]
98. Vigueras-Villaseñor, R.M.; Ojeda, I.; Gutierrez-Pérez, O.; Chavez-Saldaña, M.; Cuevas, O.; Maria, D.S.; Rojas-Castañeda, J.C. Protective effect of α-tocopherol on damage to rat testes by experimental cryptorchidism. *Int. J. Exp. Pathol.* **2011**, *92*, 131–139. [CrossRef]

99. Li, E.; Guo, Y.; Wang, G.; Chen, F.; Li, Q. Effect of resveratrol on restoring spermatogenesis in experimental cryptorchid mice and analysis of related differentially expressed proteins. *Cell Biol. Int.* **2015**, *39*, 733–740. [CrossRef]
100. Hassanin, A.M.; Ahmed, H.H.; Kaddah, A.N. A global view of the pathophysiology of varicocele. *Andrology* **2018**, *6*, 654–661. [CrossRef]
101. Cho, C.L.; Esteves, S.C.; Agarwal, A. Novel insights into the pathophysiology of varicocele and its association with reactive oxygen species and sperm DNA fragmentation. *Asian J. Androl.* **2016**, *18*, 186–193. [PubMed]
102. Mostafa, T.; Nabil, N.; Rashed, L.; Abo-Sief, A.F.; Eissa, H.H. Seminal SIRT1-oxidative stress relationship in infertile oligoasthenoteratozoospermic men with varicocele after its surgical repair. *Andrologia* **2019**, *52*, e13456. [CrossRef] [PubMed]
103. Lu, X.L.; Liu, J.J.; Li, J.T.; Yang, Q.A.; Zhang, J.M. Melatonin therapy adds extra benefit to varicecelectomy in terms of sperm parameters, hormonal profile and total antioxidant capacity: A placebo-controlled, double-blind trial. *Andrologia* **2018**, *50*, e13033. [CrossRef] [PubMed]
104. Santi, D.; Magnani, E.; Michelangeli, M.; Grassi, R.; Vecchi, B.; Pedroni, G.; Roli, L.; De Santis, M.C.; Baraldi, E.; Setti, M.; et al. Seasonal variation of semen parameters correlates with environmental temperature and air pollution: A big data analysis over 6 years. *Environ. Pollut.* **2018**, *235*, 806–813. [CrossRef]
105. Rao, M.; Zhao, X.L.; Yang, J.; Hu, S.F.; Lei, H.; Xia, W.; Zhu, C.H. Effect of transient scrotal hyperthermia on sperm parameters, seminal plasma biochemical markers, and oxidative stress in men. *Asian J. Androl.* **2015**, *17*, 668–675.
106. Houston, B.J.; Nixon, B.; Martin, J.H.; De Iuliis, G.N.; Trigg, N.A.; Bromfield, E.G.; McEwan, K.E.; Aitken, R.J. Heat exposure induces oxidative stress and DNA damage in the male germ line. *Biol. Reprod.* **2018**, *98*, 593–606. [CrossRef]
107. Samarasinghe, S.V.A.C.; Krishnan, K.; Naidu, R.; Megharaj, M.; Miller, K.; Fraser, B.; Aitken, R.J. Parabens generate reactive oxygen species in human spermatozoa. *Andrology* **2018**, *6*, 532–541. [CrossRef]
108. Sedha, S.; Kumar, S.; Shukla, S. Role of oxidative stress in male reproductive dysfunctions with reference to phthalate compounds. *Urol. J.* **2015**, *12*, 2304–2316.
109. Rahman, M.S.; Pang, M.G. Understanding the molecular mechanisms of bisphenol A action in spermatozoa. *Clin. Exp. Reprod. Med.* **2019**, *46*, 99–106. [CrossRef]
110. Rahman, M.S.; Kang, K.H.; Arifuzzaman, S.; Pang, W.K.; Ryu, D.Y.; Song, W.; Park, Y.J.; Pang, M.G. Effect of antioxidants on BPA-induced stress on sperm function in a mouse model. *Sci. Rep.* **2019**, *9*, 10584. [CrossRef]
111. Ullah, A.; Pirzada, M.; Afsar, T.; Razak, S.; Almajwal, A.; Jahan, S. Effect of bisphenol F, an analog of bisphenol A, on the reproductive functions of male rats. *Environ. Health Prev. Med.* **2019**, *24*, 41. [CrossRef] [PubMed]
112. Ullah, A.; Pirzada, M.; Jahan, S.; Ullah, H.; Khan, M.J. Bisphenol A analogues bisphenol B, bisphenol F, and bisphenol S induce oxidative stress, disrupt daily sperm production, and damage DNA in rat spermatozoa: A comparative in vitro and in vivo study. *Toxicol. Ind. Health* **2019**, *35*, 294–303. [CrossRef] [PubMed]
113. Barbonetti, A.; Castellini, C.; Di Giammarco, N.; Santilli, G.; Francavilla, S.; Francavilla, F. In vitro exposure of human spermatozoa to bisphenol A induces pro-oxidative/apoptotic mitochondrial dysfunction. *Reprod. Toxicol.* **2016**, *66*, 61–67. [CrossRef] [PubMed]
114. Lu, Y.; Lin, M.; Aitken, R.J. Exposure of spermatozoa to dibutyl phthalate induces abnormal embryonic development in a marine invertebrate *Galeolaria caespitosa* (Polychaeta: Serpulidae). *Aquat. Toxicol.* **2017**, *191*, 189–200. [CrossRef] [PubMed]
115. Leach, M.; Aitken, R.J.; Sacks, G. Sperm DNA fragmentation abnormalities in men from couples with a history of recurrent miscarriage. *Aust. N. Z. J. Obstet. Gynaecol.* **2015**, *55*, 379–383. [CrossRef]
116. Dhawan, V.; Kumar, M.; Deka, D.; Malhotra, N.; Singh, N.; Dadhwal, V.; Dada, R. Paternal factors and embryonic development: Role in recurrent pregnancy loss. *Andrologia* **2019**, *51*, e13171. [CrossRef]
117. Aitken, R.J.; Clarkson, J.S. Significance of reactive oxygen species and antioxidants in defining the efficacy of sperm preparation techniques. *J. Androl.* **1988**, *9*, 367–376. [CrossRef]
118. Ainsworth, C.; Nixon, B.; Aitken, R.J. Development of a novel electrophoretic system for the isolation of human spermatozoa. *Hum. Reprod.* **2005**, *20*, 2261–2270. [CrossRef]
119. Aitken, R.J.; Finnie, J.M.; Muscio, L.; Whiting, S.; Connaughton, H.S.; Kuczera, L.; Rothkirch, T.B.; De Iuliis, G.N. Potential importance of transition metals in the induction of DNA damage by sperm preparation media. *Hum. Reprod.* **2014**, *29*, 2136–2147. [CrossRef]

120. Muratori, M.; Tarozzi, N.; Cambi, M.; Boni, L.; Iorio, A.L.; Passaro, C.; Luppino, B.; Nadalini, M.; Marchiani, S.; Tamburrino, L.; et al. Variation of DNA fragmentation levels during density gradient sperm selection for assisted reproduction techniques: A possible new male predictive parameter of pregnancy? *Medicine* **2016**, *95*, e3624. [CrossRef]

121. Rodriguez, P.C.; Beconi, M.T. Peroxynitrite participates in mechanisms involved in capacitation of cryopreserved cattle. *Anim. Reprod. Sci.* **2009**, *110*, 96–107. [CrossRef]

122. Bize, I.; Santander, G.; Cabello, P.; Driscoll, D.; Sharpe, C. Hydrogen peroxide is involved in hamster sperm capacitation in vitro. *Biol. Reprod.* **1991**, *44*, 398–403. [CrossRef]

123. Aitken, R.J.; Ryan, A.L.; Baker, M.A.; McLaughlin, E.A. Redox activity associated with the maturation and capacitation of mammalian spermatozoa. *Free Radic. Biol. Med.* **2004**, *36*, 994–1010. [CrossRef]

124. Griveau, J.F.; Renard, P.; Le Lannou, D. An in vitro promoting role for hydrogen peroxide in human sperm capacitation. *Int. J. Androl.* **1994**, *17*, 300–307. [CrossRef]

125. Córdoba, M.; Mora, N.; Beconi, M.T. Respiratory burst and NAD(P)H oxidase activity are involved in capacitation of cryopreserved bovine spermatozoa. *Theriogenology* **2006**, *65*, 882–892. [CrossRef]

126. Aitken, R.J.; Harkiss, D.; Buckingham, D.W. Analysis of lipid peroxidation mechanisms in human spermatozoa. *Mol. Reprod. Dev.* **1993**, *35*, 302–315. [CrossRef]

127. Koppers, A.J.; De Iuliis, G.N.; Finnie, J.M.; McLaughlin, E.A.; Aitken, R.J. Significance of mitochondrial reactive oxygen species in the generation of oxidative stress in spermatozoa. *J. Clin. Endocrinol. Metab.* **2008**, *93*, 3199–3207. [CrossRef]

128. Walters, J.L.H.; De Iuliis, G.N.; Dun, M.D.; Aitken, R.J.; McLaughlin, E.A.; Nixon, B.; Bromfield, E.G. Pharmacological inhibition of arachidonate 15-lipoxygenase protects human spermatozoa against oxidative stress. *Biol. Reprod.* **2018**, *98*, 784–794. [CrossRef]

129. Aitken, R.J.; Wingate, J.K.; De Iuliis, G.N.; Koppers, A.J.; McLaughlin, E.A. Cis-unsaturated fatty acids stimulate reactive oxygen species generation and lipid peroxidation in human spermatozoa. *J. Clin. Endocrinol. Metab.* **2006**, *91*, 4154–4163. [CrossRef]

130. Koppers, A.J.; Garg, M.L.; Aitken, R.J. Stimulation of mitochondrial reactive oxygen species production by unesterified, unsaturated fatty acids in defective human spermatozoa. *Free Radic. Biol. Med.* **2010**, *48*, 112–119. [CrossRef]

131. Musset, B.; Clark, R.A.; DeCoursey, T.E.; Petheo, G.L.; Geiszt, M.; Chen, Y.; Cornell, J.E.; Eddy, C.A.; Brzyski, R.G.; El Jamali, A. NOX5 in human spermatozoa: Expression, function, and regulation. *J. Biol. Chem.* **2012**, *287*, 9376–9388. [CrossRef] [PubMed]

132. Donà, G.; Fiore, C.; Tibaldi, E.; Frezzato, F.; Andrisani, A.; Ambrosini, G.; Fiorentin, D.; Armanini, D.; Bordin, L.; Clari, G. Endogenous reactive oxygen species content and modulation of tyrosine phosphorylation during sperm capacitation. *Int. J. Androl.* **2011**, *34*, 411–419. [CrossRef] [PubMed]

133. Aitken, R.J.; Ryan, A.L.; Curry, B.J.; Baker, M.A. Multiple forms of redox activity in populations of human spermatozoa. *Mol. Hum. Reprod.* **2003**, *9*, 645–661. [CrossRef] [PubMed]

134. Aktan, G.; Doğru-Abbasoğlu, S.; Küçükgergin, C.; Kadıoğlu, A.; Ozdemirler-Erata, G.; Koçak-Toker, N. Mystery of idiopathic male infertility: Is oxidative stress an actual risk? *Fertil. Steril.* **2013**, *99*, 1211–1215. [CrossRef] [PubMed]

135. Moazamian, R.; Polhemus, A.; Connaughton, H.; Fraser, B.; Whiting, S.; Gharagozloo, P.; Aitken, R.J. Oxidative stress and human spermatozoa: Diagnostic and functional significance of aldehydes generated as a result of lipid peroxidation. *Mol. Hum. Reprod.* **2015**, *21*, 502–515. [CrossRef]

136. Aitken, R.J.; Smith, T.B.; Lord, T.; Kuczera, L.; Koppers, A.J.; Naumovski, N.; Connaughton, H.; Baker, M.A.; De Iuliis, G.N. On methods for the detection of reactive oxygen species generation by human spermatozoa: Analysis of the cellular responses to catechol oestrogen, lipid aldehyde, menadione and arachidonic acid. *Andrology* **2013**, *1*, 192–205. [CrossRef]

137. Aitken, R.J.; Baker, M.A.; O'Bryan, M. Shedding light on chemiluminescence: The application of chemiluminescence in diagnostic andrology. *J. Androl.* **2004**, *25*, 455–465. [CrossRef]

138. Gomez, E.; Buckingham, D.W.; Brindle, J.; Lanzafame, F.; Irvine, D.S.; Aitken, R.J. Development of an image analysis system to monitor the retention of residual cytoplasm by human spermatozoa: Correlation with biochemical markers of the cytoplasmic space, oxidative stress, and sperm function. *J. Androl.* **1996**, *17*, 276–287.

139. Aitken, R.J. Nitroblue tetrazolium (NBT) assay. *Reprod. Biomed. Online* **2018**, *36*, 90–91. [CrossRef]

140. Vilim, V.; Wilhelm, J. What do we measure by a luminol-dependent chemiluminescence of phagocytes? *Free Radic. Biol. Med.* **1989**, *6*, 623–629. [CrossRef]
141. Aitken, R.J.; Buckingham, D.W.; West, K.M. Reactive oxygen species and human spermatozoa: Analysis of the cellular mechanisms involved in luminol- and lucigenin-dependent chemiluminescence. *J. Cell. Physiol.* **1992**, *151*, 466–477. [CrossRef] [PubMed]

 © 2020 by the authors. Licensee MDPI, Basel, Switzerland. This article is an open access article distributed under the terms and conditions of the Creative Commons Attribution (CC BY) license (http://creativecommons.org/licenses/by/4.0/).

Review

Oxidation of Sperm Nucleus in Mammals: A Physiological Necessity to Some Extent with Adverse Impacts on Oocyte and Offspring

Joël R. Drevet [1],* and Robert John Aitken [2,3,4]

[1] Faculty of Medicine, GReD Institute, INSERM U1103—CNRS UMR6293—Université Clermont Auvergne, CRBC building, 28 place Henri Dunant, 63001 Clermont-Ferrand, France
[2] School of Environmental and Life Sciences, Priority Research Centre for Reproductive Sciences, The University of Newcastle, Callaghan, Newcastle 2308, Australia; john.aitken@newcastle.edu.au
[3] Faculty of Health and Medicine, The University of Newcastle, Callaghan, Newcastle 2308, Australia
[4] Medical Genetics, Hunter Medical Research Institute, New Lambton Heights, 13 2305 Newcastle, Australia
* Correspondence: joel.drevet@uca.fr; Tel.: +33-473407413

Received: 10 December 2019; Accepted: 21 January 2020; Published: 23 January 2020

Abstract: Sperm cells have long been known to be good producers of reactive oxygen species, while they are also known to be particularly sensitive to oxidative damage affecting their structures and functions. As with all organic cellular components, sperm nuclear components and, in particular, nucleic acids undergo oxidative alterations that have recently been shown to be commonly encountered in clinical practice. This review will attempt to provide an overview of this situation. After a brief coverage of the biological reasons why the sperm nucleus and associated DNA are sensitive to oxidative damage, a summary of the most recent results concerning the oxidation of sperm DNA in animal and human models will be presented. The study will then attempt to cover the possible consequences of sperm nuclear oxidation on male fertility and beyond.

Keywords: spermatozoa; nuclear integrity; oxidative DNA damage; putative transgenerational impacts

---

1. Oxidative Stress

Oxidative stress is inherent in the consumption of oxygen by aerobic organisms that metazoans have made their "fuel" for the production of energy via the cellular mitochondrial respiratory chain. In doing so, cells produce active oxygen derivatives commonly referred to as reactive oxygen species (ROS) which include free radicals (such as the superoxide anion $O_2\cdot^-$ and the hydroxyl radical $OH\cdot$) and non-radical molecules such as hydrogen peroxide (H_2O_2). These molecules are unstable and propagate instability by trying to capture a stabilizing electron, which leads to the oxidation of other molecules which, in turn, seek other targets [1].

Even if lipids are the most sensitive organic components to oxidation, none of them escape it, proteins, sugars, and nucleic acids also being involved. To fight against oxidative attack, cells of aerobic organisms have developed a set of countermeasures in the form of small molecules that are capable of trapping free radicals (glutathione, thioredoxin, vitamins, polyamines, polyphenols) and antioxidant enzymes (superoxide dismutase (SOD), glutathione peroxidases (GPxs), Catalase, peroxiredoxins (PRxs), glutaredoxins (GRx)) [2,3] that intervene intra- and extracellularly to regulate the presence of reactive oxygen species (ROS).

Although ROS have long been considered as aggressors leading to cell death and pathophysiology, their important physiological actions should not be neglected [4]. Indeed, ROS are regulators of cellular activity, acting essentially as second messengers and participating in the physiological oxidation of cellular components [5,6]. H_2O_2 is an ROS at the crossroads, both an intra- and extracellular signaling molecule, a powerful bactericidal agent also acting as an important player in our inflammatory/immune responses [7,8], and an essential factor in the disulfide bridging of proteins carrying thiol groups [9], whether spontaneously or mediated upon by enzymes (see Figure 1).

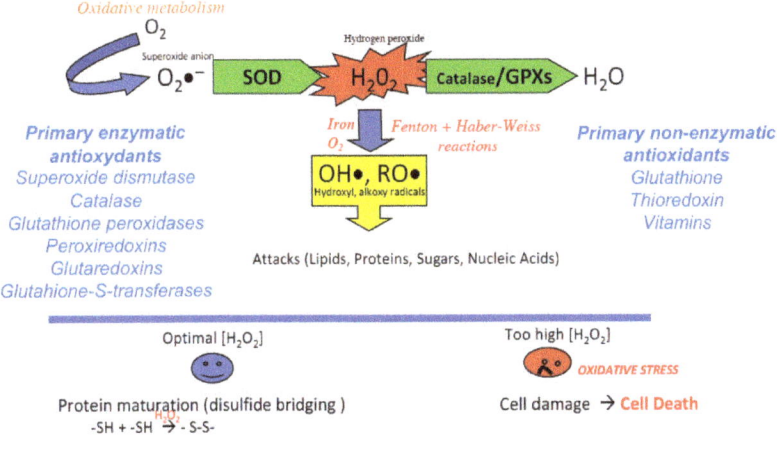

Figure 1. The classic reaction cascade for the production of reactive oxygen species of metazoans. The consumption of oxygen, which supports the production of energy by the mitochondrial respiratory chain, results in the production of the anion radical superoxide ($O_2 \cdot^-$). Although weakly reactive and not permeant the superoxide anion is readily transformed (via the action of superoxide dismutase: SOD) into an active oxygen derivative, hydrogen peroxide (H_2O_2). This molecule occupies a crossroad position having important actions in the maturation of proteins via the oxidation of thiol groups as well as serving as a stimulus and second messenger in critical signal transduction pathways. However, if an excess of H_2O_2 is produced, its strong penetration into the cellular compartments and its rapid reactivity with iron and oxygen (via the Fenton and Haber-Weiss reactions) lead to the production of very aggressive free radicals (hydroxyl and alkoxy radicals) for which there is no active recycling system. These free radicals attack all organic components (lipids, proteins, sugars, and up to nucleic acids). The multiple alterations generated if they are not sufficiently corrected can lead to cellular dysfunctions and eventually to cell death. To finely regulate the concentration of H_2O_2, both in the extracellular and intracellular compartments, metazoans have developed several enzymatic activities grouped under the classification of primary enzymatic antioxidants (GPxs: glutathione peroxidases, CAT: catalase, PRxs: peroxiredoxins, GRx: glutaredoxins, GSTs: glutathione S-transferases) to transform H_2O_2 into a neutral element, water (H_2O). In the same way, several non-enzymatic molecules (glutathione, thioredoxin, vitamins) are at work to trap free radicals.

The balance between the production and recycling of ROS is, therefore, a key element of cell homeostasis. An excess of ROS (whether due to overproduction or lack of recycling) can then lead to a situation known as "oxidative stress." This is accompanied by a set of alterations to cellular components, which affect cellular structures and functions and ultimately lead to cell death (Figure 1). Although oxidative stress is now recognized as a component of cell dysfunction, pathophysiology, and aging, it is important to note that its opposite, reductive stress (not enough ROS and/or too much antioxidants), is just as problematic for cell homeostasis. The notion of redox balance underlines this dual aspect. What are the situations in which this redox balance is called into question?

The answer is both easy and complicated in that almost all of a mammal's interactions with its environment are likely to generate both local and systemic oxidative stress (see Figure 2 for an illustration of the classic causes that can lead to overproduction of ROS). Most cells interpret this ROS signaling, whether internal and/or external, use it and/or fight it in a very refined and effective manner. However, the mature male gamete, for multiple reasons discussed below, has some gaps in its management of ROS that make it particularly vulnerable [10].

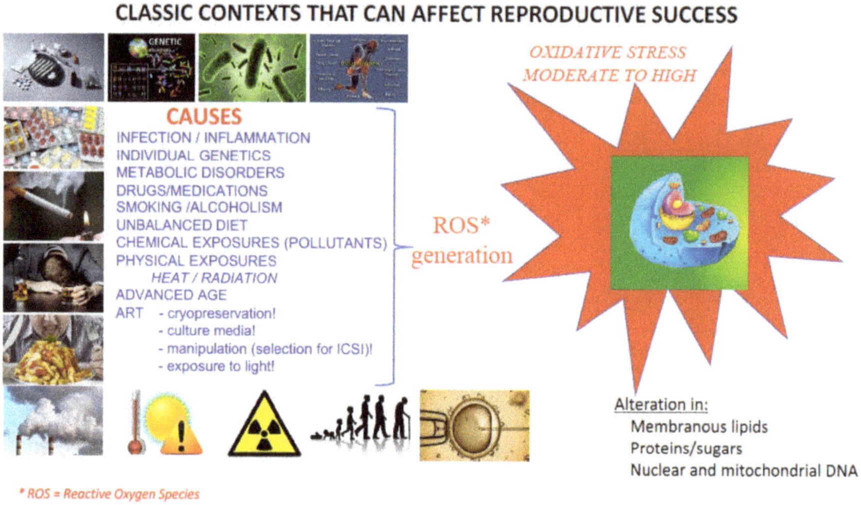

Figure 2. Classic situations promoting the generation of ROS. Situations as varied and cumulative as individual genetic predisposition, infectious/inflammatory pathophysiology or metabolic disorders, medication or addiction (drugs, alcohol, tobacco), nutritional imbalances, exposure to environmental pollutants and physical stresses (such as excessive heat, ionizing radiation) all generate systemic or local ROS that can affect all cells of the body including gametes. Aging is also a situation that promotes oxidative stress. Indeed, according to the "free radical theory of ageing", the lower efficiency of ROS recycling systems when one ages leads to an inevitable increase in ROS production. Finally, in the very specific context of assisted reproductive technologies (ART), gamete cryopreservation, culture in different artificial media, long gamete selection protocols for in vitro fertilization, intracytoplasmic sperm injection (IVF ICSI) and exposure to light alone are all sources that can lead to a pro-oxidant situation.

2. The Spermatozoon's Particular Susceptibility to Oxidative Insults

Several characteristics of spermatozoa explain their sensitivity to oxidative stress. If most cells fight oxidative stress by the presence of small molecules and the activity of antioxidant enzymes contained in their cytoplasm and/or, if necessary, by transcriptional activation of the genes corresponding to these proteins, then spermatozoa are the exception. Evolution has chosen, in internally fertilized metazoans, to produce an extremely cyto-differentiated male gamete characterized by the exclusion of most cytoplasm and the significant compaction of its genetic material, making this silent cell unable to transcribe and synthesize new proteins. The "silent" nature of this cell also explains its inability to repair the alterations that affect it, and in particular, it cannot repair the damage to its genetic material. Death by necrosis or apoptosis is the only alternative for this cell if submitted to acute stressors. As a result, spermatozoa are unable to defend themselves effectively against oxidative stress. As a consequence, in their post-testicular life, these cells largely depend on their immediate environment and their very particular organization (highly compacted nucleus) for protection. This situation of fragility in relation to oxidative stress is also aggravated by the peculiar lipid composition of the plasma membrane of the spermatozoon. Of all the differentiated cells in a mammal, spermatozoa contain the highest level of polyunsaturated fatty acids (PUFAs) in their membranes [11], which are the main targets of ROS. The oxidation of these PUFAs generates toxic aldehydes which, in a vicious cycle, amplify the production of ROS [12] and their pathological consequences. Figure 3 illustrates this in a very schematic way (as it is not the focus of the present review) capturing the known and suspected consequences of oxidative stress on sperm structures and functions. In summary, peroxidation of membrane lipids will affect sperm motility by altering the fluidity, and therefore flexibility, of the membrane, which are important factors for flagellar movements. Sperm motility will also be hampered by the loss of efficiency of mitochondria when subjected to oxidative stress. In addition, the oxidation of membrane lipids and transmembrane proteins incorporated in the lipid bilayer will affect both spermatozoa-oocyte interaction and the signaling cascades resulting from this event. This will lead to poor spermatozoa/oocyte recognition as well as altered capacitation and acrosomal reaction processes, which are crucial steps for successful fertilization [13–15].

IMPACTS OF ROS ON SPERM STRUCTURES AND FUNCTIONS

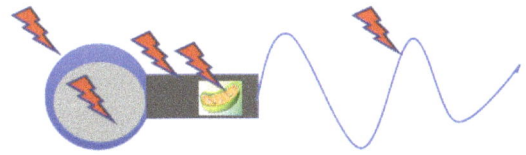

- plasma membrane fluidity loss → **Motility affected**

- mitochondrial function → **Motility affected**

- oxidation of lipids, sugars and membrane proteins → **gamete recognition affected**

- interference with H_2O_2-mediated cell signalling → **compromised fertilization**

- sperm DNA attacks → **mutagenic risk inheritable if not corrected by the oocyte**
        -alteration in embryonic development
        -increase in pathologies in the offspring
        -genetic weakening

Figure 3. Known impacts of ROS on sperm structures and functions. ROS (red arrowheads above and within the schematized spermatozoa) have detrimental effects on sperm cells, resulting in changes in their structure and function. Peroxidation of membrane lipids affects the fluidity of the lipid bilayer, resulting in changes in sperm motility. Similarly, oxidative damage to enzyme complexes in the mitochondrial respiratory chain leads to changes in sperm motility. In an entirely different register, damage to the plasma membrane, whether from lipids, sugars and/or transmembrane proteins incorporated into the lipid bilayer, can alter the spermatozoa's ability to interact with their target, the oocyte. At the same time, ROS (especially H_2O_2) can disrupt the terminal signaling pathways of capacitation and acrosomal reaction, thereby disrupting the fertilization stage. Finally, by attacking nucleic acids, ROS induce a mutagenic risk that can be transmitted to the embryo and future generations if these alterations in the genetic and epigenetic information contained within the paternal nucleus/DNA are not effectively corrected by oocyte repair systems. These de novo mutations, created in the female germline but originating in the male, then increase the risk of abnormal development, the appearance of pathologies in the offspring and, in the long term, may lead to the genetic impoverishment of the species. Please refer to articles [13–15] for appropriate literature covering these aspects.

Last, but not least, one of the major consequences of oxidative stress on sperm cells involves damage to the paternal genetic material. Figure 4 shows some of the multiple ways in which oxidative stress can damage the cell nucleus and its contents. Depending on the intensity of the oxidative stress, it can range from simple oxidation of bases (guanosine and adenosine being the most sensitive bases to oxidative stress) to DNA fragmentation (by single- or double-strand breaks). In between, other DNA oxidative events can be found, including the generation of abasic sites and DNA-protein cross-linking. As mentioned above, as mature sperm cells lack a fully functional DNA repair system, they will need to rely on oocyte DNA repair systems (mainly the post-fertilization oocyte base excision repair pathway, BER) to correct these oxidative alterations. Even in situations of moderate to low oxidative stress, which will not cause DNA breakdown, oxidation of the bases will occur and must be corrected (i.e., each oxidized base must be replaced by a non-oxidized base).

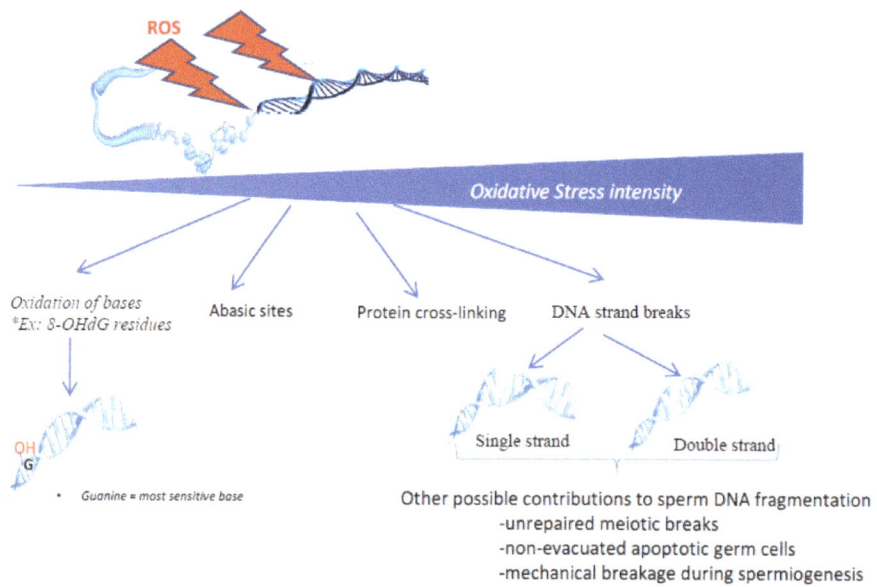

Figure 4. The oxidative damage of DNA has many faces. Depending on the intensity of oxidative stress, different types of nuclear/DNA alterations can be observed. In the case of low/light oxidative stress, the first type of damage concerns the oxidation of DNA bases (in particular guanosine and adenosine) leading to oxidized residues (such as 8-oxo-guanosine, the so-called 8-OHdG residue). When the level of oxidative stress increases, other alterations can occur, including the generation of abasic sites, cross-link of nuclear proteins and rupture of DNA strands, either single or double. Although DNA strand breaks may have an oxidative origin, it should be kept in mind that the fragmentation of sperm DNA may have other origins, including unrepaired meiotic breaks, non-processed apoptotic germ cells or mechanical shearing during late spermatogenesis (spermiogenesis) when protamines replace sperm nuclear histones.

3. Oxidation Processes are Required for Optimal Maturation of Sperm

Although, as mentioned above, sperm structures and functions are easily threatened by oxidative alterations, oxidative processes contribute to the production of fully mature sperm cells. In the late 1980s, it was elegantly reported that during post-testicular maturation of sperm cells (i.e., during epididymal maturation), sperm proteins undergo high disulfide bridging activity [16]. In short, if we look at the level of thiol groups carried by sperm proteins in the caput epididymis, we see that it consists mainly of free thiols, whereas in spermatozoa from the caudal epididymis, most free thiols have been converted into disulfide bridges. Disulfide isomerases are at work in the epididymis to create these bridges by using hydrogen peroxide as an oxidizing agent. Many sperm proteins feature disulfide bridges, whether they are part of the plasma membrane or more internal; regardless of their location in the head, midpiece, or tail segment of the spermatozoon. It is assumed that these intra- and/or inter-protein interactions are created for mechanical purposes, stiffening key structures within the spermatozoon and making the latter more resistant to attack. It is also suspected that these finely tuned protein-protein interactions involving disulfide bridges could determine optimal flagellar movement and motility of sperm cells. In the nucleus, this disulfide bridging activity is also at play and concerns cysteine-rich protamines. We and others have helped to show that an enzymatic activity contained in the sperm nucleus (snGPx4, for sperm nucleus glutathione peroxidase 4) was actually a disulfide isomerase using luminal hydrogen peroxide in the caput epididymis to make inter- and intra-protamine disulfide bridges that further condense the sperm nucleus and lock it in an optimally compacted state during epididymal transit [17–19]. This was further reinforced by the observation that when the luminal hydrogen peroxide concentration of the epididymis was modified [19,20], it immediately resulted in a higher transient condensation of the sperm nucleus in the caput epididymis [20,21], followed rapidly by excessive oxidative alterations of the DNA due to both the ability of H_2O_2 to cross the plasma membrane and the difficulty spermatozoa experience in defending themselves against such attacks. Finely controlled post-testicular oxidation processes are therefore at work to define an optimal state of nuclear condensation of the sperm nucleus. This is of paramount importance as the high compaction of paternal DNA is one of the means chosen by evolution to protect sperm DNA from mutagenic alterations. The process is finely balanced because not all free thiols of cysteine-rich protamines are affected by disulfide bridging. Some thiols are associated with zinc, thus preventing a number of them from being involved in disulfide bridges. It is therefore assumed that it is important not to over-condense the already highly compacted sperm nucleus so that the oocyte does not have too much difficulty post-fertilization, in creating the male pronucleus, a process that has recently been shown to be controlled by thiol-reducing activity [22]. This beneficial post-testicular oxidation process illustrates very well the Jeckyl and Hyde act played by ROS on sperm structures and functions and in particular on the nucleus.

The fact that it has long been reported that sperm cells are themselves very good producers of ROS [23], a logical consequence of their high mitochondrial activity when they are motile, is ambiguous. Could evolution have had the opportunity to do otherwise? Certainly not! Sexual reproduction requires the physical union of male and female gametes, so one or both cells must be capable of movement to achieve fertilization. Movement requires energy, which is supplied to aerobic organisms by oxygen consumption via the mitochondrial respiratory chain, inevitably leading to the generation of ROS. Thus, ROS are an inevitable consequence of sperm movement. However, evolution has been smart enough to dampen the harmful effects of ROS by grouping mitochondria into a well-defined subcellular compartment, the sperm midpiece, where some of the free radicals can be neutralized before they have a chance to enter the nucleus.

4. Sperm DNA Oxidation is Conditioned by the Chromatin Organization

We have already mentioned above, that whenever ROS homeostasis is modified around sperm cells, there is a risk of excessive oxidation that can affect its structures and functions. In this section, we will focus on the oxidation of sperm DNA and, in particular, where it occurs in the mammalian sperm nucleus from recent data coming from animal and human studies [21,24–26]. We have shown that the peripheral regions of the mouse sperm nucleus are more sensitive to oxidative DNA alterations [24]. We also showed that the basal region of the sperm nucleus in the immediate vicinity of the sperm midpiece, the internal source of ROS, was another area preferably affected by oxidative damage [24]. Given the notion of chromosomal territories (CT) within spermatozoa (referring to the fact that within the sperm nucleus, chromosomes are not randomly organized but occupy specific positions identical from one sperm cell to another [27]), it was logical to find that the chromosomes that were most sensitive to oxidative damage were those located in the nuclear periphery and at the base of the sperm nucleus [25]. In the case of the mouse sperm nucleus, this referred to the small autosomes (Chr19, Chr18 and Chr17), all three of which are located near the neck of the cell, closer to the midpiece (see Figure 5). This also concerned the Y chromosome, which occupies a particular position in the murine sperm nucleus near the thin, hook-shaped apical head region [21,25]. In addition, it was very logical to find that regions of low compaction within chromosomes (corresponding to regions maintained in a nucleosomal organization, i.e., still associated with persisting histones that were not replaced by protamines during spermiogenesis) were particularly sensitive to oxidative attack [25]. This conclusion was strongly supported by confocal microscopic images showing that the oxidized regions of the murine sperm nucleus fully corresponded to the nuclear domains enriched with persistent histones [24]. In addition, we further observed that the small DNA regions (about 1 kb long) connecting one protamine toroid to the next one (the interlinker regions) were systematically more sensitive to oxidation along each chromosome than the domains associated with protamines [25]. This has been illustrated by the rhythmic presence of oxidized DNA domains on the chromosomes at about 50 kb intervals, which corresponds exactly to the length of DNA associated with each protamine toroidal ring [25]. Therefore, paternal DNA regions of easy access, lower condensation and near ROS sources (external/internal) are those that will preferably undergo oxidative alterations.

Identical investigations conducted on human spermatozoa confirmed these observations [26]. In the nucleus of human spermatozoa, the chromosome regions sensitive to oxidation concern the peripheral nuclear territories and the chromosome domains associated with histones [26]. The same rhythmic pattern of oxidized regions occurring approximately every 50 kb on the chromosomes and corresponding to the inter-toroid DNA linkers has also been observed in the human sperm nucleus [26]. The only difference between the chromosome regions of the mouse and human sperm nucleus sensitive to oxidation comes from the observation that in the human sperm nucleus, almost all chromosomes were fairly equally affected by oxidative damage [26]. This led to the observation of a linear relationship between the number of oxidized regions and the length of each chromosome. This was not the case in the highly compacted mouse sperm nucleus where small chromosomes because of their localization at the nuclear periphery were most affected by oxidative damage [25]. This difference between the two species is easily explained by the fact that the human sperm nucleus retains a very high proportion of persistent histones, compared to the mouse sperm nucleus, which makes it much less condensed [28]. Naturally, this structural feature is likely to facilitate oxidative damage along the length of each human sperm chromosome. Following the same logic, the number of regions sensitive to oxidation in the weakly condensed chromatin of human spermatozoa was much higher than that of a mouse sperm cell (see Figure 5). However, this comparison is of limited value and may be purely fortuitous as it is difficult to compare one model with the other. Indeed, it is unlikely that the basal oxidation level of spermatozoa from a normozoospermic human donor can be considered equivalent to the basal oxidation level of wild-type (WT) mouse sperm cell. Despite the more homogeneous distribution of oxidized regions on the chromosomes of the human sperm nucleus, some chromosome domains have nevertheless proved to be particularly sensitive to DNA oxidation [26]. Such an example is the q11–q14 domain of chromosome 15 of the human sperm nucleus where 3 regions of susceptibility to oxidation have been found. Interestingly, one of these regions (q13–q14) overlaps a locus in which are located genes involved in syndromes whose frequency in offspring has been associated with the age of the father and sperm DNA lesions. These syndromes were also associated with poor DNA repair activities in the oocyte, probably related to maternal age [26]. Contrary to what might have been expected by looking at the repetitive and G-rich signature of the telomeres (5'-TTAGGG-3'), we did not observe any particular sensitivity to telomere oxidation in any of the models studied (human [26] or mouse spermatozoa [24]). Although this may have been due to the technical limitations associated with sequencing regions full of repetitive sequences, this observation is not conducive to a strong impact of oxidation on the length of spermatozoa telomeres. Despite reports showing that short telomeres are associated with sperm DNA alteration (mainly fragmentation) and male infertility [29,30], it is not yet clear that DNA oxidation directly influences spermatozoa telomere length [31,32]. A recent report suggests, however, that ROS may inhibit telomerase activity [33] thus indirectly explaining why male infertility cases may be found characterized by spermatozoa with shorter telomeres.

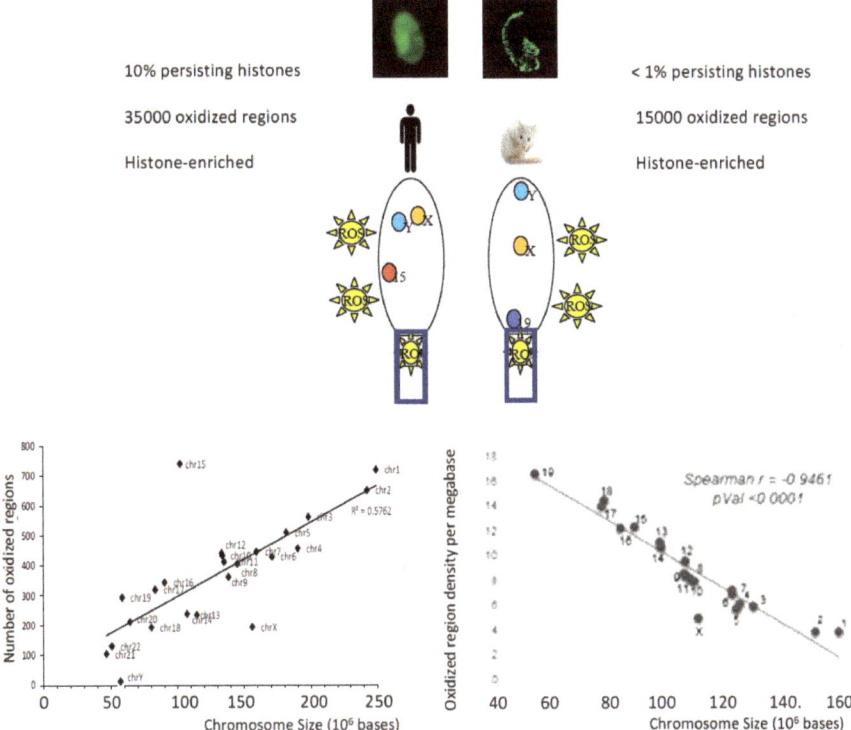

Figure 5. The nuclear regions sensitive to oxidation depend on the chromatin organization of the sperm cells, which is species-specific. In humans and mice, DNA oxidative alterations of spermatozoa revealed by the presence of 8-OHdG residues by fluorescent confocal microscopy show distinct patterns. The low nuclear condensation state of human spermatozoa in which a large part of the chromatin is still in a nucleosomal organization (i.e., associated with persistent histones) explains why 8-OHdG residues are found throughout the sperm nucleus and why there are so many oxidized domains. This low state of nuclear compaction is also illustrated by the observation that the number of oxidized regions follows a linear relationship with the length of the chromosomes. In the more compacted mouse sperm nucleus containing few histone-bound nucleosomes, the regions sensitive to oxidation are more peripheral and less numerous. In this less accessible context, the linear relationship between chromosome length and the number of oxidized regions is no longer valid and only the most exposed (peripheral) chromosomes are affected by oxidative alterations (as illustrated by autosome 19 and chromosome Y [21,25]). In both species, the more chromosomes are located at the periphery of the sperm nucleus the more sensitive they will be to oxidative DNA damage [21,25,26]. In both species, oxidized nuclear regions are regions enriched with persistent histones [21,25,26].

5. Consequences of Sperm DNA Oxidation

In the transgenic $gpx5^{-/-}$ mouse model (characterized by a light oxidative epididymal luminal environment [20]), it is estimated that more than one million oxidized guanosine residues (so-called 8-oxodG residues) distributed in 16,000 regions on the chromosomes have occurred, of which 1000 are particularly oxidized (hot spots [25]). Considering that two of the four deoxynucleotides (guanosine and adenosine) are sensitive to oxidation, this gives an idea of the extent of oxidative DNA damage that a sperm nucleus can undergo. One of the interesting aspects of this mouse model is that the oxidation of sperm DNA is not associated with DNA fragmentation [20], two conditions that are widely confused in both the clinical and scientific communities. Indeed, while it is true that massive oxidation of sperm DNA can lead to DNA fragmentation, fragmentation of sperm DNA can also result from unrepaired meiotic arrests, poor elimination of apoptotic germ cells and mechanical shearing of the sperm nucleus during the protamination process of spermiogenesis. Thus, DNA fragmentation should not automatically result in the oxidation of sperm DNA. Conversely, the absence of DNA fragmentation does not necessarily mean the absence of DNA oxidation as moderate oxidation does not create single and/or double-strand breaks. This finding, which has not yet percolated clinically, is very well illustrated by the $gpx5^{-/-}$ mouse model, whose phenotype is only a slight oxidation of the sperm DNA without an increase in DNA fragmentation [20]. Nevertheless, when knockout (KO) males were mated with WT females of proven fertility, an increase in abortions, abnormal developments and perinatal mortality were recorded compared to WT/WT crosses [20]. This clearly shows that even a mild oxidation of the sperm DNA alone is sufficient to trigger reproductive failures. As the fertilizing capacity of these spermatozoa with an oxidized nucleus is not affected in any way, reproductive failures could only result from an aberrant repair of paternal DNA by the oocyte. It is the oocyte's task to ensure that the oxidized residues of the paternal nucleus are removed during the decondensation and de-protamination steps following fertilization. This will bring the paternal nucleus into a pronuclear state facilitating syngamy and the restoration of diploidy.

It has been shown that the sperm nucleus contains only the first step of the base excision repair pathway (BER) represented by the activity of 8-oxoguanine DNA glycosylase 1 (OGG1 [34]). It appears that sperm do not have the other components of the BER pathway, which includes *apurinic endonuclease 1* (APE1) and *X-ray repair complementing defective repair in Chinese hamster cells 1* (XRCC1) activities [34]. Only the oocyte BER system can complete the repair process using its XRCC1/APE1 activities to engineer the replacement of oxidized bases with non-oxidized residues. This oocyte-driven DNA repair process underlines the importance of oocyte quality in preventing transmission of paternal DNA alterations into the embryo. Whenever the oocyte's ability to repair is reduced (e.g., due to maternal age or non-physiological pressure linked to forced oocyte production following hormonal stimulation during the course of ART IVF procedures [35]), there is a risk that unrepaired oxidized residues remain in the paternal pronucleus. If the oocyte BER pathway is not fully effective, and/or if the level of oxidized bases in the sperm nucleus is too high to be treated properly, this will result in nonrepair and/or false repair leading to mainly transversion-type mutations (following Hoogsteen base pairing between 8-OHdG and adenine [36]) that will ultimately be transmitted to the developing embryo and future generations [37]. Depending on where this occurs, it may call into question the completion and/or normality of embryonic development as well as the quality of life of the future individual and beyond. In addition, unrepaired 8-OHdG in paternal pronucleus could, especially if they occur on CpG islands, have an impact on the reprogramming of the methyl epigenetic mark of adjacent cytosines. Indeed, it has been shown that oxidized guanine suppresses the methylation of an adjacent cytosine [38–40]. This could lead to aberrant DNA methylation in the embryo as remethylation of the paternal nucleus occurs after fertilization [41]. Aberrant methylation of embryonic DNA may explain abnormal development, altered gene expression, genomic instability and the susceptibility of offspring to disease [42].

6. Sperm Nuclear Oxidation May Go Well Beyond Base Alterations

If sperm DNA is sensitive to oxidative attacks, this is also the case for the other components of the nuclear compartment, i.e., the nuclear proteins and the recently characterized nuclear complement of non-coding RNA (ncRNAs). Together with methylation of cytosine residues, nuclear protein modifications and the ncRNA profile represent the three levels of epigenetic information carried by the spermatozoon.

Spermatozoa cytosine hypomethylation has been associated with infertile patients with oxidative DNA damage (mainly DNA fragmentation) and elevated seminal ROS; a situation that has been corrected by antioxidant supplementation [43]. It has been suggested that glutathione synthesis and homocysteine recycling via the single carbon cycle are the pathways linking oxidative stress and cytosine hypomethylation [44]. The oxidation of DNMTs (DNA methyltransferases) decreasing their activity and, as indicated above, the lower cytosine methylation in oxidized CpG regions, are other pathways by which oxidative stress can influence sperm DNA methylation [45]. Modification of the sperm cytosine methylation profile by oxidation is an important issue that deserves the attention of the clinical and scientific communities, as it may be closely related to environmental exposures and ART [46,47].

In addition to the impact of an oxidized G residue on the methylation process, there is a second question to consider theoretically if a post-testicular oxidative stress situation occurs. It concerns the oxidation of methylcytosine (meC) residues carried by the spermatozoa to create hydroxymethylcytosine residues (hmeC). It is interesting to note that the generation of hmeC is the first step in an enzymatically-mediated oxidation process (via the TET enzymes: Ten of Eleven Translocases) which is used to remove meC marks [48] during the post-fertilization reprogramming of the male pronucleus. This is particularly important for the male pronucleus because the sperm nucleus has been highly methylated during spermatogenesis. However, it has been shown that some regions of the male pronucleus must escape this meC erasure process and are therefore maintained in a silent transcriptional state [49]. If these regions are not properly hydroxymethylated in an oxidation process independent of TET, this could lead to the demethylation of paternal genomic regions that should normally be methylated. Such events could lead to significant changes in the embryonic epigenetic fingerprint later in development. Experiments are underway to test this hypothesis. Preliminary data suggest that a post-testicular pro-oxidant environment alters both meC and hmeC distribution within the sperm nucleus [50] In the genetic contexts of WT and $gpx5^{-/-}$ mice, we are presently carrying out sperm chromatin immunoprecipitation experiments using antibodies specific for meC and hmeC to identify chromatin regions subjected to differential methylation/hydroxymethylation of cytosine residues.

Sperm nuclear proteins can also be affected by oxidative alterations as is the case for any protein. Protein oxidation essentially results in protein carbonylation [51] and redox thiol modification. Protein carbonylation is defined as the covalent and irreversible modification of the side chains of the amino acids cysteine, histidine and lysine by peroxidized lipid intermediates such as 4-hydroxy, 4-oxoneonenal (4-HNE) [52]. As noted above, there is considerable evidence that oxidation of sperm nuclear proteins containing thiols (including protamines) affects the structure and function of sperm cells. Besides thiol-oxidation, the carbonylation of sperm protamines can occur in a pro-oxidant environment because the protamines are rich in cysteine, histidine and lysine residues. It is not expected to be particularly damaging to the embryo as the nuclear protamines in the sperm are quickly removed after fertilization and replaced by histones. It is, however, possible that protamine carbonylation, advanced glycation end products (AGE) and other sperm nuclear protein-protein cross-linking events that are facilitated upon protein oxidation might modify the kinetics of this protamine replacement process, which was recently shown to be redox-mediated in the oocyte [22]. More important is probably the oxidative alteration of the persistent nuclear histones in spermatozoa. To date, it is understood that these paternal histones will not be replaced in the oocyte and will, therefore, be transmitted to the developing embryo. If oxidative alterations occur in these histones, this could create unsuspected problems in the developing embryo, as these histones will be part of the zygote histone code. To our knowledge, this particular area has not yet been investigated.

Over the past decade, there has been considerable evidence that sperm cells provide the oocyte with a complex and highly dynamic load of non-coding RNA (ncRNA), which represents another aspect of the paternal epigenetic heritage. Two recent studies have shown that environmental constraints such as a particular diet or exposure to behavioral stress modify the profile of sperm RNA [53,54].

These studies also showed unequivocally that these different sperm ncRNA contents were responsible for the transmission of the paternal phenotype to the offspring. In addition, it has been shown that the apocrine secretory activity of the epididymis (i.e., via epididymosomes) is the source of these changes. We have very recently contributed to this area by showing that the distribution of ncRNAs of the epididymal epithelium is significantly altered when WT mice are compared to the pro-oxidant situation experienced by $gpx5^{-/-}$ KO mice [55]. As the epididymis is a provider of ncRNA during sperm transit, it is therefore expected that if the ncRNA profile of the epididymis changes, the sperm ncRNA profile will also change. We are currently comparing the ncRNA content of sperm from WT mice versus $gpx5^{-/-}$ Not surprisingly, our preliminary results [56] confirm this hypothesis. How and to what extent these different sperm ncRNAs can affect the embryo development program and the health of the offspring are investigations that must now be conducted.

7. Conclusions

Whether they are unrepaired or falsely repaired oxidized bases promoting the creation of de novo mutations in the embryo, multiple epigenetic alterations (whether due to changes in the methylation/hydroxymethylation status of sperm cytosines, the paternally-associated histone code and/or the sperm ncRNA profile), it is obvious that oxidative stress will have a profound impact on the sperm structures and functions as well as on the messages it carries in the oocyte and embryo. These are important questions that still need to be better assessed, especially with regard to the clinical consequences of ART. As sperm DNA oxidation is much more common in infertile patients than sperm DNA fragmentation, a better understanding of its consequences on the embryo and the future individual is needed. In particular, artificial oxidative situations generated in the context of ART should be seriously evaluated, as they may partly explain the low success rate associated with this technology, which has not improved significantly over the past 25 years.

Author Contributions: J.R.D. wrote the manuscript which was then edited and revised by R.J.A. All authors have read and agreed to the published version of the manuscript.

Funding: This research received no external funding.

Conflicts of Interest: The two authors are scientific advisors to an American biotechnology company (CellOxess LLC, New Jersey, USA) specializing in the commercialization of antioxidant supplements for the treatment of oxidative stress.

References

1. Davies, K.J. Oxidative stress: The paradox of aerobic life. *Biochem. Soc. Symp.* **1995**, *61*, 1–31.
2. O'Flaherty, C. The Enzymatic Antioxidant System of Human Spermatozoa. *Adv. Androl.* **2014**, *2014*, 626374. [CrossRef]
3. Hanschmann, E.M.; Godoy, J.R.; Berndt, C.; Hudemann, C.; Lillig, C.H. Thioredoxins, Glutaredoxins, and Peroxiredoxins—Molecular Mechanisms and Health Significance: From Cofactors to Antioxidants to Redox Signaling. *Antioxid. Redox Signal.* **2013**, *19*, 1539–1605. [CrossRef]
4. Sies, H.; Berndt, C.; Jones, D.P. Oxidative stress. *Annu. Rev. Biochem.* **2017**, *86*, 715–748. [CrossRef] [PubMed]
5. Patel, R.; Rinker, L.; Peng, J.; Chilian, W.M. *Reactive Oxygen Species (ROS) in Living Cells*; Intechopen Publisher: London, UK, 2018. [CrossRef]
6. Sies, H. Role of metabolic H_2O_2 generation: Redox signaling and oxidative stress. *J. Biol. Chem.* **2014**, *289*, 8735–8741. [CrossRef] [PubMed]
7. Clifford, D.; Repine, J. Hydrogen peroxide mediated killing of bacteria. *Mol. Cell. Biochem.* **1982**, *49*, 143–149. [CrossRef] [PubMed]
8. Reth, M. Hydrogen peroxide as second messenger in lymphocyte activation. *Nat. Immunol.* **2002**, *3*, 1129–1134. [CrossRef]
9. Winterbourn, C.C. Biological Production, Detection, and Fate of Hydrogen Peroxide. *Antioxid. Redox Signal.* **2018**, *29*, 541–551. [CrossRef]

10. Aitken, R.J.; Jones, K.T.; Robertson, S.A. Reactive Oxygen Species and Sperm Function–In Sickness and In Health. *J. Androl.* **2012**, *33*, 1096–1106. [CrossRef]
11. Wathes, D.C.; Abayasekara, D.R.E.; Aitken, R.J. Polyunsaturated Fatty Acids in Male and Female Reproduction. *Biol. Reprod.* **2007**, *77*, 190–201. [CrossRef]
12. Moazamian, R.; Polhemus, A.; Connaughton, H.; Fraser, B.; Whiting, S.; Gharagozloo, P.; Aitken, R.J. Oxidative stress and human spermatozoa: Diagnostic and functional significance of aldehydes generated as a result of lipid peroxidation. *Mol. Hum. Reprod.* **2015**, *21*, 502–515. [CrossRef] [PubMed]
13. Aitken, R.J. Reactive oxygen species as mediators of sperm capacitation and pathological damage. *Mol. Reprod. Dev.* **2017**, *84*, 1039–1052. [CrossRef] [PubMed]
14. O'flaherty, C. Redox regulation of mammalian sperm capacitation. *Asian J. Androl.* **2015**, *17*, 583–590. [CrossRef] [PubMed]
15. Aitken, R.J.; Gibb, Z.; Baker, M.A.; Drevet, J.; Gharagozloo, P. Causes and consequences of oxidative stress in spermatozoa. *Reprod. Fertil. Dev.* **2016**, *28*, 1–10. [CrossRef] [PubMed]
16. Shalgi, R. Dynamics of the thiol status of rat spermatozoa during maturation: Analysis with the fluorescent labeling agent monobromobimane. *Biol. Reprod.* **1989**, *40*, 1037–1045. [CrossRef] [PubMed]
17. Conrad, M.; Moreno, S.G.; Sinowatz, F.; Ursini, F.; Kölle, S.; Roveri, A.; Brielmeier, M.; Wurst, W.; Maiorino, M.; Bornkamm, G.W. The Nuclear Form of Phospholipid Hydroperoxide Glutathione Peroxidase Is a Protein Thiol Peroxidase Contributing to Sperm Chromatin Stability. *Mol. Cell. Biol.* **2005**, *25*, 7637–7644. [CrossRef]
18. Noblanc, A.; Peltier, M.; Damon-Soubeyrand, C.; Kerchkove, N.; Chabory, E.; Vernet, P.; Saez, F.; Cadet, R.; Janny, L.; Pons-Rejraji, H.; et al. Epididymis response partly compensates for spermatozoa oxidative defects in snGPx4 and GPx5 double mutant mice. *PLoS ONE* **2012**, *7*, e38565. [CrossRef]
19. Noblanc, A.; Kocer, A.; Chabory, E.; Vernet, P.; Saez, F.; Cadet, R.; Conrad, M.; Drevet, J.R. Glutathione peroxidases at work on epididymal spermatozoa: An example of the dual effect of reactive oxygen species on mammalian sperm fertilizing ability. *J. Androl.* **2011**, *32*, 641–650. [CrossRef]
20. Chabory, E.; Damon, C.; Lenoir, A.; Kauselmann, G.; Kern, H.; Zevnik, B.; Garrel, C.; Saez, F.; Cadet, R.; Henry-Berger, J.; et al. Epididymis seleno-independent glutathione peroxidase 5 maintains sperm DNA integrity in mice. *J. Clin. Investig.* **2009**, *119*, 2074–2085. [CrossRef]
21. Champroux, A.; Damon-Soubeyrand, C.; Goubely, C.; Bravard, S.; Henry-Berger, J.; Guiton, R.; Saez, F.; Drevet, J.; Kocer, A. Nuclear Integrity but Not Topology of Mouse Sperm Chromosome is Affected by Oxidative DNA Damage. *Genes* **2018**, *9*, 501. [CrossRef]
22. Tirmarche, S.; Kimura, S.; Dubruille, R.; Horard, B.; Loppin, B. Unlocking sperm chromatin at fertilization requires a dedicated egg thioredoxin in Drosophila. *Nat. Commun.* **2016**, *7*, 13539. [CrossRef] [PubMed]
23. Tosic, J.; Walton, A. Formation of hydrogen peroxide by spermatozoa and its inhibitory effect of respiration. *Nature* **1946**, *158*, 485. [CrossRef] [PubMed]
24. Noblanc, A.; Damon-Soubeyrand, C.; Karrich, B.; Henry-Berger, J.; Cadet, R.; Saez, F.; Guiton, R.; Janny, L.; Pons-Rejraji, H.; Alvarez, J.G.; et al. DNA oxidative damage in mammalian spermatozoa: Where and why is the male nucleus affected? *Free Radic. Biol. Med.* **2013**, *65*, 719–723. [CrossRef] [PubMed]
25. Kocer, A.; Henry-Berger, J.; Noblanc, A.; Champroux, A.; Pogorelcnik, R.; Guiton, R.; Janny, L.; Pons-Rejraji, H.; Saez, F.; Johnson, G.D.; et al. Oxidative DNA damage in mouse sperm chromosomes: Size matters. *Free Radic. Biol. Med.* **2015**, *89*, 993–1002. [CrossRef] [PubMed]
26. Xavier, M.J.; Nixon, B.; Roman, S.D.; Scott, R.J.; Drevet, J.R.; Aitken, R.J. Paternal impacts on development: Identification of genomic regions vulnerable to oxidative DNA damage in human spermatozoa. *Hum. Reprod.* **2019**, *34*, 1876–1890. [CrossRef]
27. Zalensky, A.; Zalenskaya, I. Organization of chromosomes in spermatozoa: An additional layer of epigenetic information? *Biochem. Soc. Trans.* **2007**, *35*, 609–611. [CrossRef]
28. Johnson, G.D.; Lalancette, C.; Linneman, A.K.; Leduc, F.; Boissonneault, G.; Krawetz, S.A. The sperm nucleus: Chromatin, RNA, and the nuclear matrix. *Reproduction* **2011**, *141*, 21–36. [CrossRef]
29. Rodríguez, S.; Goyanes, V.; Segrelles, E.; Blasco, M.A.; Gosálvez, J.; Fernández, J.L. Critically short telomeres are associated with sperm DNA fragmentation. *Fertil. Steril.* **2005**, *84*, 843–845. [CrossRef]
30. Tahamtan, S.; Tavalaee, M.; Izadi, T.; Barikrow, N.; Zakeri, Z.; Lockshin, R.A.; Abbasi, H.; Esfahani, M.H.N. Reduced sperm telomere length in individuals with varicocele is associated with reduced genomic integrity. *Sci. Rep.* **2019**, *9*, 4336. [CrossRef]

31. Balmori, C.; Varela, E. Should we consider telomere length and telomerase activity in male factor infertility? *Curr. Opin. Obstet. Gynecol.* **2018**, *30*, 197–202. [CrossRef]
32. Rocca, M.S.; Foresta, C.; Ferlin, A. Telomere length: Lights and shadows on their role in human reproduction. *Biol. Reprod.* **2019**, *100*, 305–317. [CrossRef] [PubMed]
33. Ahmed, W.; Lingner, J. PRDX1 and MHT1 cooperate to prevent ROS-mediated inhibition of telomerase. *Genes Dev.* **2018**, *32*, 658–669. [CrossRef] [PubMed]
34. Smith, T.B.; Dun, M.D.; Smith, N.D.; Curry, B.J.; Connaughton, H.S.; Aitken, R.J. The presence of a truncated base excision repair pathway in human spermatozoa that is mediated by OGG1. *J. Cell Sci.* **2013**, *126*, 1488–1497. [CrossRef] [PubMed]
35. Bosch, E.; Labarta, E.; Kolibianakis, E.; Rosen, M.; Meldrum, D. Regimen of ovarian stimulation affects oocyte and therefore embryo quality. *Fertil. Steril.* **2016**, *105*, 560–570. [CrossRef] [PubMed]
36. Ghossal, G.; Muniyappa, K. Hoogsteen base-pairing revisited: Resolving a role in normal biological processes ad human diseases. *Biochem. Biophys. Res. Commun.* **2006**, *343*, 1–7. [CrossRef] [PubMed]
37. Ohno, M.; Sakumi, K.; Fukumura, R.; Furuichi, M.; Iwasaki, Y.; Hokama, M.; Ikemura, T.; Tsuzuki, T.; Gondo, Y.; Nakabeppu, Y. 8-oxoguanine causes spontaneous de novo germline mutations in mice. *Sci. Rep.* **2014**, *4*, 4689. [CrossRef] [PubMed]
38. Wachsman, J.T. DNA methylation and the association between genetic and epigenetic changes: Relation to carcinogenesis. *Mutat. Res. Mol. Mech. Mutagenes.* **1997**, *375*, 1–8. [CrossRef]
39. Franco, R.; Schoneveld, O.; Georgakilas, A.G.; Panayiotidis, M.I. Oxidative stress, DNA methylation and carcinogenesis. *Cancer Lett.* **2008**, *266*, 6–11. [CrossRef]
40. Wu, Q.; Ni, X. ROS-mediated DNA methylation pattern alterations in carcinogenesis. *Curr. Drug Targets* **2015**, *16*, 13–19. [CrossRef]
41. Ma, X.; Wang, X.; Qin, L.; Song, C.; Lin, F.; Song, J.; Zhu, C.; Liu, H. De novo DNA methylation of the paternal genome in 2-cell mouse embryos. *Genet. Mol. Res.* **2014**, *13*, 8632–8639. [CrossRef]
42. Ziech, D.; Franco, R.; Pappa, A.; Panayiotidis, M.I. Reactive Oxygen Species (ROS)—Induced genetic and epigenetic alterations in human carcinogenesis. *Mutat. Res. Mol. Mech. Mutagenes.* **2011**, *711*, 167–173. [CrossRef] [PubMed]
43. Tunc, O.; Tremellen, K. Oxidative DNA damage impairs global sperm DNA methylation in infertile men. *J. Assist. Reprod. Genet.* **2009**, *26*, 537–544. [CrossRef] [PubMed]
44. Menezo, Y.; Silvestris, E.; Dale, B.; Elder, K. Oxidative stress and alterations in DNA methylation: Twin sides of the same coin. *Reprod. Biomed. Online* **2016**, *33*, 668–683. [CrossRef] [PubMed]
45. Weitzman, S.A.; Turk, P.W.; Milkowski, D.H.; Kozlowski, K. Free radical adducts induce alterations in DNA cytosine methylation. *Proc. Natl. Acad. Sci. USA* **1994**, *91*, 1261–1264. [CrossRef] [PubMed]
46. Menezo, Y.J.; Dale, B.; Elder, K. The negative impact of the environment on methylation/epigenetic marking in gametes and embryos: A plea for action to protect the fertility of future generations. *Mol. Reprod. Dev.* **2019**, *86*, 1273–1282. [CrossRef]
47. Menezo, Y.; Clement, P.; Dale, B. DNA methylation patterns in the early human embryo and the epigenetic/imprinting problems: A plea for a more carful approach to human assisted reproductive technology (ART). *Int. J. Mol. Sci.* **2019**, *20*, 1342. [CrossRef]
48. Wu, X.; Zhang, Y. TET-mediated active DNA demethylation: Mechanism, function and beyond. *Nat. Rev. Genet.* **2017**, *18*, 517–534. [CrossRef]
49. Salvaing, J.; Aguirre-Lavin, T.; Boulesteix, C.; Lehmann, G.; Debey, P.; Beaujean, N. 5-Methylcytosine and 5-Hydroxymethylcytosine Spatiotemporal Profiles in the Mouse Zygote. *PLoS ONE* **2012**, *7*, e38156. [CrossRef]
50. Shaygania, E. Université Clermont Auvergne, Clermont-Ferrand, France. Unpublished work. 2019.
51. Lone, S.A.; Mohanty, T.K.; Baithalu, R.K.; Yadav, H.P. Sperm protein carbonylation. *Andrologia* **2019**, *51*, e13233. [CrossRef]
52. Nayak, J.; Jena, S.R.; Samanta, L. Oxidative stress and sperm dysfunction: An insight into dynamics of semen proteome. In *Oxidants, Antioxidants and Impact of the Oxidative Status in Male Reproduction*; Academic Press: Cambridge, MA, USA; Elsevier: Cambridge, MA, USA, 2018; pp. 261–272.
53. Sharma, U.; Conine, C.C.; Shea, J.M.; Boskovic, A.; Derr, A.G.; Bing, X.Y.; Belleannee, C.; Kucukural, A.; serra, R.W.; Sun, F.; et al. Biogenesis and function of tRNA fragments during sperm maturation and fertilization in mammals. *Science* **2016**, *351*, 391–396. [CrossRef]

54. Chen, Q.; Yan, M.; Cao, Z.; Li, X.; Zhang, Y.; Shi, J.; Feng, G.H.; Peng, H.; Zhang, X.; Zhang, Y.; et al. Sperm tsRNAs contribute to intergenerational inheritance of an acquired metabolic disorder. *Science* **2016**, *351*, 397–400. [CrossRef] [PubMed]
55. Chu, C.; Henry-Berger, J.; Ru, Y.; Kocer, A.; Champroux, A.; Li, Z.T.; He, M.; Xie, S.; Ma, W.; Ni, M.; et al. Knockout of glutathione peroxidase 5 down-regulates the piRNAs in the caput epididymis of aged mice. *Asian J. Androl.* **2019**. In press.
56. Chu, C. University of Chinese Academy of Sciences, Shanghai, China & Université Clermont Auvergne, Clermont-Ferrand, France. Unpublished work. 2018.

© 2020 by the authors. Licensee MDPI, Basel, Switzerland. This article is an open access article distributed under the terms and conditions of the Creative Commons Attribution (CC BY) license (http://creativecommons.org/licenses/by/4.0/).

Review

Redox Regulation and Oxidative Stress: The Particular Case of the Stallion Spermatozoa

Fernando J. Peña [1,*], Cristian O'Flaherty [2], José M. Ortiz Rodríguez [1], Francisco E. Martín Cano [1], Gemma L. Gaitskell-Phillips [1], María C. Gil [1] and Cristina Ortega Ferrusola [1]

[1] Laboratory of Equine Reproduction and Equine Spermatology, Veterinary Teaching Hospital, University of Extremadura, 10003 Cáceres, Spain; jmortizro@gmail.com (J.M.O.R.); femartincano@gmail.com (F.E.M.C.); gemmagaitskell@hotmail.com (G.L.G.-P.); crgil@unex.es (M.C.G.); cristinaof@unex.es (C.O.F.)
[2] Departments of Surgery (Urology Division) and Pharmacology and Therapeutics, Faculty of Medicine, McGill University, Montréal, QC H4A 3J1, Canada; cristian.oflaherty@mcgill.ca
* Correspondence: fjuanpvega@unex.es; Tel.: +34-927-257-167

Received: 29 September 2019; Accepted: 15 November 2019; Published: 19 November 2019

Abstract: Redox regulation and oxidative stress have become areas of major interest in spermatology. Alteration of redox homeostasis is recognized as a significant cause of male factor infertility and is behind the damage that spermatozoa experience after freezing and thawing or conservation in a liquid state. While for a long time, oxidative stress was just considered an overproduction of reactive oxygen species, nowadays it is considered as a consequence of redox deregulation. Many essential aspects of spermatozoa functionality are redox regulated, with reversible oxidation of thiols in cysteine residues of key proteins acting as an "on–off" switch controlling sperm function. However, if deregulation occurs, these residues may experience irreversible oxidation and oxidative stress, leading to malfunction and ultimately death of the spermatozoa. Stallion spermatozoa are "professional producers" of reactive oxygen species due to their intense mitochondrial activity, and thus sophisticated systems to control redox homeostasis are also characteristic of the spermatozoa in the horse. As a result, and combined with the fact that embryos can easily be collected in this species, horses are a good model for the study of redox biology in the spermatozoa and its impact on the embryo.

Keywords: horses; spermatozoa; reactive oxygen species (ROS); oxidative stress; redox regulation; equine

1. Introduction

The male gamete, the spermatozoon, is generated in the germinal epithelium of the testes in a process called spermatogenesis. This epithelium consists of germ cells in different stages of development, intermingled with Sertoli cells that provide structural support and nursing, protecting the germ cells. Spermatogenesis is initiated by the differentiation of spermatogonia from a stem cell pool. These cells initiate a proliferative phase entering a continuous process of mitotic division, dramatically increasing spermatogonial numbers. This process is usually termed spermatocytogenesis. In the next step, cells enter a meiotic phase that includes duplication and exchange of genetic information and two meiotic divisions which reduce the chromosome complement to form round haploid spermatids. During the spermiogenesis phase, round spermatids experience a dramatic transformation that includes compaction and silencing of DNA and elongation of the nucleus, development of specific structures such as the sperm tail and acrosome, relocation of the mitochondria in the midpiece, in addition to the loss of other organelles and most of the cytoplasm. Fully developed spermatozoa are released in the lumen of the seminiferous tubules in a process termed spermiation. Recent reviews on this topic can be found elsewhere [1–4]. Chemically, oxidation is the loss of an electron, while reduction is the gain of an electron. This nomenclature reflects the tendency of oxygen, a highly electronegative atom, to partially or fully steal an electron from other molecules. Reactive oxygen species (ROS) [5,6] are atoms or molecules with a single unpaired electron, including, among others, superoxide ($O_2\bullet^-$), the hydroxyl radical (HO•) and the lipid peroxide radical (LOO•). Although hydrogen peroxide (H_2O_2) is not a free radical, it is a precursor of HO•. UV radiation and the presence of metal ions (Fe^{2+}, Fe^{3+} or Cu^{2+}) generate HO•. All aerobic organisms depend on the generation of ATP from electrochemical energy generated in the four electron reduction of molecular oxygen into water. During this process the mitochondrial transport chain may lose electrons, leading to the formation of ROS.

Moreover, mitochondrial dysfunction may exacerbate the loss of electrons and thus increase the production of ROS to toxic levels disrupting redox homeostasis [6]. This particular effect is especially critical in horses. The stallion spermatozoon is characterized by an unusually intense mitochondrial activity in comparison with other mammals [7–11].

Spermatozoa were the first cells known to be capable of generating ROS [12]. This early report demonstrated that bovine spermatozoa produce H_2O_2 as a consequence of cellular respiration. It also showed that the production of H_2O_2 inhibits respiration and concluded that bovine spermatozoa must be equipped with a mechanism for the elimination of H_2O_2 at a low rate, to keep it at physiological levels. For a long time, the production of ROS was considered solely as a toxic byproduct of sperm metabolism; however, nowadays, extensive evidence indicates that crucial functions of the spermatozoa are redox regulated, and redox regulation has become a major area of research in sperm biology [13–20]. Since the discovery of ROS production by the spermatozoa, the concept of oxidative stress has evolved, and enormous research interest in this topic has developed in the last decade. As an example, a recent search in PubMed retrieved 215842 entries using the term oxidative stress, when this term was combined with spermatozoa 2777 entries were obtained (https://www.ncbi.nlm.nih.gov/pubmed/, accessed September, 1 2019). Under aerobic conditions, production of ROS is unavoidable. However, organisms have evolved to develop complex mechanisms to maintain the production of ROS at physiological levels (oxidative eustress) and the redox signaling dependent on ROS regulated [21–23]. Interestingly, the ability to respond to ROS appeared very early in the course of evolution, well before the increase of atmospheric oxygen, probably in response to low ozone levels, since U.V. radiation splits water into ROS [24].

2. Sources of ROS in the Spermatozoa

Several pathways lead to the generation of ROS, including the production of $O_2^-\bullet$, H_2O_2, reactive nitrogen species (RNS), and OH• [25]. The superoxide anion is generated from the coupling of O_2 with an electron (e^-). The electron donor is usually NADH or NADPH, and the reaction is catalyzed by various oxidases; NADPH oxidases, xanthine oxidase and complex I/II/III/IV from the mitochondria [25]. The generation of H_2O_2 occurs after the dismutation of $O_2^-\bullet$, mostly catalyzed by superoxide dismutases (SODs), although a small percentage occurs spontaneously. Some oxidases also have dismutase activity and may contribute to direct production of peroxide from superoxide. The reaction of $O_2^-\bullet$ with reduced transition metals may lead to formation of H_2O_2 [25]. Most of the OH• is generated from H_2O_2 and $O_2^-\bullet$ in a reaction catalyzed by a metal ion (iron or cupper). This is known as the Habor–Weiss reaction. This reaction occurs in two steps; in the first step, $O_2^-\bullet$ reduces Fe^{3+} to Fe^{2+} (Fe^{3+} $O_2^-\bullet \rightarrow Fe^{2+} + O_2$), and the second step is the Fenton reaction where Fe^{2+} reacts with H_2O_2 to generate OH• and OH^- ($Fe^{2+} + H_2O_2 \rightarrow Fe^{3+} + OH\bullet + OH^-$) [25]. Nitric oxide and $ONOO^-$ (form by the combination of NO and $O_2^-\bullet$) are the most important RNS in spermatozoa [25].

Several potential sources can be responsible for ROS production in the spermatozoa, including the spermatozoa itself and contaminating cells in the ejaculate. Dead spermatozoa are a major source of ROS, frequently overlooked in reproductive technologies [26]. L-amino oxidase (LAAO) is present in stallion spermatozoa being able to generate significant amounts of ROS; aromatic amino acids are substrates for this enzyme, producing substantial amounts of ROS, especially in the presence of dead spermatozoa [26]. Interestingly, cryopreservation media contain sufficient amounts of aromatic amino acids to activate this enzyme. Ongoing proteomic studies in our laboratory have also confirmed the presence of this enzyme in stallion spermatozoa. A NADP oxidoreductase system has been detected in the membrane [27], however nowadays it is considered that the main source of reactive oxygen species is electron leakage in the mitochondrial electron transport chain (ETC) [7,8,10,28–31]. In particular, defective mitochondria may represent a hallmark of male infertility. Evidences of mitophagy in human sperm were described in our laboratory, suggesting that activation of mitophagy is a mechanism that maintains proper sperm function [32]. The sources of reactive oxygen species in the electron transport chain of the stallion spermatozoa have also recently been investigated in our laboratory [9,10], confirming the role of the ETC as a main source of ROS in stallion spermatozoa.

3. Redox Regulation and Signaling

Although initially, oxidative stress was defined as a disturbance in the pro-oxidant-antioxidant balance in favor of the former, current knowledge has evolved and oxidative stress is better defined in terms of regulation of redox signaling. Numerous processes are redox regulated in biological systems. Redox regulation is similar to pH regulation, the pH varies in different cellular compartments, also the redox state is not an overall redox state and vary in different compartments of the spermatozoa [33]. Redox reactions consist of the transfer of electrons (e^-) from one molecule (oxidation) to another molecule (reduction). Thus, reduction implies a decrease in overall charge (more e^-) of the molecule, while oxidation implies an increase in overall charge (fewer e^-). Reactive oxygen species, such as the superoxide anion $O_2^-\bullet$, are low molecular weight compounds that are chemically unstable, particularly in biological systems [21]. The hydroxyl radical is the most reactive and oxidizes virtually any closer molecule. The reactivity of $HO\bullet$ is 7×10^9 L mol^{-1} s^{-1}, while the rate constant for $O_2^-\bullet$ is <0.3 and is 2×10^{-2} L mol^{-1} s^{-1} for H_2O_2 [33]. Another electronically excited state of interest in spermatology is singlet molecular oxygen, generated by photoexcitation mainly by ultraviolet A and B light rays, but even infrared and visible light may also generate photobiological responses. This is the rationale of the customary procedure of avoiding light exposure during semen processing [33]. Other species include alkoxyl and peroxyl radicals, non-radical species such as hypochlorite, peroxynitrite, singlet oxygen and lipid peroxydes, among others [34]. To understand the basis of redox signaling it is important to bear in mind the characteristics of different ROS. As previously mentioned the $HO\bullet$ is the most reactive, and has the shortest half-life (10^{-15} s.) [24]. The $HO\bullet$, is considered to be the most harmful oxidant, with no signaling functions. Although $O_2^-\bullet$ may have difficulty diffusing through membranes due to its anionic charge, it may use specific channels in some tissues [35–37]. Hydrogen peroxide is a stable compound and in addition is a nonpolar molecule that can easily diffuse through membranes, and is also transported through aquaporin channels [24,38–40]; all of which make H_2O_2 a suitable molecule for redox signaling. The primary target of hydrogen peroxide is the thiol group of the amino acid cysteine, which is oxidized in a reversible fashion. The presence of glutathione (GSH) and other thiols in spermatozoa is well known [41], also the role of oxidative regulation in significant biological processes occurs in very early stages of development. For example, studies in sea urchin, show an oxidative burst that occurs at the time of fertilization preventing polyspermy through the activation of a dual oxidase (Udx1), that induces cross linking of surface proteins on the egg surface [42,43]. Also, oxidation reduction processes of sulfhydryl groups of protamines are critical for chromatin condensation during spermatogenesis [44].

Nitric oxide is a ubiquitous free radical generated from the oxidation of L-arginine to L-citruline by three isoforms of reduced nicotinamide adenine dinucleotide phosphate (NADPH)-dependent NO-synthases (NOS) [45]. Among other functions, NO is relevant for spermatogenesis, penile erection, folliculogenesis, and ovulation [46]. In spermatozoa, NO appeared to play a major role in the regulation of sperm motility and capacitation [47–49]. Studies in our laboratory have identified the presence of NOS in stallion spermatozoa, its role in sperm functionality and, interestingly, we also showed the effect of egg yolk present in freezing extenders scavenging NO [50]. While the NO produced by NOS is a messenger molecule, it may react with $O_2^-\bullet$ to form peroxynitrite ($ONOO^-$) [33], an oxidant that may induce 3-nitrotyrosine residues in proteins, affecting mitochondrial functions and triggering cell death via oxidation and nitration reactions [51]; however, due to the high content of SOD (1000 times more than intracellular NO levels), the production of ONOOO$^-$ is prevented by the rapid dismutation of $O_2^-\bullet$ [25].

Many cellular processes are redox regulated. In spermatozoa, redox regulation has been extensively studied in relation to capacitation [13,15,52–57]. Capacitation is the maturational process that sensitizes spermatozoa to recognize and fertilize the oocyte. Capacitation involves, removal of cholesterol from the plasma membrane, removal of coating materials from the membrane, a rise in intracellular Ca^{2+}, an increase in intracellular cAMP, and a dramatic increase in tyrosine phosphorylation.

Removal of cholesterol from the membrane is preceded by its oxidation, stimulated by bicarbonate, and the formation oxysterols [58–60] that are depleted from the sperm membrane by albumin. Different aspects are worth mentioning in the context of the present review; one is the fact that bovine studies have demonstrated that after freezing and thawing this oxidative mechanism is altered, offering an explanation of the reduced fertility of cryopreserved spermatozoa [61]. The stallion spermatozoa present difficulties to capacitate in vitro, explaining the poor results of conventional IVF in this species. This issue has been the subject of an excellent recent review [62], and the reader is referred to it for detailed information in the topic; however, the possibility that this may relate to the specific redox regulation in spermatozoa is an intriguing possibility that warrants to be further explored; interestingly, intracellular glutathione (GSH) is much higher in horses than in other domestic species. Also, during capacitation the sperm plasma membrane potential (E(m)) hyperpolarizes [56,63,64], and spermatozoa experience alkalinization. Detailed reviews on the molecular aspects of capacitation can be found elsewhere [17]. Interestingly, only a subpopulation of spermatozoa is able to experience capacitation [52,56]. Tyrosine phosphorylation is a redox regulated process [17,20,54,65–70]. Other functions of the spermatozoa, such as activated motility may also be redox regulated [17,71], in relation to tyrosine phosphatases (PTPs), which are intracellular targets for ROS [72]. The activity of PTPs depends on a conserved cysteine (Cys) residue, where oxidation results in the inactivation of the enzyme [22,73]. On the other hand, ROS can also activate kinases. In addition to hydrogen peroxide, other species such as reduced glutathione (GSSG), hydrogen sulphide and lipid peroxides (LPO) can inactivate PTPs [74]. Reversible oxidation of target cysteine residues in specific proteins modulates its activity [22]. In order to function in a reversible manner oxidized cysteine (Cyss) residues need to be reduced. This reversibility depends on adequate availability of reducing molecules including the peroxiredoxin (PRDX) family of antioxidant enzymes [22]. Peroxirredoxins have been described in spermatozoa [13–15,75] and play a major role in sperm function, stressing the importance of redox signaling in these highly specialized cells. Reversing the oxidized Cys residue in this family of pathways involves thiorredoxin or GSH. Reduction of the higher oxidation state (sulphinic acid SO_2H) may require sulfiredoxin or sestrins [22,76]. This reversible sequential oxidation of PRDXs allows a tight regulation of the function of these proteins in a regulation described as a "floodgate" model [77,78]. Spermatozoa are rich in thiols [41], with the majority of thiol groups associated with proteins, which may suggest that redox regulation is an important regulatory mechanism in these cells. Spermatozoa are transcriptionally silent cells whose regulation depends on post transcriptional modification of proteins. One interesting example, since mitophagy has been recently described in spermatozoa [32], of proteins regulated by reversible oxidation of Cys residues, is the large family of Cys-dependent proteases [22]. In particular, the cysteine protease HsAtg4 is a direct target for oxidation by H_2O_2, specifically a residue located near the protein's catalytic site [79]. The presence of a similar mechanism in spermatozoa is an intriguing possibility and deserves further research [32]. Other functions in the spermatozoa that are redox regulated, include control of motility [71], and binding to the oviductal epithelium to form the sperm reservoir [80–82].

4. Modern Concept of Oxidative Stress Applied to Spermatozoa

Since redox regulation is being unveiled as a major mechanism regulating sperm function, probably at the same level as tyrosine phosphorylation and other post translational modifications of sperm proteins, sophisticated mechanisms must be present to maintain redox status under physiological control. Both seminal plasma and the spermatozoa itself contain enzymatic and non-enzymatic systems that contribute to maintenance of oxidative eustress. Recent research from our laboratory shows that in stallion spermatozoa seminal plasma plays a major role in regulating redox status. The steady state redox potential (E_h) can be estimated using the Nerst Equation: $Eh = Eo + RT/Ln$ [oxidized molecule/reduced molecule], where Eo is the standard reduction potential, R = gas constant, T is the absolute temperature, n = number of electrons transferred and F is the Faraday constant [23]. Recently, a system to easily measure the steady state in semen has become available and is being introduced into reproductive medicine and clinics. Using this system, E_h is provided as the static oxidation reduction potential (sORP) and is expressed as millivolts per million spermatozoa. E_h in raw semen (seminal plasma present) was measured and was found to be 1.62 ± 0.06 mV/10^6 spermatozoa, when seminal plasma was removed, it was 7.9 ± 0.79 mV/10^6 spermatozoa, thus showing a much higher overall oxidation status [83]. This finding suggests that regulation of the extracellular medium may also be of great importance as is the case in other cells [83], from this viewpoint it is well recognized that equine seminal plasma is rich in antioxidants [84–89]. On the other hand, it is important to consider that once the semen is deposited in the mare's uterus or is processed, the antioxidants in seminal plasma are removed from close contact with the spermatozoa, meaning the importance of intrinsic antioxidant defenses in the spermatozoa become critical [13,15,90–92].

The spermatozoa itself also has antioxidant defenses, including glutathione, and other enzymatic antioxidant defenses such as the paraoxonase [93–97], thioredoxin [15,98–104] and peroxiredoxin [13,14, 51,75,90,91,105,106] families of proteins. Ongoing proteomic studies in our laboratory have identified peroxiredoxins 5 and 6, and thioredoxin reductase in stallion spermatozoa. Interestingly, and as previously indicated, the concentration of intracellular GSH in the horse spermatozoa is higher than in most domestic species. A recent study in our laboratory revealed that the mean concentration of GSH in stallions was 8.2 ± 2.1 μM/10^9 spermatozoa [107], while values reported in other species are in the nanomolar ranges per billion spermatozoa [41]. These high levels of GSH in stallion spermatozoa, may be linked to the intense mitochondrial activity of the spermatozoa in this species. Intense mitochondrial activity causes increased ROS production, and thus sophisticated mechanisms to maintain redox homeostasis may have evolved differently between species with spermatozoa less dependent on oxidative phosphorylation for ATP production. In relation to this, evidence of the presence and activity of the Cystine antiporter SLC7A11 in stallion spermatozoa has been discovered [83]. This antiporter exchanges extracellular cystine (oxidized form of cysteine) for intracellular glutamate. Once in the cell, cystine is reduced and used for GSH synthesis. Indirect evidence of the presence of a system exporting glutamate in spermatozoa were reported as early as in 1959 [108]. Evidence of GSH synthesis in stallion spermatozoa [107], include the presence of the enzymes glutathione synthetase (GSS) and gamma glutamylcysteine synthetase (GCLC). In addition, functional studies indicate their activity; the use of the specific inhibitor L-Buthioninine sulfoximide (BSO) reduced GSH synthesis from cysteine. In this particular experiment, mass spectrometry (MS) was used to specifically identify GSH and avoid interference with other thiols. Overall these results point to a sophisticated redox regulation in stallion spermatozoa. It is considered that most extracellular cysteine is present in the disulfide form (cystine), thus the presence of the xCT/SLCTA11 antiporter may be a major mechanism of cystine incorporation in the spermatozoa. This antiporter is present and active in stallion spermatozoa [83]. In addition to its role in the incorporation of cysteine for GSH synthesis, a potential role in an active Cys/Cyss redox node in the spermatozoa must be considered. Overall, these recent findings support the hypothesis of a complex redox regulation in the spermatozoa. Oxidative stress is thus better defined as the fail in the regulation of redox signaling due either to overproduction of ROS, or exhaustion of regulatory mechanisms. This latter point has recently been addressed, and functionality of the stallion spermatozoa is linked to thiol content. When thiols are exhausted stallion spermatozoa rapidly enters senescence, which is characterized by increased production of lipid peroxides, activation of caspase 3, loss of motility and death [109,110]. Remarkably, this senescence is triggered by ROS as is capacitation. It has been proposed that both processes are linked. Only one capacitated spermatozoa will fertilize the oocyte, while the redundant spermatozoa finally succumb in a truncated apoptotic cascade, characterized by enhanced mitochondrial ROS production, lipid peroxidation, caspase activation, loss of motility and phosphatidylserine externalization, representing a signal to phagocytic cells for the elimination of redundant spermatozoa without significant inflammatory reaction [111].

The stallion spermatozoa is a paradigm of this sophisticated redox regulation; recent research has shown apparently paradoxical results, in this regard more fertile spermatozoa show increased ROS production [8], further underlining the concept that a tightly controlled redox regulation occurs in stallion spermatozoa.

5. The Mitochondria in Redox Signaling

Electrons can be prematurely leaked to oxygen in the ETC or associated to catabolism of substrates [112,113]. Depending of the number of electrons being leaked, different outcomes are possible. If leaked one by one they generate superoxide radicals, if in pairs they generate hydrogen peroxide. When they are properly transferred four at a time, they generate water and drive OXPHOS at complex IV of the ETC. A growing body of scientific evidence is stressing the role of proper mitochondrial function in sperm physiology [7,9–11,28,31,32,114–118]; moreover, definition of oxidative stress as the result of mitochondrial malfunction, states that it is the result of "a dysfunction of electron transfer reactions leading to oxidant/antioxidant imbalance and oxidative damage to macromolecules" [119]. This theory states that $O_2^-\bullet$ does not accidentally leak from the ECT, but instead is a signaling molecule [119]. Recent research in our laboratory with an aryl hydrocarbon receptor deficient (AhR$^{-/-}$) mouse strain, showing males of unusually high fertility (also in terms of number of pups born) revealed that this strain was characterized by higher mitochondrial activity [120]. Other reports also link mitochondrial activity with fertility in humans and equines [7,8,28,31,116,121,122]. Interestingly, the mitochondria are the more sensitive structure in the spermatozoa to stress induced by different biotechnologies, and have been proposed as a sensitive marker of sperm quality and fertilization ability [120]. Mitochondrial roles in the spermatozoa may include Ca_2^+ storage and signaling, production of ATP, control of sperm lifespan and activation of a specific form of apoptosis for silent, non-inflammatory elimination of redundant spermatozoa after insemination, and potentially control of redox signaling. Numerous evidences point to mitochondria as the hallmark of fertile spermatozoa. However, proper evaluation of mitochondrial function in spermatozoa is still elusive, and rarely performed under clinical settings. Fluorescent probes and flow cytometry represent the method of choice to study mitochondrial function in spermatozoa, with the potential for analysis of thousands of spermatozoa and simultaneous functions in every single spermatozoon, together with the recent development of computational methods [29] to study sperm subpopulations makes this technique the gold standard. However, technical difficulties preclude its wider use in reproductive medicine. These difficulties relate to special characteristics of commonly used probes, such as the JC-1. This dye is difficult to compensate using the 488 nm excitation laser due to the spectral characteristics of the fluorochrome, and the dual excitation depending on the formation of monomers (low mitochondrial membrane potential) of aggregates (high mitochondrial membrane potential). This particular issue can be addressed using dual excitation; monomers with the blue 488 nm laser, and aggregates with the 561 nm yellow laser. The application of computational methods to the analysis of data also improves the identification of specific spermatic subpopulations. The production of hydrogen peroxide in stallion mitochondria have been investigated in our laboratory [10], inhibition of complex I of the ETC increased the production of mitochondrial superoxide and hydrogen peroxide, suggesting that mitochondrial malfunction is a potential source of redox deregulation in stallion spermatozoa. The inhibition of complex III also caused increased ROS production. In addition, the above-mentioned study underpinned the importance of cautious selection of probes to assess ROS in spermatozoa. However, mitochondrial dysfunction may lead to either reduced or increased production of ROS [112] depending on the cause of the dysfunction and caution interpreting the results of the analysis of ROS production in spermatozoa is always advised. Specific antioxidant defenses in the mitochondria of the stallion spermatozoa include mitochondrial GSH, peroxiredoxin 5 and manganese-dependent superoxide dismutase (Mn-SOD). Mitochondrial ROS have been implicated in numerous signaling pathways in somatic cells [112] and is also likely that these species may participate in signaling in spermatozoa. Together with its importance in sperm regulation, the special characteristics of the spermatozoa, a cell devoid of most organelles and a very limited cytoplasm, may also mean this cell is a suitable model for the study of mitochondrial function.

6. Redox Regulation and Sperm Metabolism

Together with mitochondria, in recent years stallion sperm metabolism have been of increased interest for scientists focused in equine reproduction [11,117]. Mitochondria play major roles in cellular metabolism, being the energetic power-house of the cell [123]. Oxidative phosphorylation (OXPHOS) and the tricarboxylic acid cycle (TCA cycle) are well known mitochondrial functions. Recent specific research in horses has underlined the importance of mitochondria as a provider of energy in the form of ATP, and the consequences it has for sperm physiology and the functional evaluation of the spermatozoa [10]. Early studies suggested that spermatozoa were glycolytic cells, however the participation of oxidative phosphorylation in production of energy is now acknowledged [9]. Early studies also suggested that ATPs produced by mitochondrial respiration could not reach distal parts of the flagellum. To solve this problem, shuttle systems and/or glycolysis ought to be present [124]. Also, species specific strategies occur in the predominance of one energy source. Recent proteomic studies indicate that the spermatozoa can use different substrates for energy, possessing the ability to oxidize fatty acids [125,126]. The stallion spermatozoa is considered to predominantly use OXPHOS for the generation of energy [7,8,11,117]. The adenine nucleotide translocator (ANT) catalyzes the transmembrane exchange of ATP, generated by oxidative phosphorylation, for cytosolic ADP [127]. Inhibition of this protein leads to reduced sperm motility suggesting that ATP produced by OXPHOS in the mitochondria plays an important role in spermatic motility in horses. Further studies aimed to clarify the role of mitochondrial ATP in stallion sperm motility. Inhibition of OXPHOS reduced spermatic motility and ATP content in stallion but not in human spermatozoa suggesting species specific differences in energetic metabolism [8]. Moreover, this study showed paradoxical relations between fertility and oxidative stress, fertile stallions were characterized by spermatozoa showing increased levels of 8-hydroxiguanidine and $O_2^-\bullet$. These increased levels were attributed to increased mitochondrial activity in the spermatozoa of fertile stallions [8]. The relation between increased mitochondrial activity and ROS production has also been confirmed in independent studies [11]. In addition, and in line with these findings, a dramatic decrease in sperm ATP content after mitochondrial uncoupling and inhibition of mitochondrial respiration was reported [9]. Reduction of ATP was accompanied by low motilities and velocities, and interestingly, inhibition of mitochondrial respiration at the ATP synthase complex collapsed sperm membranes. This may relate to the high ATP consumption necessary to maintain the activity of the Na^+-K^+ ATPase pump in the spermatozoa [128]. The relation between ROS production and mitochondrial activity was also confirmed. Despite the predominance of OXPHOS, glycolysis and other sources of energy are also present in the spermatozoa. OXPHOS takes place in the mitochondria located in the sperm midpiece, while glycolysis occurs mainly in the flagellum in which the fibrous sheath is rich in glycolytic enzymes where they are anchored [129–131]. The substrate for glycolysis is glucose, which is incorporated into the spermatozoa through diverse glucose transporters (GLUTs) [132]. Oxidative phosphorylation uses diverse sources of substrates derived from the metabolism of carbohydrates, lipids and amino acids. While for a long time a debate has existed among spermatologists regarding the main source of energy in spermatozoa, the existence of different bioenergetic strategies in different species is now becoming clear [133], and thanks to the introduction of the "omics" technologies into spermatology, the spermatozoa is being unveiled as a cell with a much higher bioenergetic plasticity that previously assumed [126,134]. In this regard, recent proteomic studies in horses and humans reveal that beta oxidation of fatty acids plays an important role in providing energy for the spermatozoa [126,135]. The pentose phosphate cycle pathway (PPP) is also present in spermatozoa [133,136–141]. NADPH produced by the PPP is important for the re-activation of 2-CysPRDXS. [90] In human spermatozoa, the pentose phosphate pathway can respond dynamically to oxidative stress [142] and the inhibition of glutathione reductase impairs the ability of sperm to resist oxidative stress and lipid peroxidation [140]. Also, NADPH may play a role in relation to the activity of an NADPH oxidase which plays a role in capacitation [137]. The glutathione peroxidase-glutathione reductase-pentose phosphate pathway system is functional and provides an effective antioxidant defense in normal human spermatozoa [140,143]. Overall, current knowledge on sperm metabolism suggests species specific differences and a great metabolic plasticity in the spermatozoa, which are able to adapt their metabolism to the changing environments that they are exposed to, on their travel to fertilize the oocyte. Recent research using the strategy of intervention on the metabolic flexibility of stallion spermatozoa seems promising [7,11,26,117,144], both in the development of new extenders for long time liquid storage, and as an intervention for the development of thawing extenders. In this particular aspect, current extenders in use for stallion spermatozoa contain high concentrations of glucose, around 270–300 mM, these concentrations are far from being physiological, and may preclude long term preservation of liquid semen. It is well known that supraphysiological concentrations of glucose may lead to cell death [145] due to accumulation of advanced glycation end products (AGEs) [146–149]. The discovery of endocrine features in the spermatozoa also underlines the complex metabolism of these cells that represent an area of great interest for research in the coming decade [138,150]. Finally, amino-acid metabolism ought to be considered, this has been reported in fish spermatozoa, and anecdotal reports in mammals using amino-acids as semen additives support this possibility [151,152]. Additionally, indirect evidence of the role of the amino acid glutamine in stallion spermatozoa has been recently reported by our laboratory. Inhibition of the xCT antiporter, and thus increased intracellular glutamate improved sperm function in fresh extended stallion spermatozoa, but not in frozen thawed samples [83]. The amino-acid glutamine may enter the Krebs cycle and improve mitochondrial function under some circumstances [153]. Glutamine metabolism can provide considerable amounts of NADPH, through the pentose phosphate pathway, and can occur in parallel with aerobic glycolysis depending on glucose-6-phosphate availability [154]. The increase in sperm functionality after using the xCT antiporter inhibitor sulfasalazine can be explained through this mechanism.

7. Consequences of Redox Deregulation

In accordance with the current biochemical literature, redox regulation is tightly regulated in the spermatozoa, with interactions between spermatic metabolism, mitochondrial production and scavenging of ROS. A summary of current knowledge on redox regulation in stallion spermatozoa is presented in Figure 1. Many factors can deregulate this complex network in humans and other animals, including aging, exposure to toxins, particularly alcohol and tobacco in humans, poor diet, lack of physical activity and systemic diseases including obesity and diabetes [30,155–158]. Also, current sperm biotechnologies such as cryopreservation cause redox deregulation of spermatozoa, mainly through a severe mitochondrial osmotic stress [110,118,128,159,160]. Deregulation of redox homeostasis has a profound impact on sperm physiology and fertility, all spermatic compartments and functions may be affected. Moreover, impacts on the embryo and the offspring may also occur.

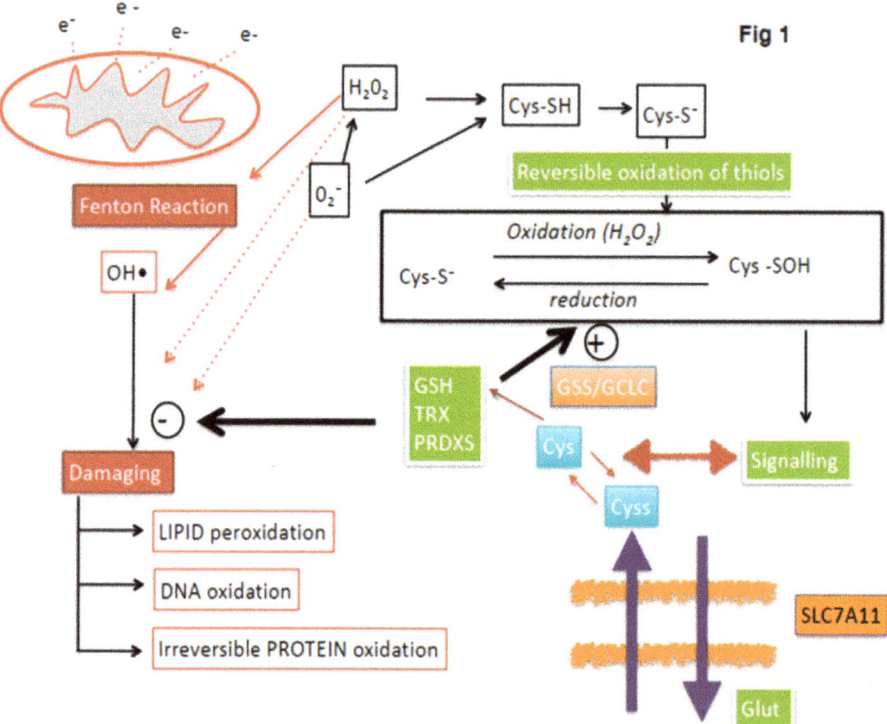

Figure 1. Overview of redox regulation in stallion spermatozoa. Electron (e^-) leakage at the mitochondria is one of the main sources of ROS. Mechanisms to maintain redox homeostasis include thioredoxin (TRX) and peroxiredoxin (PRDX) systems and gluthatione (GSH) (green boxes). The stallion spermatozoa can incorporate cystine (cyss) (blue boxes), through the SlC7A11 x-CT antiporter by exchange for intracellular glutamate (Glut). Cystine is reduced in the cytoplasm to Cysteine and contribute to the intracellular GSH pool by the action of the enzymes involved in the synthesis of GHS, Glutathion syntethase (GSS) and glutamate cysteine ligase (GCLC); this mechanism has been described only in horses. Controlled levels of ROS regulate sperm functionality through reversible oxidation of thiols in cysteine containing proteins (blank boxes). If redox regulation is lost, irreversible oxidation of thiols and oxidative attack to lipids DNA and proteins occurs leading to sperm malfunction and finally death (red boxes). The hydroxyl radical (OH•) is the most damaging ROS, produced by the Habor–Weiss/Fenton reaction.

8. Effects on Lipids

Lipid peroxidation is well recognized as a consequence of redox deregulation and loss of redox homeostasis in spermatozoa. In the stallion model, lipid peroxidation occurs as a consequence of aging (Figure 2) and sperm biotechnologies such as cryopreservation and chromosomal sex sorting [89,109,110,161–164]. Deregulation of redox signalling, aging and cell senescence is well documented, and aged stallions show increased peroxidation of the lipids in the sperm membranes. Cryopreservation leads to a paradoxical situation, while osmotic induced damage in the mitochondria may lead to reduced production of ROS, lipid peroxidation increases after freezing and thawing. On the other hand, spermatozoa that withstands cryopreservation better is also characterized by increased production of ROS [31]. Lipid peroxidation (LPO) occurs after the oxidative attack of lipids, mainly the phospholipids and cholesterol of the membranes. Interestingly, LPO induces changes in the permeability and fluidity of the membranes that can be easily monitored using probes like

YoPro-1 [165,166]. LPO results in the production of lipid hydroperoxides, which are unstable and decompose to more stable and less reactive secondary compounds [167–169]. Lipid peroxidation occurs in three phases, in the *initiation* phase abstraction of H• from a lipid chain (LH) gives a lipid radical (L•). Formation of L• is favored in the membrane of the horse spermatozoa due to their abundance in PUFAs [170,171], in this type of lipid the resulting radical is resonance stabilized [167]. Following *initiation* the *propagation* phase continues and the lipid radical reacts with oxygen to generate a lipoperoxyl radical (LOO•), that reacts with a lipid to yield a L• and a lipid hydroperoxyde (LOOH), these are unstable molecules that generate new peroxyl and alkoxyl radicals and decompose to form secondary products [168]. Finally the reaction ends when it gives a non-radical, or non-propagating species [169]. Among the secondary products formed upon lipid peroxidation of the polyunsaturated fatty acids (PUFAs) of the sperm membranes, aldehydes have received special attention due to their toxicity to spermatozoa [109,110,172–179]. Depending on the oxidation of different PUFAs, distinct compounds can originate, malondialdehyde originates from the oxidation of PUFAs containing at least three double bonds, like arachidonic acid. 4 hydroxy-2(E)-nonenal (4-HNE) originates from the oxidation of $\omega 6$ fatty acids. The composition of the sperm membrane, suggests that 4-HNE should be the prevalent compound upon LPO, since docosopentanoic acid (C22: 5ω6) is the predominant PUFA in the phospholipids of stallion spermatozoa [170]. Interestingly, recently, seasonal variation in the lipid composition of the sperm membranes has been reported [180]. It should also be noted that 4-HNE, while triggered by an initial oxidative step, can later continue independent of oxidative stress and continues providing a source of ω-6 fatty acids is available [181]. 4-hydroxynonenal reacts with GSH by Michael addition to form GSH conjugates, and although this reaction can happen spontaneously it occurs much faster in the presence of glutathione-S-transferases. Also, the aldehyde function of 4-HNE can be reduced into alcohol or oxidized into acid, with the participation of alcohol dehydrogenase and aldehyde dehydrogenase, forming 1,4-dihydroxynonene and 4-hydroxynonenoic acid, which can undergo beta oxidation [167]. The role of GSH and aldehyde dehydrogenase has recently been investigated in stallion spermatozoa in relation to oxidative stress [107,109,110,175], suggesting that these mechanisms for 4-HNE detoxification are of pivotal importance for spermatic function. The relation between GSH and 4-HNE in cryopreserved stallion spermatozoa suggest that GSH is effectively a major mechanism for detoxifying 4-HNE [110]. Also, aldehyde dehydrogenase has proven to be a major detoxifying mechanism for 4-HNE in stallion spermatozoa [175]. Lipid peroxidation has been traditionally detected using BODIPY dyes [89,182]; however, its dual fluorescence and its lipid binding can make this dye difficult to interpret upon flow cytometry analysis. More recently, lipid peroxidation is being detected using antibodies against 4-hydroxynonenal (4-HNE) [110,175,183]. The availability of secondary antibodies marked with different probes makes this technique suitable for multicolor panels, and to study the relation between increased levels of 4-HNE and sperm functionality using multiparametric analysis. Mass spectrometry is also a suitable tool for the study of lipid peroxidation induced changes in the spermatozoa and has recently been used in our laboratory to monitor GSH [107].

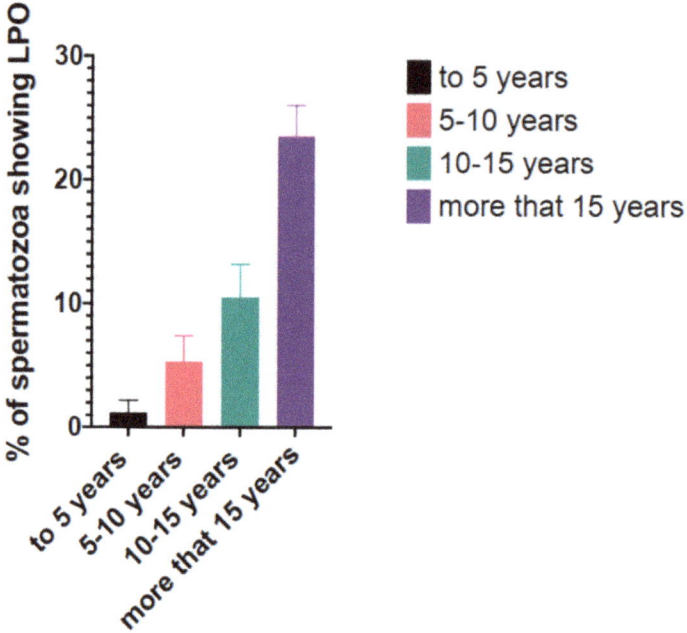

Figure 2. Effect of stallion age in the peroxidation of sperm membranes, semen was collected from stallions of different ages (to 5 years old, 5–10, 10–15 and more than 15 years old) and lipid peroxidation was assessed flow cytometrically after BODIPY 581/591 C11, as seen in the figure, lipid peroxidation increases with age.

9. Effects on Proteins

Oxidative modifications of structural and functional proteins are one of the major factors involved in protein dysfunction. Protein carbonyl content is a commonly used biomarker of oxidative damage of proteins. Toxic adducts derived from LPO can diffuse through membranes allowing the reactive aldehydes to covalently modify proteins [173,174,184,185]. In addition to advanced lipid peroxidation end products (ALEs), products derived from the glycoxidation of carbohydrates, that will form advanced glycation end products (AGEs) can also induce protein carbonylation [169]. There is an excellent recent review of this particular topic focused on the spermatozoa [51] and the reader is referred to it for complete details.

10. Oxidative DNA Damage

Spermatozoa harbor the haploid paternal genome and also important epigenetic information with regulatory roles for early embryo development [186]. Recently, it has been reported that biotechnologies such as cryopreservation damage sperm genes with important roles in fertilization and early embryo development, even in the absence of detectable DNA fragmentation [187,188]. Cryopreservation can also damage the sperm epigenome [189]. Many assays have been developed to investigate DNA integrity in the spermatozoa [190,191]. It is considered that most of the DNA damage is caused by an oxidative mechanism. Oxidation of nucleotides can cause abasic pairs in DNA, increasing the risk of replication errors. Loss of a base in DNA, i.e., creation of an abasic site leaving a deoxyribose residue in the strand, is a frequent lesion that may occur spontaneously, or under the action of radiation or alkylating agents, or enzymatically as an intermediate in the repair of modified or abnormal bases. The abasic site lesion is mutagenic or lethal if not repaired. From a chemical view point, the abasic site is an alkali-labile residue that leads to strand breakage through beta- and delta- elimination [192,193]. More recently, multiple consequences of the electrophilic nature of abasic lesions have been revealed [194], and oxidized abasic sites are nowadays considered irreparable, leading to the most deleterious form of DNA damage, inter-strand cross links and double strand breaks [195,196]. Detection of oxidized nucleotides in sperm with flow cytometry has been reported using a specific antibody against the oxidative derivative of guanosine, 8-hydroxyguanosine [109,197], and threshold values for fertility have recently been reported in humans [198]. Another newly developed flow-cytometry-based assay, for evaluation of oxidative stress in sperm DNA, is the γHA2AX assay [199]. Although most histones are replaced by protamines, a small fraction remain in the nucleosome (5-15% in humans). This fraction contains the H2AH histone that is phosphorylated in Ser139 when under oxidative stress. The detection of γHA2AX (the phosphorylated form of the histone) has proven to be more sensitive than the TUNEL assay to detect DNA fragmentation, and also to be better correlated with pregnancy outcome in humans [200].

11. Impact of Early Embryo Development (EED)

Fecundation of the egg by spermatozoa with compromised redox regulation or experiencing non-lethal oxidative stress has important consequences with regard to embryo viability and the health and well-being of the offspring [201]. Assisted reproductive technologies such as in vitro fertilization and ICSI are associated with an increased incidence of birth defects in offspring [202]. Animal studies indicate that fecundation with spermatozoa experiencing oxidative stress may cause embryonic death [203], an effect that has been linked to oxidative damage in the spermatozoa [204]. Recent research from our laboratory has compared the effect of cryopreservation on the transcriptome of early equine embryos [205]. Using the same ejaculate, half processed as fresh sperm and the other half frozen and thawed, we obtained embryos from the same mare and stallion after artificial insemination with the aliquot of fresh sperm and, in the mare's next cycle using the frozen thawed semen aliquot. The transcriptional profile of embryos obtained with frozen thawed spermatozoa differed significantly from that of embryos obtained with the fresh sperm aliquot of the same ejaculate. Significant downregulation of genes involved in biological pathways related to the gene ontology (GO) terms *oxidative phosphorylation, DNA binding, DNA replication,* and *immune response*. Interestingly, many genes with reduced expression were orthologs of genes in which knockouts are embryonic lethal in mice [205]. While the exact mechanism behind these changes remains to be elucidated, redox deregulation and oxidative stress in the spermatozoa seem to be an important factor. The spermatozoa is known to carry proteins [201], and numerous ncRNAs [206] to the oocyte, with important functions in early embryogenesis. However, it has recently been reported that caput epidydimal mouse sperm, which has not yet incorporated RNAs, can support full development [207]. The impact of redox deregulation on sperm proteins is well recognized and has recently been reviewed [51,208], so it is not unlikely that oxidized proteins can be incorporated by the embryo impacting its development. Recently, preimplantation proteins in the human embryo with potential sperm origin have been identified [201]. In particular, 93 different proteins have been proposed as related to zygote and early embryo development before implantation in humans, moreover up to 560 sperm proteins with known roles in the regulation of gene expression in other cells or tissues have also been identified [201]. Even though further investigation is needed in this field, oxidative damage to sperm proteins with important functions during early embryo development may occur. Further supporting this hypothesis is the fact that biological processes such as *DNA binding and replication,* and *Histone Acetylation* were downregulated in embryos obtained with cryopreserved spermatozoa [205], and many of the proteins mentioned above have roles in these processes [201].

12. Concluding Remarks

Redox regulation plays a major role in controlling sperm functionality, recent research is unveiling the existence of sophisticated redox regulation systems that may constitute targets for the treatment of the male factor subfertility. In addition, the interaction between metabolism and redox regulation may offer alternatives to traditional methods of sperm conservation. The increasing use of proteomic techniques in research in spermatology will provide significant advances in the understanding of redox regulation in the spermatozoa in coming years.

Funding: The authors received financial support for their studies from the Ministerio de Economía y Competitividad-FEDER, Madrid, Spain, grant AGL2017-83149-R and the Junta de Extremadura-FEDER (IB16030 and GR18008).

Conflicts of Interest: The authors declare that there is no conflict of interest that may affect the impartiality of the information presented in this paper. "The funders had no role in the design of the study; in the collection, analyses, or interpretation of data; in the writing of the manuscript, or in the decision to publish the results".

References

1. Staub, C.; Johnson, L. Review: Spermatogenesis in the bull. *Animal* **2018**, *12*, s27–s35. [CrossRef]
2. Bose, R.; Sheng, K.; Moawad, A.R.; Manku, G.; O'Flaherty, C.; Taketo, T.; Culty, M.; Fok, K.L.; Wing, S.S. Ubiquitin Ligase Huwe1 Modulates Spermatogenesis by Regulating Spermatogonial Differentiation and Entry into Meiosis. *Sci. Rep.* **2017**, *7*, 17759. [CrossRef]
3. Gervasi, M.G.; Visconti, P.E. Molecular changes and signaling events occurring in spermatozoa during epididymal maturation. *Andrology* **2017**, *5*, 204–218. [CrossRef] [PubMed]
4. Shiraishi, K.; Matsuyama, H. Gonadotoropin actions on spermatogenesis and hormonal therapies for spermatogenic disorders. *Endocr. J.* **2017**, *64*, 123–131. [CrossRef] [PubMed]
5. Kalyanaraman, B.; Cheng, G.; Hardy, M.; Ouari, O.; Bennett, B.; Zielonka, J. Teaching the basics of reactive oxygen species and their relevance to cancer biology: Mitochondrial reactive oxygen species detection, redox signaling, and targeted therapies. *Redox Biol.* **2018**, *15*, 347–362. [CrossRef] [PubMed]
6. Kalyanaraman, B. Teaching the basics of redox biology to medical and graduate students: Oxidants, antioxidants and disease mechanisms. *Redox Biol.* **2013**, *1*, 244–257. [CrossRef]
7. Swegen, A.; Lambourne, S.R.; Aitken, R.J.; Gibb, Z. Rosiglitazone Improves Stallion Sperm Motility, ATP Content, and Mitochondrial Function. *Biol. Reprod.* **2016**, *95*, 107. [CrossRef]
8. Gibb, Z.; Lambourne, S.R.; Aitken, R.J. The paradoxical relationship between stallion fertility and oxidative stress. *Biol. Reprod.* **2014**, *91*, 77. [CrossRef]
9. Davila, M.P.; Munoz, P.M.; Bolanos, J.M.; Stout, T.A.; Gadella, B.M.; Tapia, J.A.; da Silva, C.B.; Ferrusola, C.O.; Pena, F.J. Mitochondrial ATP is required for the maintenance of membrane integrity in stallion spermatozoa, whereas motility requires both glycolysis and oxidative phosphorylation. *Reproduction* **2016**, *152*, 683–694. [CrossRef]
10. Plaza Davila, M.; Martin Munoz, P.; Tapia, J.A.; Ortega Ferrusola, C.; Balao da Silva, C.C.; Pena, F.J. Inhibition of Mitochondrial Complex I Leads to Decreased Motility and Membrane Integrity Related to Increased Hydrogen Peroxide and Reduced ATP Production, while the Inhibition of Glycolysis Has Less Impact on Sperm Motility. *PLoS ONE* **2015**, *10*, e0138777. [CrossRef]
11. Darr, C.R.; Varner, D.D.; Teague, S.; Cortopassi, G.A.; Datta, S.; Meyers, S.A. Lactate and Pyruvate Are Major Sources of Energy for Stallion Sperm with Dose Effects on Mitochondrial Function, Motility, and ROS Production. *Biol. Reprod.* **2016**, *95*, 34. [CrossRef] [PubMed]
12. Tosic, J.; Walton, A. Formation of hydrogen peroxide by spermatozoa and its inhibitory effect of respiration. *Nature* **1946**, *158*, 485. [CrossRef] [PubMed]
13. Lee, D.; Moawad, A.R.; Morielli, T.; Fernandez, M.C.; O'Flaherty, C. Peroxiredoxins prevent oxidative stress during human sperm capacitation. *Mol. Hum. Reprod.* **2017**, *23*, 106–115. [CrossRef] [PubMed]
14. Liu, Y.; O'Flaherty, C. In vivo oxidative stress alters thiol redox status of peroxiredoxin 1 and 6 and impairs rat sperm quality. *Asian J. Androl.* **2017**, *19*, 73–79. [CrossRef] [PubMed]
15. O'Flaherty, C. Redox regulation of mammalian sperm capacitation. *Asian J. Androl.* **2015**, *17*, 583–590. [CrossRef]
16. O'Flaherty, C.; de Souza, A.R. Hydrogen peroxide modifies human sperm peroxiredoxins in a dose-dependent manner. *Biol. Reprod.* **2011**, *84*, 238–247. [CrossRef]
17. de Lamirande, E.; O'Flaherty, C. Sperm activation: Role of reactive oxygen species and kinases. *Biochim. Biophys. Acta* **2008**, *1784*, 106–115. [CrossRef]

18. O'Flaherty, C.; de Lamirande, E.; Gagnon, C. Positive role of reactive oxygen species in mammalian sperm capacitation: Triggering and modulation of phosphorylation events. *Free Radic. Biol. Med.* **2006**, *41*, 528–540. [CrossRef]
19. O'Flaherty, C.; de Lamirande, E.; Gagnon, C. Reactive oxygen species and protein kinases modulate the level of phospho-MEK-like proteins during human sperm capacitation. *Biol. Reprod.* **2005**, *73*, 94–105. [CrossRef]
20. O'Flaherty, C.M.; Beorlegui, N.B.; Beconi, M.T. Reactive oxygen species requirements for bovine sperm capacitation and acrosome reaction. *Theriogenology* **1999**, *52*, 289–301. [CrossRef]
21. Fujii, S.; Sawa, T.; Nishida, M.; Ihara, H.; Ida, T.; Motohashi, H.; Akaike, T. Redox signaling regulated by an electrophilic cyclic nucleotide and reactive cysteine persulfides. *Arch. Biochem. Biophys.* **2016**, *595*, 140–146. [CrossRef] [PubMed]
22. Holmstrom, K.M.; Finkel, T. Cellular mechanisms and physiological consequences of redox-dependent signalling. *Nat. Rev. Mol. Cell Biol.* **2014**, *15*, 411–421. [CrossRef] [PubMed]
23. Go, Y.M.; Jones, D.P. The redox proteome. *J. Biol. Chem.* **2013**, *288*, 26512–26520. [CrossRef] [PubMed]
24. Briehl, M.M. Oxygen in human health from life to death—An approach to teaching redox biology and signaling to graduate and medical students. *Redox Biol.* **2015**, *5*, 124–139. [CrossRef]
25. Zhang, L.; Wang, X.; Cueto, R.; Effi, C.; Zhang, Y.; Tan, H.; Qin, X.; Ji, Y.; Yang, X.; Wang, H. Biochemical basis and metabolic interplay of redox regulation. *Redox Biol.* **2019**, *26*, 101284. [CrossRef]
26. Aitken, J.B.; Naumovski, N.; Curry, B.; Grupen, C.G.; Gibb, Z.; Aitken, R.J. Characterization of an L-amino acid oxidase in equine spermatozoa. *Biol. Reprod.* **2015**, *92*, 125. [CrossRef]
27. Vernet, P.; Fulton, N.; Wallace, C.; Aitken, R.J. Analysis of reactive oxygen species generating systems in rat epididymal spermatozoa. *Biol. Reprod.* **2001**, *65*, 1102–1113. [CrossRef]
28. Cueto, R.; Zhang, L.; Shan, H.M.; Huang, X.; Li, X.; Li, Y.F.; Lopez, J.; Yang, W.Y.; Lavallee, M.; Yu, C.; et al. Identification of homocysteine-suppressive mitochondrial ETC complex genes and tissue expression profile—Novel hypothesis establishment. *Redox Biol.* **2018**, *17*, 70–88. [CrossRef]
29. Ortega-Ferrusola, C.; Anel-Lopez, L.; Martin-Munoz, P.; Ortiz-Rodriguez, J.M.; Gil, M.C.; Alvarez, M.; de Paz, P.; Ezquerra, L.J.; Masot, A.J.; Redondo, E.; et al. Computational flow cytometry reveals that cryopreservation induces spermptosis but subpopulations of spermatozoa may experience capacitation-like changes. *Reproduction* **2017**, *153*, 293–304. [CrossRef]
30. Aitken, R.J.; Gibb, Z.; Baker, M.A.; Drevet, J.; Gharagozloo, P. Causes and consequences of oxidative stress in spermatozoa. *Reprod. Fertil. Dev.* **2016**, *28*, 1–10. [CrossRef]
31. Yeste, M.; Estrada, E.; Rocha, L.G.; Marin, H.; Rodriguez-Gil, J.E.; Miro, J. Cryotolerance of stallion spermatozoa is related to ROS production and mitochondrial membrane potential rather than to the integrity of sperm nucleus. *Andrology* **2015**, *3*, 395–407. [CrossRef] [PubMed]
32. Aparicio, I.M.; Espino, J.; Bejarano, I.; Gallardo-Soler, A.; Campo, M.L.; Salido, G.M.; Pariente, J.A.; Pena, F.J.; Tapia, J.A. Autophagy-related proteins are functionally active in human spermatozoa and may be involved in the regulation of cell survival and motility. *Sci. Rep.* **2016**, *6*, 33647. [CrossRef] [PubMed]
33. Sies, H.; Berndt, C.; Jones, D.P. Oxidative Stress. *Annu. Rev. Biochem.* **2017**. [CrossRef] [PubMed]
34. Schmidt, H.H.; Stocker, R.; Vollbracht, C.; Paulsen, G.; Riley, D.; Daiber, A.; Cuadrado, A. Antioxidants in Translational Medicine. *Antioxid. Redox Signal.* **2015**, *23*, 1130–1143. [CrossRef] [PubMed]
35. Han, D.; Antunes, F.; Canali, R.; Rettori, D.; Cadenas, E. Voltage-dependent anion channels control the release of the superoxide anion from mitochondria to cytosol. *J. Biol. Chem.* **2003**, *278*, 5557–5563. [CrossRef] [PubMed]
36. Han, D.; Antunes, F.; Daneri, F.; Cadenas, E. Mitochondrial superoxide anion production and release into intermembrane space. *Methods Enzymol.* **2002**, *349*, 271–280. [PubMed]
37. Han, D.; Williams, E.; Cadenas, E. Mitochondrial respiratory chain-dependent generation of superoxide anion and its release into the intermembrane space. *Biochem. J.* **2001**, *353*, 411–416. [CrossRef]
38. Vieceli Dalla Sega, F.; Zambonin, L.; Fiorentini, D.; Rizzo, B.; Caliceti, C.; Landi, L.; Hrelia, S.; Prata, C. Specific aquaporins facilitate Nox-produced hydrogen peroxide transport through plasma membrane in leukaemia cells. *Biochim. Biophys. Acta* **2014**, *1843*, 806–814. [CrossRef]
39. Mubarakshina Borisova, M.M.; Kozuleva, M.A.; Rudenko, N.N.; Naydov, I.A.; Klenina, I.B.; Ivanov, B.N. Photosynthetic electron flow to oxygen and diffusion of hydrogen peroxide through the chloroplast envelope via aquaporins. *Biochim. Biophys. Acta* **2012**, *1817*, 1314–1321. [CrossRef]

40. Bienert, G.P.; Moller, A.L.; Kristiansen, K.A.; Schulz, A.; Moller, I.M.; Schjoerring, J.K.; Jahn, T.P. Specific aquaporins facilitate the diffusion of hydrogen peroxide across membranes. *J. Biol. Chem.* **2007**, *282*, 1183–1192. [CrossRef]
41. Li, T.K. The glutathione and thiol content of mammalian spermatozoa and seminal plasma. *Biol. Reprod.* **1975**, *12*, 641–646. [CrossRef] [PubMed]
42. Wong, J.L.; Creton, R.; Wessel, G.M. The oxidative burst at fertilization is dependent upon activation of the dual oxidase Udx1. *Dev. Cell* **2004**, *7*, 801–814. [CrossRef] [PubMed]
43. Wong, J.L.; Wessel, G.M. Free-radical crosslinking of specific proteins alters the function of the egg extracellular matrix at fertilization. *Development* **2008**, *135*, 431–440. [CrossRef] [PubMed]
44. Chapman, J.C.; Michael, S.D. Proposed mechanism for sperm chromatin condensation/decondensation in the male rat. *Reprod. Biol. Endocrinol.* **2003**, *1*, 20. [CrossRef] [PubMed]
45. Herrero, M.B.; de Lamirande, E.; Gagnon, C. Nitric oxide is a signaling molecule in spermatozoa. *Curr. Pharm. Des.* **2003**, *9*, 419–425. [CrossRef]
46. Roselli, M.; Buonomo, O.; Piazza, A.; Guadagni, F.; Vecchione, A.; Brunetti, E.; Cipriani, C.; Amadei, G.; Nieroda, C.; Greiner, J.W.; et al. Novel clinical approaches in monoclonal antibody-based management in colorectal cancer patients: Radioimmunoguided surgery and antigen augmentation. *Semin. Surg. Oncol.* **1998**, *15*, 254–262. [CrossRef]
47. Maciel, V.L., Jr.; Caldas-Bussiere, M.C.; Marin, D.F.D.; Paes de Carvalho, C.S.; Quirino, C.R.; Leal, A. Nitric oxide impacts bovine sperm capacitation in a cGMP-dependent and cGMP-independent manner. *Reprod. Domest. Anim.* **2019**. [CrossRef]
48. Maciel, V.L., Jr.; Caldas-Bussiere, M.C.; Silveira, V.; Reis, R.S.; Rios, A.F.L.; Paes de Carvalho, C.S. L-arginine alters the proteome of frozen-thawed bovine sperm during in vitro capacitation. *Theriogenology* **2018**, *119*, 1–9. [CrossRef]
49. Staicu, F.D.; Lopez-Ubeda, R.; Romero-Aguirregomezcorta, J.; Martinez-Soto, J.C.; Matas Parra, C. Regulation of boar sperm functionality by the nitric oxide synthase/nitric oxide system. *J. Assist. Reprod. Genet.* **2019**, *36*, 1721–1736. [CrossRef]
50. Ortega Ferrusola, C.; Gonzalez Fernandez, L.; Macias Garcia, B.; Salazar-Sandoval, C.; Morillo Rodriguez, A.; Rodriguez Martinez, H.; Tapia, J.A.; Pena, F.J. Effect of cryopreservation on nitric oxide production by stallion spermatozoa. *Biol. Reprod.* **2009**, *81*, 1106–1111. [CrossRef]
51. O'Flaherty, C.; Matsushita-Fournier, D. Reactive oxygen species and protein modifications in spermatozoa. *Biol. Reprod.* **2017**, *97*, 577–585. [CrossRef] [PubMed]
52. Luque, G.M.; Dalotto-Moreno, T.; Martin-Hidalgo, D.; Ritagliati, C.; Puga Molina, L.C.; Romarowski, A.; Balestrini, P.A.; Schiavi-Ehrenhaus, L.J.; Gilio, N.; Krapf, D.; et al. Only a subpopulation of mouse sperm displays a rapid increase in intracellular calcium during capacitation. *J. Cell. Physiol.* **2018**, *233*, 9685–9700. [CrossRef] [PubMed]
53. Alvau, A.; Battistone, M.A.; Gervasi, M.G.; Navarrete, F.A.; Xu, X.; Sanchez-Cardenas, C.; De la Vega-Beltran, J.L.; Da Ros, V.G.; Greer, P.A.; Darszon, A.; et al. The tyrosine kinase FER is responsible for the capacitation-associated increase in tyrosine phosphorylation in murine sperm. *Development* **2016**, *143*, 2325–2333. [CrossRef] [PubMed]
54. Stival, C.; Puga Molina Ldel, C.; Paudel, B.; Buffone, M.G.; Visconti, P.E.; Krapf, D. Sperm Capacitation and Acrosome Reaction in Mammalian Sperm. *Adv. Anat. Embryol. Cell Biol.* **2016**, *220*, 93–106. [CrossRef] [PubMed]
55. Stival, C.; La Spina, F.A.; Baro Graf, C.; Arcelay, E.; Arranz, S.E.; Ferreira, J.J.; Le Grand, S.; Dzikunu, V.A.; Santi, C.M.; Visconti, P.E.; et al. Src Kinase Is the Connecting Player between Protein Kinase A (PKA) Activation and Hyperpolarization through SLO3 Potassium Channel Regulation in Mouse Sperm. *J. Biol. Chem.* **2015**, *290*, 18855–18864. [CrossRef] [PubMed]
56. Escoffier, J.; Navarrete, F.; Haddad, D.; Santi, C.M.; Darszon, A.; Visconti, P.E. Flow cytometry analysis reveals that only a subpopulation of mouse sperm undergoes hyperpolarization during capacitation. *Biol. Reprod.* **2015**, *92*, 121. [CrossRef] [PubMed]
57. Visconti, P.E.; Krapf, D.; de la Vega-Beltran, J.L.; Acevedo, J.J.; Darszon, A. Ion channels, phosphorylation and mammalian sperm capacitation. *Asian J. Androl.* **2011**, *13*, 395–405. [CrossRef]

58. Boerke, A.; Brouwers, J.F.; Olkkonen, V.M.; van de Lest, C.H.; Sostaric, E.; Schoevers, E.J.; Helms, J.B.; Gadella, B.M. Involvement of bicarbonate-induced radical signaling in oxysterol formation and sterol depletion of capacitating mammalian sperm during in vitro fertilization. *Biol. Reprod.* **2013**, *88*, 21. [CrossRef]
59. Aitken, R.J. The capacitation-apoptosis highway: Oxysterols and mammalian sperm function. *Biol. Reprod.* **2011**, *85*, 9–12. [CrossRef]
60. Zerbinati, C.; Caponecchia, L.; Puca, R.; Ciacciarelli, M.; Salacone, P.; Sebastianelli, A.; Pastore, A.; Palleschi, G.; Petrozza, V.; Porta, N.; et al. Mass spectrometry profiling of oxysterols in human sperm identifies 25-hydroxycholesterol as a marker of sperm function. *Redox Biol.* **2017**, *11*, 111–117. [CrossRef]
61. Brouwers, J.F.; Boerke, A.; Silva, P.F.; Garcia-Gil, N.; van Gestel, R.A.; Helms, J.B.; van de Lest, C.H.; Gadella, B.M. Mass spectrometric detection of cholesterol oxidation in bovine sperm. *Biol. Reprod.* **2011**, *85*, 128–136. [CrossRef] [PubMed]
62. Leemans, B.; Stout, T.A.E.; De Schauwer, C.; Heras, S.; Nelis, H.; Hoogewijs, M.; Van Soom, A.; Gadella, B.M. Update on mammalian sperm capacitation: How much does the horse differ from other species? *Reproduction* **2019**. [CrossRef] [PubMed]
63. Battistone, M.A.; Da Ros, V.G.; Salicioni, A.M.; Navarrete, F.A.; Krapf, D.; Visconti, P.E.; Cuasnicu, P.S. Functional human sperm capacitation requires both bicarbonate-dependent PKA activation and down-regulation of Ser/Thr phosphatases by Src family kinases. *Mol. Hum. Reprod.* **2013**, *19*, 570–580. [CrossRef] [PubMed]
64. Chavez, J.C.; Hernandez-Gonzalez, E.O.; Wertheimer, E.; Visconti, P.E.; Darszon, A.; Trevino, C.L. Participation of the Cl^-/HCO_3^- exchangers SLC26A3 and SLC26A6, the Cl^- channel CFTR, and the regulatory factor SLC9A3R1 in mouse sperm capacitation. *Biol. Reprod.* **2012**, *86*, 1–14. [CrossRef]
65. Salicioni, A.M.; Platt, M.D.; Wertheimer, E.V.; Arcelay, E.; Allaire, A.; Sosnik, J.; Visconti, P.E. Signalling pathways involved in sperm capacitation. *Soc. Reprod. Fertil. Suppl.* **2007**, *65*, 245–259.
66. Hernandez-Gonzalez, E.O.; Sosnik, J.; Edwards, J.; Acevedo, J.J.; Mendoza-Lujambio, I.; Lopez-Gonzalez, I.; Demarco, I.; Wertheimer, E.; Darszon, A.; Visconti, P.E. Sodium and epithelial sodium channels participate in the regulation of the capacitation-associated hyperpolarization in mouse sperm. *J. Biol. Chem.* **2006**, *281*, 5623–5633. [CrossRef]
67. Lefievre, L.; Jha, K.N.; de Lamirande, E.; Visconti, P.E.; Gagnon, C. Activation of protein kinase A during human sperm capacitation and acrosome reaction. *J. Androl.* **2002**, *23*, 709–716.
68. Visconti, P.E.; Stewart-Savage, J.; Blasco, A.; Battaglia, L.; Miranda, P.; Kopf, G.S.; Tezon, J.G. Roles of bicarbonate, cAMP, and protein tyrosine phosphorylation on capacitation and the spontaneous acrosome reaction of hamster sperm. *Biol. Reprod.* **1999**, *61*, 76–84. [CrossRef]
69. O'Flaherty, C.; de Lamirande, E.; Gagnon, C. Phosphorylation of the Arginine-X-X-(Serine/Threonine) motif in human sperm proteins during capacitation: Modulation and protein kinase A dependency. *Mol. Hum. Reprod.* **2004**, *10*, 355–363. [CrossRef]
70. O'Flaherty, C.; Beorlegui, N.; Beconi, M.T. Participation of superoxide anion in the capacitation of cryopreserved bovine sperm. *Int. J. Androl.* **2003**, *26*, 109–114. [CrossRef]
71. Freitas, M.J.; Vijayaraghavan, S.; Fardilha, M. Signaling mechanisms in mammalian sperm motility. *Biol. Reprod.* **2017**, *96*, 2–12. [CrossRef] [PubMed]
72. Gonzalez-Fernandez, L.; Ortega-Ferrusola, C.; Macias-Garcia, B.; Salido, G.M.; Pena, F.J.; Tapia, J.A. Identification of protein tyrosine phosphatases and dual-specificity phosphatases in mammalian spermatozoa and their role in sperm motility and protein tyrosine phosphorylation. *Biol. Reprod.* **2009**, *80*, 1239–1252. [CrossRef] [PubMed]
73. Denu, J.M.; Tanner, K.G. Specific and reversible inactivation of protein tyrosine phosphatases by hydrogen peroxide: Evidence for a sulfenic acid intermediate and implications for redox regulation. *Biochemistry* **1998**, *37*, 5633–5642. [CrossRef] [PubMed]
74. Frijhoff, J.; Dagnell, M.; Godfrey, R.; Ostman, A. Regulation of protein tyrosine phosphatase oxidation in cell adhesion and migration. *Antioxid. Redox Signal.* **2014**, *20*, 1994–2010. [CrossRef] [PubMed]
75. Ozkosem, B.; Feinstein, S.I.; Fisher, A.B.; O'Flaherty, C. Advancing age increases sperm chromatin damage and impairs fertility in peroxiredoxin 6 null mice. *Redox Biol.* **2015**, *5*, 15–23. [CrossRef] [PubMed]
76. Jeong, W.; Bae, S.H.; Toledano, M.B.; Rhee, S.G. Role of sulfiredoxin as a regulator of peroxiredoxin function and regulation of its expression. *Free Radic. Biol. Med.* **2012**, *53*, 447–456. [CrossRef]

77. Wood, Z.A.; Schroder, E.; Robin Harris, J.; Poole, L.B. Structure, mechanism and regulation of peroxiredoxins. *Trends Biochem. Sci.* **2003**, *28*, 32–40. [CrossRef]
78. Wood, Z.A.; Poole, L.B.; Karplus, P.A. Peroxiredoxin evolution and the regulation of hydrogen peroxide signaling. *Science* **2003**, *300*, 650–653. [CrossRef]
79. Scherz-Shouval, R.; Shvets, E.; Fass, E.; Shorer, H.; Gil, L.; Elazar, Z. Reactive oxygen species are essential for autophagy and specifically regulate the activity of Atg4. *EMBO J.* **2007**, *26*, 1749–1760. [CrossRef]
80. Gualtieri, R.; Mollo, V.; Duma, G.; Talevi, R. Redox control of surface protein sulphhydryls in bovine spermatozoa reversibly modulates sperm adhesion to the oviductal epithelium and capacitation. *Reproduction* **2009**, *138*, 33–43. [CrossRef]
81. Gualtieri, R.; Iaccarino, M.; Mollo, V.; Prisco, M.; Iaccarino, S.; Talevi, R. Slow cooling of human oocytes: Ultrastructural injuries and apoptotic status. *Fertil. Steril.* **2009**, *91*, 1023–1034. [CrossRef] [PubMed]
82. Talevi, R.; Zagami, M.; Castaldo, M.; Gualtieri, R. Redox regulation of sperm surface thiols modulates adhesion to the fallopian tube epithelium. *Biol. Reprod.* **2007**, *76*, 728–735. [CrossRef] [PubMed]
83. Ortiz-Rodriguez, J.M.; Martin-Cano, F.E.; Ortega-Ferrusola, C.; Masot, J.; Redondo, E.; Gazquez, A.; Gil, M.C.; Aparicio, I.M.; Rojo-Dominguez, P.; Tapia, J.A.; et al. The incorporation of cystine by the soluble carrier family 7 member 11 (SLC7A11) is a component of the redox regulatory mechanism in stallion spermatozoa. *Biol. Reprod.* **2019**. [CrossRef] [PubMed]
84. Ball, B.A.; Gravance, C.G.; Medina, V.; Baumber, J.; Liu, I.K. Catalase activity in equine semen. *Am. J. Vet. Res.* **2000**, *61*, 1026–1030. [CrossRef]
85. Baumber, J.; Ball, B.A. Determination of glutathione peroxidase and superoxide dismutase-like activities in equine spermatozoa, seminal plasma, and reproductive tissues. *Am. J. Vet. Res.* **2005**, *66*, 1415–1419. [CrossRef]
86. Brummer, M.; Hayes, S.; Dawson, K.A.; Lawrence, L.M. Measures of antioxidant status of the horse in response to selenium depletion and repletion. *J. Anim. Sci* **2013**, *91*, 2158–2168. [CrossRef]
87. Leone, E. Ergothioneine in the equine ampullar secretion. *Nature* **1954**, *174*, 404–405. [CrossRef]
88. Mann, T. Biochemistry of stallion semen. *J. Reprod. Fertil. Suppl.* **1975**, *23*, 47–52.
89. Ortega Ferrusola, C.; Gonzalez Fernandez, L.; Morrell, J.M.; Salazar Sandoval, C.; Macias Garcia, B.; Rodriguez-Martinez, H.; Tapia, J.A.; Pena, F.J. Lipid peroxidation, assessed with BODIPY-C11, increases after cryopreservation of stallion spermatozoa, is stallion-dependent and is related to apoptotic-like changes. *Reproduction* **2009**, *138*, 55–63. [CrossRef]
90. Fernandez, M.C.; O'Flaherty, C. Peroxiredoxin 6 is the primary antioxidant enzyme for the maintenance of viability and DNA integrity in human spermatozoa. *Hum. Reprod.* **2018**. [CrossRef]
91. Moawad, A.R.; Fernandez, M.C.; Scarlata, E.; Dodia, C.; Feinstein, S.I.; Fisher, A.B.; O'Flaherty, C. Deficiency of peroxiredoxin 6 or inhibition of its phospholipase A2 activity impair the in vitro sperm fertilizing competence in mice. *Sci. Rep.* **2017**, *7*, 12994. [CrossRef] [PubMed]
92. Neagu, V.R.; Garcia, B.M.; Sandoval, C.S.; Rodriguez, A.M.; Ferrusola, C.O.; Fernandez, L.G.; Tapia, J.A.; Pena, F.J. Freezing dog semen in presence of the antioxidant butylated hydroxytoluene improves postthaw sperm membrane integrity. *Theriogenology* **2010**, *73*, 645–650. [CrossRef] [PubMed]
93. Efrat, M.; Stein, A.; Pinkas, H.; Breitbart, H.; Unger, R.; Birk, R. Paraoxonase 1 (PON1) attenuates sperm hyperactivity and spontaneous acrosome reaction. *Andrology* **2018**. [CrossRef] [PubMed]
94. Barranco, I.; Tvarijonaviciute, A.; Perez-Patino, C.; Alkmin, D.V.; Ceron, J.J.; Martinez, E.A.; Rodriguez-Martinez, H.; Roca, J. The activity of paraoxonase type 1 (PON-1) in boar seminal plasma and its relationship with sperm quality, functionality, and in vivo fertility. *Andrology* **2015**, *3*, 315–320. [CrossRef]
95. Barranco, I.; Roca, J.; Tvarijonaviciute, A.; Ruber, M.; Vicente-Carrillo, A.; Atikuzzaman, M.; Ceron, J.J.; Martinez, E.A.; Rodriguez-Martinez, H. Measurement of activity and concentration of paraoxonase 1 (PON-1) in seminal plasma and identification of PON-2 in the sperm of boar ejaculates. *Mol. Reprod. Dev.* **2015**, *82*, 58–65. [CrossRef]
96. Lazaros, L.A.; Xita, N.V.; Hatzi, E.G.; Kaponis, A.I.; Stefos, T.J.; Plachouras, N.I.; Makrydimas, G.V.; Sofikitis, N.V.; Zikopoulos, K.A.; Georgiou, I.A. Association of paraoxonase gene polymorphisms with sperm parameters. *J. Androl.* **2011**, *32*, 394–401. [CrossRef]
97. Verit, F.F.; Verit, A.; Ciftci, H.; Erel, O.; Celik, H. Paraoxonase-1 activity in subfertile men and relationship to sperm parameters. *J. Androl.* **2009**, *30*, 183–189. [CrossRef]

98. Moradi, M.N.; Karimi, J.; Khodadadi, I.; Amiri, I.; Karami, M.; Saidijam, M.; Vatannejad, A.; Tavilani, H. Evaluation of the p53 and Thioredoxin reductase in sperm from asthenozoospermic males in comparison to normozoospermic males. *Free Radic. Biol. Med.* **2018**, *116*, 123–128. [CrossRef]
99. Emelyanov, A.V.; Fyodorov, D.V. Thioredoxin-dependent disulfide bond reduction is required for protamine eviction from sperm chromatin. *Genes Dev.* **2016**, *30*, 2651–2656. [CrossRef]
100. Tirmarche, S.; Kimura, S.; Dubruille, R.; Horard, B.; Loppin, B. Unlocking sperm chromatin at fertilization requires a dedicated egg thioredoxin in Drosophila. *Nat. Commun.* **2016**, *7*, 13539. [CrossRef]
101. Su, D.; Novoselov, S.V.; Sun, Q.A.; Moustafa, M.E.; Zhou, Y.; Oko, R.; Hatfield, D.L.; Gladyshev, V.N. Mammalian selenoprotein thioredoxin-glutathione reductase. Roles in disulfide bond formation and sperm maturation. *J. Biol. Chem.* **2005**, *280*, 26491–26498. [CrossRef] [PubMed]
102. Miranda-Vizuete, A.; Tsang, K.; Yu, Y.; Jimenez, A.; Pelto-Huikko, M.; Flickinger, C.J.; Sutovsky, P.; Oko, R. Cloning and developmental analysis of murid spermatid-specific thioredoxin-2 (SPTRX-2), a novel sperm fibrous sheath protein and autoantigen. *J. Biol. Chem.* **2003**, *278*, 44874–44885. [CrossRef] [PubMed]
103. Yu, Y.; Oko, R.; Miranda-Vizuete, A. Developmental expression of spermatid-specific thioredoxin-1 protein: Transient association to the longitudinal columns of the fibrous sheath during sperm tail formation. *Biol. Reprod.* **2002**, *67*, 1546–1554. [CrossRef] [PubMed]
104. Kuribayashi, Y.; Gagnon, C. Effect of catalase and thioredoxin addition to sperm incubation medium before in vitro fertilization on sperm capacity to support embryo development. *Fertil. Steril.* **1996**, *66*, 1012–1017. [CrossRef]
105. Ozkosem, B.; Feinstein, S.I.; Fisher, A.B.; O'Flaherty, C. Absence of Peroxiredoxin 6 Amplifies the Effect of Oxidant Stress on Mobility and SCSA/CMA3 Defined Chromatin Quality and Impairs Fertilizing Ability of Mouse Spermatozoa. *Biol. Reprod.* **2016**, *94*, 68. [CrossRef]
106. O'Flaherty, C. Peroxiredoxins: Hidden players in the antioxidant defence of human spermatozoa. *Basic Clin. Androl.* **2014**, *24*, 4. [CrossRef]
107. Ortega Ferrusola, C.; Martin Munoz, P.; Ortiz-Rodriguez, J.M.; Anel-Lopez, L.; Balao da Silva, C.; Alvarez, M.; de Paz, P.; Tapia, J.A.; Anel, L.; Silva-Rodriguez, A.; et al. Depletion of thiols leads to redox deregulation, production of 4-hydroxinonenal and sperm senescence: A possible role for GSH regulation in spermatozoa. *Biol. Reprod.* **2018**. [CrossRef]
108. Keil, M.; Wetterauer, U.; Heite, H.J. Glutamic acid concentration in human semen–Its origin and significance. *Andrologia* **1979**, *11*, 385–391. [CrossRef]
109. Munoz, P.M.; Ferrusola, C.O.; Lopez, L.A.; Del Petre, C.; Garcia, M.A.; de Paz Cabello, P.; Anel, L.; Pena, F.J. Caspase 3 Activity and Lipoperoxidative Status in Raw Semen Predict the Outcome of Cryopreservation of Stallion Spermatozoa. *Biol. Reprod.* **2016**, *95*, 53. [CrossRef]
110. Martin Munoz, P.; Ortega Ferrusola, C.; Vizuete, G.; Plaza Davila, M.; Rodriguez Martinez, H.; Pena, F.J. Depletion of Intracellular Thiols and Increased Production of 4-Hydroxynonenal that Occur During Cryopreservation of Stallion Spermatozoa Lead to Caspase Activation, Loss of Motility, and Cell Death. *Biol. Reprod.* **2015**, *93*, 143. [CrossRef]
111. Aitken, R.J.; Baker, M.A.; Nixon, B. Are sperm capacitation and apoptosis the opposite ends of a continuum driven by oxidative stress? *Asian J. Androl.* **2015**, *17*, 633–639. [CrossRef] [PubMed]
112. Brand, M.D. Mitochondrial generation of superoxide and hydrogen peroxide as the source of mitochondrial redox signaling. *Free Radic. Biol. Med.* **2016**, *100*, 14–31. [CrossRef] [PubMed]
113. Goncalves, R.L.; Bunik, V.I.; Brand, M.D. Production of superoxide/hydrogen peroxide by the mitochondrial 2-oxoadipate dehydrogenase complex. *Free Radic. Biol. Med.* **2016**, *91*, 247–255. [CrossRef] [PubMed]
114. Samanta, L.; Agarwal, A.; Swain, N.; Sharma, R.; Gopalan, B.; Esteves, S.C.; Durairajanayagam, D.; Sabanegh, E. Proteomic Signatures of Sperm Mitochondria in Varicocele: Clinical Use as Biomarkers of Varicocele Associated Infertility. *J. Urol.* **2018**, *200*, 414–422. [CrossRef] [PubMed]
115. Lu, X.; Zhang, Y.; Bai, H.; Liu, J.; Li, J.; Wu, B. Mitochondria-targeted antioxidant MitoTEMPO improves the post-thaw sperm quality. *Cryobiology* **2018**, *80*, 26–29. [CrossRef]
116. Amaral, S.; S Tavares, R.; Baptista, M.; Sousa, M.I.; Silva, A.; Escada-Rebelo, S.; Paiva, C.P.; Ramalho-Santos, J. Mitochondrial Functionality and Chemical Compound Action on Sperm Function. *Curr. Med. Chem.* **2016**, *23*, 3575–3606. [CrossRef]
117. Gibb, Z.; Lambourne, S.R.; Quadrelli, J.; Smith, N.D.; Aitken, R.J. L-carnitine and pyruvate are prosurvival factors during the storage of stallion spermatozoa at room temperature. *Biol. Reprod.* **2015**, *93*, 104. [CrossRef]

118. Pena, F.J.; Plaza Davila, M.; Ball, B.A.; Squires, E.L.; Martin Munoz, P.; Ortega Ferrusola, C.; Balao da Silva, C. The Impact of Reproductive Technologies on Stallion Mitochondrial Function. *Reprod. Domest. Anim.* **2015**, *50*, 529–537. [CrossRef]
119. Jones, D.P. Disruption of mitochondrial redox circuitry in oxidative stress. *Chem. Biol. Interact.* **2006**, *163*, 38–53. [CrossRef]
120. Rico-Leo, E.M.; Moreno-Marin, N.; Gonzalez-Rico, F.J.; Barrasa, E.; Ortega-Ferrusola, C.; Martin-Munoz, P.; Sanchez-Guardado, L.O.; Llano, E.; Alvarez-Barrientos, A.; Infante-Campos, A.; et al. piRNA-associated proteins and retrotransposons are differentially expressed in murine testis and ovary of aryl hydrocarbon receptor deficient mice. *Open Biol.* **2016**, *6*, 160186. [CrossRef]
121. Losano, J.D.A.; Angrimani, D.S.R.; Ferreira Leite, R.; Simoes da Silva, B.D.C.; Barnabe, V.H.; Nichi, M. Spermatic mitochondria: Role in oxidative homeostasis, sperm function and possible tools for their assessment. *Zygote* **2018**, *26*, 251–260. [CrossRef] [PubMed]
122. Amaral, A.; Lourenco, B.; Marques, M.; Ramalho-Santos, J. Mitochondria functionality and sperm quality. *Reproduction* **2013**, *146*, R163–R174. [CrossRef]
123. Vakifahmetoglu-Norberg, H.; Ouchida, A.T.; Norberg, E. The role of mitochondria in metabolism and cell death. *Biochem. Biophys. Res. Commun.* **2017**, *482*, 426–431. [CrossRef] [PubMed]
124. du Plessis, S.S.; Agarwal, A.; Mohanty, G.; van der Linde, M. Oxidative phosphorylation versus glycolysis: What fuel do spermatozoa use? *Asian J. Androl.* **2015**, *17*, 230–235. [CrossRef] [PubMed]
125. Amaral, A.; Paiva, C.; Attardo Parrinello, C.; Estanyol, J.M.; Ballesca, J.L.; Ramalho-Santos, J.; Oliva, R. Identification of proteins involved in human sperm motility using high-throughput differential proteomics. *J. Proteome Res.* **2014**, *13*, 5670–5684. [CrossRef]
126. Amaral, A.; Castillo, J.; Estanyol, J.M.; Ballesca, J.L.; Ramalho-Santos, J.; Oliva, R. Human sperm tail proteome suggests new endogenous metabolic pathways. *Mol. Cell. Proteom.* **2013**, *12*, 330–342. [CrossRef]
127. Klingenberg, M. The ADP and ATP transport in mitochondria and its carrier. *Biochim. Biophys. Acta* **2008**, *1778*, 1978–2021. [CrossRef]
128. Ortega Ferrusola, C.; Anel-Lopez, L.; Ortiz-Rodriguez, J.M.; Martin Munoz, P.; Alvarez, M.; de Paz, P.; Masot, J.; Redondo, E.; Balao da Silva, C.; Morrell, J.M.; et al. Stallion spermatozoa surviving freezing and thawing experience membrane depolarization and increased intracellular Na^+. *Andrology* **2017**, *5*, 1174–1182. [CrossRef]
129. Moscatelli, N.; Lunetti, P.; Braccia, C.; Armirotti, A.; Pisanello, F.; De Vittorio, M.; Zara, V.; Ferramosca, A. Comparative Proteomic Analysis of Proteins Involved in Bioenergetics Pathways Associated with Human Sperm Motility. *Int. J. Mol. Sci.* **2019**, *20*, 3000. [CrossRef]
130. Ferramosca, A.; Zara, V. Bioenergetics of mammalian sperm capacitation. *Biomed. Res. Int.* **2014**, *2014*, 902953. [CrossRef]
131. Piomboni, P.; Focarelli, R.; Stendardi, A.; Ferramosca, A.; Zara, V. The role of mitochondria in energy production for human sperm motility. *Int. J. Androl.* **2012**, *35*, 109–124. [CrossRef] [PubMed]
132. Bucci, D.; Rodriguez-Gil, J.E.; Vallorani, C.; Spinaci, M.; Galeati, G.; Tamanini, C. GLUTs and mammalian sperm metabolism. *J. Androl.* **2011**, *32*, 348–355. [CrossRef] [PubMed]
133. Marin, S.; Chiang, K.; Bassilian, S.; Lee, W.N.; Boros, L.G.; Fernandez-Novell, J.M.; Centelles, J.J.; Medrano, A.; Rodriguez-Gil, J.E.; Cascante, M. Metabolic strategy of boar spermatozoa revealed by a metabolomic characterization. *FEBS Lett.* **2003**, *554*, 342–346. [CrossRef]
134. Asghari, A.; Marashi, S.A.; Ansari-Pour, N. A sperm-specific proteome-scale metabolic network model identifies non-glycolytic genes for energy deficiency in asthenozoospermia. *Syst. Biol. Reprod. Med.* **2017**, *63*, 100–112. [CrossRef]
135. Swegen, A.; Curry, B.J.; Gibb, Z.; Lambourne, S.R.; Smith, N.D.; Aitken, R.J. Investigation of the stallion sperm proteome by mass spectrometry. *Reproduction* **2015**, *149*, 235–244. [CrossRef] [PubMed]
136. Qiu, J.H.; Li, Y.W.; Xie, H.L.; Li, Q.; Dong, H.B.; Sun, M.J.; Gao, W.Q.; Tan, J.H. Effects of glucose metabolism pathways on sperm motility and oxidative status during long-term liquid storage of goat semen. *Theriogenology* **2016**, *86*, 839–849. [CrossRef]
137. Miraglia, E.; Lussiana, C.; Viarisio, D.; Racca, C.; Cipriani, A.; Gazzano, E.; Bosia, A.; Revelli, A.; Ghigo, D. The pentose phosphate pathway plays an essential role in supporting human sperm capacitation. *Fertil. Steril.* **2010**, *93*, 2437–2440. [CrossRef]

138. Ando, S.; Aquila, S. Arguments raised by the recent discovery that insulin and leptin are expressed in and secreted by human ejaculated spermatozoa. *Mol. Cell. Endocrinol.* **2005**, *245*, 1–6. [CrossRef]
139. Urner, F.; Sakkas, D. Involvement of the pentose phosphate pathway and redox regulation in fertilization in the mouse. *Mol. Reprod. Dev.* **2005**, *70*, 494–503. [CrossRef]
140. Williams, A.C.; Ford, W.C. Functional significance of the pentose phosphate pathway and glutathione reductase in the antioxidant defenses of human sperm. *Biol. Reprod.* **2004**, *71*, 1309–1316. [CrossRef]
141. Urner, F.; Sakkas, D. A possible role for the pentose phosphate pathway of spermatozoa in gamete fusion in the mouse. *Biol. Reprod.* **1999**, *60*, 733–739. [CrossRef] [PubMed]
142. Evdokimov, V.V.; Barinova, K.V.; Turovetskii, V.B.; Muronetz, V.I.; Schmalhausen, E.V. Low Concentrations of Hydrogen Peroxide Activate the Antioxidant Defense System in Human Sperm Cells. *Biochemistry* **2015**, *80*, 1178–1185. [CrossRef] [PubMed]
143. Ford, W.C.; Whittington, K.; Williams, A.C. Reactive oxygen species in human sperm suspensions: Production by leukocytes and the generation of NADPH to protect sperm against their effects. *Int. J. Androl.* **1997**, *20* (Suppl. 3), 44–49.
144. Ortiz-Rodriguez, J.M.; Balao da Silva, C.; Masot, J.; Redondo, E.; Gazquez, A.; Tapia, J.A.; Gil, C.; Ortega-Ferrusola, C.; Pena, F.J. Rosiglitazone in the thawing medium improves mitochondrial function in stallion spermatozoa through regulating Akt phosphorylation and reduction of caspase 3. *PLoS ONE* **2019**, *14*, e0211994. [CrossRef] [PubMed]
145. Volpe, C.M.O.; Villar-Delfino, P.H.; Dos Anjos, P.M.F.; Nogueira-Machado, J.A. Cellular death, reactive oxygen species (ROS) and diabetic complications. *Cell Death Dis.* **2018**, *9*, 119. [CrossRef] [PubMed]
146. Allaman, I.; Belanger, M.; Magistretti, P.J. Methylglyoxal, the dark side of glycolysis. *Front. Neurosci.* **2015**, *9*, 23. [CrossRef] [PubMed]
147. Nevin, C.; McNeil, L.; Ahmed, N.; Murgatroyd, C.; Brison, D.; Carroll, M. Investigating the Glycating Effects of Glucose, Glyoxal and Methylglyoxal on Human Sperm. *Sci. Rep.* **2018**, *8*, 9002. [CrossRef]
148. Chen, M.C.; Lin, J.A.; Lin, H.T.; Chen, S.Y.; Yen, G.C. Potential effect of advanced glycation end products (AGEs) on spermatogenesis and sperm quality in rodents. *Food Funct.* **2019**, *10*, 3324–3333. [CrossRef]
149. Karimi, J.; Goodarzi, M.T.; Tavilani, H.; Khodadadi, I.; Amiri, I. Relationship between advanced glycation end products and increased lipid peroxidation in semen of diabetic men. *Diabetes Res. Clin. Pract.* **2011**, *91*, 61–66. [CrossRef]
150. Aquila, S.; Gentile, M.; Middea, E.; Catalano, S.; Ando, S. Autocrine regulation of insulin secretion in human ejaculated spermatozoa. *Endocrinology* **2005**, *146*, 552–557. [CrossRef]
151. Sangeeta, S.; Arangasamy, A.; Kulkarni, S.; Selvaraju, S. Role of amino acids as additives on sperm motility, plasma membrane integrity and lipid peroxidation levels at pre-freeze and post-thawed ram semen. *Anim. Reprod. Sci.* **2015**, *161*, 82–88. [CrossRef] [PubMed]
152. Lahnsteiner, F. The role of free amino acids in semen of rainbow trout Oncorhynchus mykiss and carp Cyprinus carpio. *J. Fish. Biol.* **2009**, *75*, 816–833. [CrossRef] [PubMed]
153. Koppula, P.; Zhang, Y.; Zhuang, L.; Gan, B. Amino acid transporter SLC7A11/xCT at the crossroads of regulating redox homeostasis and nutrient dependency of cancer. *Cancer Commun. (Lond.)* **2018**, *38*, 12. [CrossRef] [PubMed]
154. Breda, C.N.S.; Davanzo, G.G.; Basso, P.J.; Saraiva Camara, N.O.; Moraes-Vieira, P.M.M. Mitochondria as central hub of the immune system. *Redox Biol.* **2019**, *26*, 101255. [CrossRef]
155. Bromfield, E.G.; Aitken, R.J.; Anderson, A.L.; McLaughlin, E.A.; Nixon, B. The impact of oxidative stress on chaperone-mediated human sperm-egg interaction. *Hum. Reprod.* **2015**, *30*, 2597–2613. [CrossRef]
156. Gharagozloo, P.; Aitken, R.J. The role of sperm oxidative stress in male infertility and the significance of oral antioxidant therapy. *Hum. Reprod.* **2011**, *26*, 1628–1640. [CrossRef]
157. Aitken, R.J.; De Iuliis, G.N.; Finnie, J.M.; Hedges, A.; McLachlan, R.I. Analysis of the relationships between oxidative stress, DNA damage and sperm vitality in a patient population: Development of diagnostic criteria. *Hum. Reprod.* **2010**, *25*, 2415–2426. [CrossRef]
158. Aitken, R.J.; Baker, M.A. Oxidative stress, sperm survival and fertility control. *Mol. Cell. Endocrinol.* **2006**, *250*, 66–69. [CrossRef]
159. Thomson, L.K.; Fleming, S.D.; Aitken, R.J.; De Iuliis, G.N.; Zieschang, J.A.; Clark, A.M. Cryopreservation-induced human sperm DNA damage is predominantly mediated by oxidative stress rather than apoptosis. *Hum. Reprod.* **2009**, *24*, 2061–2070. [CrossRef]

160. Garcia, B.M.; Moran, A.M.; Fernandez, L.G.; Ferrusola, C.O.; Rodriguez, A.M.; Bolanos, J.M.; da Silva, C.M.; Martinez, H.R.; Tapia, J.A.; Pena, F.J. The mitochondria of stallion spermatozoa are more sensitive than the plasmalemma to osmotic-induced stress: Role of c-Jun N-terminal kinase (JNK) pathway. *J. Androl.* **2012**, *33*, 105–113. [CrossRef]

161. Pena, F.J.; Ball, B.A.; Squires, E.L. A new method for evaluating stallion sperm viability and mitochondrial membrane potential in fixed semen samples. *Cytom. B Clin. Cytom.* **2016**. [CrossRef]

162. Balao da Silva, C.M.; Ortega Ferrusola, C.; Morillo Rodriguez, A.; Gallardo Bolanos, J.M.; Plaza Davila, M.; Morrell, J.M.; Rodriguez Martinez, H.; Tapia, J.A.; Aparicio, I.M.; Pena, F.J. Sex sorting increases the permeability of the membrane of stallion spermatozoa. *Anim. Reprod. Sci.* **2013**, *138*, 241–251. [CrossRef] [PubMed]

163. Rodriguez, A.M.; Ferrusola, C.O.; Garcia, B.M.; Morrell, J.M.; Martinez, H.R.; Tapia, J.A.; Pena, F.J. Freezing stallion semen with the new Caceres extender improves post thaw sperm quality and diminishes stallion-to-stallion variability. *Anim. Reprod. Sci.* **2011**, *127*, 78–83. [CrossRef]

164. Ortega Ferrusola, C.; Gonzalez Fernandez, L.; Salazar Sandoval, C.; Macias Garcia, B.; Rodriguez Martinez, H.; Tapia, J.A.; Pena, F.J. Inhibition of the mitochondrial permeability transition pore reduces "apoptosis like" changes during cryopreservation of stallion spermatozoa. *Theriogenology* **2010**, *74*, 458–465. [CrossRef] [PubMed]

165. Ortega-Ferrusola, C.; Gil, M.C.; Rodriguez-Martinez, H.; Anel, L.; Pena, F.J.; Martin-Munoz, P. Flow cytometry in Spermatology: A bright future ahead. *Reprod. Domest. Anim.* **2017**, *52*, 921–931. [CrossRef] [PubMed]

166. Gallardo Bolanos, J.M.; Miro Moran, A.; Balao da Silva, C.M.; Morillo Rodriguez, A.; Plaza Davila, M.; Aparicio, I.M.; Tapia, J.A.; Ortega Ferrusola, C.; Pena, F.J. Autophagy and apoptosis have a role in the survival or death of stallion spermatozoa during conservation in refrigeration. *PLoS ONE* **2012**, *7*, e30688. [CrossRef]

167. Gueraud, F.; Atalay, M.; Bresgen, N.; Cipak, A.; Eckl, P.M.; Huc, L.; Jouanin, I.; Siems, W.; Uchida, K. Chemistry and biochemistry of lipid peroxidation products. *Free Radic. Res.* **2010**, *44*, 1098–1124. [CrossRef]

168. Uchida, K. Lipid peroxidation and redox-sensitive signaling pathways. *Curr. Atheroscler Rep.* **2007**, *9*, 216–221. [CrossRef]

169. Uchida, K. Cellular response to bioactive lipid peroxidation products. *Free Radic. Res.* **2000**, *33*, 731–737. [CrossRef]

170. Macias Garcia, B.; Gonzalez Fernandez, L.; Ortega Ferrusola, C.; Morillo Rodriguez, A.; Gallardo Bolanos, J.M.; Rodriguez Martinez, H.; Tapia, J.A.; Morcuende, D.; Pena, F.J. Fatty acids and plasmalogens of the phospholipids of the sperm membranes and their relation with the post-thaw quality of stallion spermatozoa. *Theriogenology* **2011**, *75*, 811–818. [CrossRef]

171. Garcia, B.M.; Fernandez, L.G.; Ferrusola, C.O.; Salazar-Sandoval, C.; Rodriguez, A.M.; Martinez, H.R.; Tapia, J.A.; Morcuende, D.; Pena, F.J. Membrane lipids of the stallion spermatozoon in relation to sperm quality and susceptibility to lipid peroxidation. *Reprod. Domest. Anim.* **2011**, *46*, 141–148. [CrossRef] [PubMed]

172. Martin Munoz, P.; Anel-Lopez, L.; Ortiz-Rodriguez, J.M.; Alvarez, M.; de Paz, P.; Balao da Silva, C.; Rodriguez Martinez, H.; Gil, M.C.; Anel, L.; Pena, F.J.; et al. Redox cycling induces spermptosis and necrosis in stallion spermatozoa while the hydroxyl radical (OH*) only induces spermptosis. *Reprod. Domest. Anim.* **2018**, *53*, 54–67. [CrossRef] [PubMed]

173. Hall, S.E.; Aitken, R.J.; Nixon, B.; Smith, N.D.; Gibb, Z. Electrophilic aldehyde products of lipid peroxidation selectively adduct to heat shock protein 90 and arylsulfatase A in stallion spermatozoa. *Biol. Reprod.* **2017**, *96*, 107–121. [CrossRef] [PubMed]

174. Bromfield, E.G.; Aitken, R.J.; McLaughlin, E.A.; Nixon, B. Proteolytic degradation of heat shock protein A2 occurs in response to oxidative stress in male germ cells of the mouse. *Mol. Hum. Reprod.* **2017**, *23*, 91–105. [CrossRef]

175. Gibb, Z.; Lambourne, S.R.; Curry, B.J.; Hall, S.E.; Aitken, R.J. Aldehyde Dehydrogenase Plays a Pivotal Role in the Maintenance of Stallion Sperm Motility. *Biol. Reprod.* **2016**, *94*, 133. [CrossRef]

176. Moazamian, R.; Polhemus, A.; Connaughton, H.; Fraser, B.; Whiting, S.; Gharagozloo, P.; Aitken, R.J. Oxidative stress and human spermatozoa: Diagnostic and functional significance of aldehydes generated as a result of lipid peroxidation. *Mol. Hum. Reprod.* **2015**, *21*, 502–515. [CrossRef]

177. Baker, M.A.; Weinberg, A.; Hetherington, L.; Villaverde, A.I.; Velkov, T.; Baell, J.; Gordon, C.P. Defining the mechanisms by which the reactive oxygen species by-product, 4-hydroxynonenal, affects human sperm cell function. *Biol. Reprod.* **2015**, *92*, 108. [CrossRef]
178. Aitken, R.J.; Baker, M.A. Causes and consequences of apoptosis in spermatozoa; contributions to infertility and impacts on development. *Int. J. Dev. Biol.* **2013**, *57*, 265–272. [CrossRef]
179. Aitken, R.J.; Smith, T.B.; Lord, T.; Kuczera, L.; Koppers, A.J.; Naumovski, N.; Connaughton, H.; Baker, M.A.; De Iuliis, G.N. On methods for the detection of reactive oxygen species generation by human spermatozoa: Analysis of the cellular responses to catechol oestrogen, lipid aldehyde, menadione and arachidonic acid. *Andrology* **2013**, *1*, 192–205. [CrossRef]
180. Aurich, C.; Ortega Ferrusola, C.; Pena Vega, F.J.; Schrammel, N.; Morcuende, D.; Aurich, J. Seasonal changes in the sperm fatty acid composition of Shetland pony stallions. *Theriogenology* **2018**, *107*, 149–153. [CrossRef]
181. Zimniak, P. Relationship of electrophilic stress to aging. *Free Radic. Biol. Med.* **2011**, *51*, 1087–1105. [CrossRef] [PubMed]
182. Martinez-Pastor, F.; Mata-Campuzano, M.; Alvarez-Rodriguez, M.; Alvarez, M.; Anel, L.; de Paz, P. Probes and techniques for sperm evaluation by flow cytometry. *Reprod. Domest. Anim.* **2010**, *45* (Suppl. 2), 67–78. [CrossRef]
183. Aitken, R.J.; Gibb, Z.; Mitchell, L.A.; Lambourne, S.R.; Connaughton, H.S.; De Iuliis, G.N. Sperm motility is lost in vitro as a consequence of mitochondrial free radical production and the generation of electrophilic aldehydes but can be significantly rescued by the presence of nucleophilic thiols. *Biol. Reprod.* **2012**, *87*, 110. [CrossRef] [PubMed]
184. Bromfield, E.G.; McLaughlin, E.A.; Aitken, R.J.; Nixon, B. Heat Shock Protein member A2 forms a stable complex with angiotensin converting enzyme and protein disulfide isomerase A6 in human spermatozoa. *Mol. Hum. Reprod.* **2016**, *22*, 93–109. [CrossRef] [PubMed]
185. Aitken, R.J.; Flanagan, H.M.; Connaughton, H.; Whiting, S.; Hedges, A.; Baker, M.A. Involvement of homocysteine, homocysteine thiolactone, and paraoxonase type 1 (PON-1) in the etiology of defective human sperm function. *Andrology* **2016**, *4*, 345–360. [CrossRef] [PubMed]
186. Teperek, M.; Simeone, A.; Gaggioli, V.; Miyamoto, K.; Allen, G.E.; Erkek, S.; Kwon, T.; Marcotte, E.M.; Zegerman, P.; Bradshaw, C.R.; et al. Sperm is epigenetically programmed to regulate gene transcription in embryos. *Genome Res.* **2016**, *26*, 1034–1046. [CrossRef]
187. Valcarce, D.G.; Carton-Garcia, F.; Riesco, M.F.; Herraez, M.P.; Robles, V. Analysis of DNA damage after human sperm cryopreservation in genes crucial for fertilization and early embryo development. *Andrology* **2013**, *1*, 723–730. [CrossRef]
188. Valcarce, D.G.; Carton-Garcia, F.; Herraez, M.P.; Robles, V. Effect of cryopreservation on human sperm messenger RNAs crucial for fertilization and early embryo development. *Cryobiology* **2013**, *67*, 84–90. [CrossRef]
189. Kopeika, J.; Thornhill, A.; Khalaf, Y. The effect of cryopreservation on the genome of gametes and embryos: Principles of cryobiology and critical appraisal of the evidence. *Hum. Reprod. Update* **2015**, *21*, 209–227. [CrossRef]
190. Rex, A.S.; Aagaard, J.; Fedder, J. DNA fragmentation in spermatozoa: A historical review. *Andrology* **2017**, *5*, 622–630. [CrossRef]
191. Evenson, D.P.; Kasperson, K.; Wixon, R.L. Analysis of sperm DNA fragmentation using flow cytometry and other techniques. *Soc. Reprod. Fertil. Suppl.* **2007**, *65*, 93–113. [PubMed]
192. Lhomme, J.; Constant, J.F.; Demeunynck, M. Abasic DNA structure, reactivity, and recognition. *Biopolymers* **1999**, *52*, 65–83. [CrossRef]
193. Belmont, P.; Jourdan, M.; Demeunynck, M.; Constant, J.F.; Garcia, J.; Lhomme, J.; Carez, D.; Croisy, A. Abasic site recognition in DNA as a new strategy to potentiate the action of anticancer alkylating drugs? *J. Med. Chem.* **1999**, *42*, 5153–5159. [CrossRef]
194. Greenberg, M.M. Looking beneath the surface to determine what makes DNA damage deleterious. *Curr. Opin Chem. Biol.* **2014**, *21*, 48–55. [CrossRef] [PubMed]
195. San Pedro, J.M.; Greenberg, M.M. 5,6-Dihydropyrimidine peroxyl radical reactivity in DNA. *J. Am. Chem. Soc.* **2014**, *136*, 3928–3936. [CrossRef] [PubMed]
196. Greenberg, M.M. Abasic and oxidized abasic site reactivity in DNA: Enzyme inhibition, cross-linking, and nucleosome catalyzed reactions. *Acc. Chem. Res.* **2014**, *47*, 646–655. [CrossRef]

197. Balao da Silva, C.M.; Ortega-Ferrusola, C.; Morrell, J.M.; Rodriguez Martinez, H.; Pena, F.J. Flow Cytometric Chromosomal Sex Sorting of Stallion Spermatozoa Induces Oxidative Stress on Mitochondria and Genomic DNA. *Reprod. Domest. Anim.* **2016**, *51*, 18–25. [CrossRef]
198. Vorilhon, S.; Brugnon, F.; Kocer, A.; Dollet, S.; Bourgne, C.; Berger, M.; Janny, L.; Pereira, B.; Aitken, R.J.; Moazamian, A.; et al. Accuracy of human sperm DNA oxidation quantification and threshold determination using an 8-OHdG immuno-detection assay. *Hum. Reprod.* **2018**, *33*, 553–562. [CrossRef]
199. Li, Z.; Yang, J.; Huang, H. Oxidative stress induces H2AX phosphorylation in human spermatozoa. *FEBS Lett.* **2006**, *580*, 6161–6168. [CrossRef]
200. Garolla, A.; Cosci, I.; Bertoldo, A.; Sartini, B.; Boudjema, E.; Foresta, C. DNA double strand breaks in human spermatozoa can be predictive for assisted reproductive outcome. *Reprod. Biomed. Online* **2015**, *31*, 100–107. [CrossRef]
201. Castillo, J.; Jodar, M.; Oliva, R. The contribution of human sperm proteins to the development and epigenome of the preimplantation embryo. *Hum. Reprod. Update* **2018**, *24*, 535–555. [CrossRef] [PubMed]
202. Aitken, R.J.; Curry, B.J. Redox regulation of human sperm function: From the physiological control of sperm capacitation to the etiology of infertility and DNA damage in the germ line. *Antioxid. Redox Signal.* **2011**, *14*, 367–381. [CrossRef] [PubMed]
203. Burruel, V.; Klooster, K.L.; Chitwood, J.; Ross, P.J.; Meyers, S.A. Oxidative damage to rhesus macaque spermatozoa results in mitotic arrest and transcript abundance changes in early embryos. *Biol. Reprod.* **2013**, *89*, 72. [CrossRef] [PubMed]
204. McCarthy, M.J.; Baumber, J.; Kass, P.H.; Meyers, S.A. Osmotic stress induces oxidative cell damage to rhesus macaque spermatozoa. *Biol. Reprod.* **2010**, *82*, 644–651. [CrossRef] [PubMed]
205. Ortiz-Rodriguez, J.M.; Ortega-Ferrusola, C.; Gil, M.C.; Martin-Cano, F.E.; Gaitskell-Phillips, G.; Rodriguez-Martinez, H.; Hinrichs, K.; Alvarez-Barrientos, A.; Roman, A.; Pena, F.J. Transcriptome analysis reveals that fertilization with cryopreserved sperm downregulates genes relevant for early embryo development in the horse. *PLoS ONE* **2019**, *14*, e0213420. [CrossRef] [PubMed]
206. Jodar, M. Sperm and seminal plasma RNAs: What roles do they play beyond fertilization? *Reproduction* **2019**. [CrossRef]
207. Zhou, D.; Suzuki, T.; Asami, M.; Perry, A.C.F. Caput Epididymidal Mouse Sperm Support Full Development. *Dev. Cell* **2019**, *50*, 5–6. [CrossRef]
208. Morielli, T.; O'Flaherty, C. Oxidative stress impairs function and increases redox protein modifications in human spermatozoa. *Reproduction* **2015**, *149*, 113–123. [CrossRef]

 © 2019 by the authors. Licensee MDPI, Basel, Switzerland. This article is an open access article distributed under the terms and conditions of the Creative Commons Attribution (CC BY) license (http://creativecommons.org/licenses/by/4.0/).

Review

Antioxidants and Male Fertility: From Molecular Studies to Clinical Evidence

David Martin-Hidalgo [1,2,*], Maria Julia Bragado [2], Ana R. Batista [3], Pedro F. Oliveira [1,4,5] and Marco G. Alves [2,*]

1. Unit for Multidisciplinary Research in Biomedicine (UMIB), Laboratory of Cell Biology, Department of Microscopy, Institute of Biomedical Sciences Abel Salazar (ICBAS), University of Porto, 4050-313 Porto, Portugal; pfobox@gmail.com
2. Research Group of Intracellular Signaling and Technology of Reproduction (SINTREP), Institute of Biotechnology in Agriculture and Livestock (INBIO G+C), University of Extremadura, 10004 Cáceres, Spain; jbragado@unex.es
3. Merck S.A., 1495-190 Algés, Portugal; ana-rita.batista@merckgroup.com
4. i3S-Instituto de Investigação e Inovação em Saúde, University of Porto, 4200-135 Porto, Portugal
5. Faculty of Medicine, University of Porto, 4200-319 Porto, Portugal
* Correspondence: davidmh@unex.es (D.M.-H.); alvesmarc@gmail.com (M.G.A.)

Received: 14 March 2019; Accepted: 3 April 2019; Published: 5 April 2019

Abstract: Spermatozoa are physiologically exposed to reactive oxygen species (ROS) that play a pivotal role on several sperm functions through activation of different intracellular mechanisms involved in physiological functions such as sperm capacitation associated-events. However, ROS overproduction depletes sperm antioxidant system, which leads to a condition of oxidative stress (OS). Subfertile and infertile men are known to present higher amount of ROS in the reproductive tract which causes sperm DNA damage and results in lower fertility and pregnancy rates. Thus, there is a growing number of couples seeking fertility treatment and assisted reproductive technologies (ART) due to OS-related problems in the male partner. Interestingly, although ART can be successfully used, it is also related with an increase in ROS production. This has led to a debate if antioxidants should be proposed as part of a fertility treatment in an attempt to decrease non-physiological elevated levels of ROS. However, the rationale behind oral antioxidants intake and positive effects on male reproduction outcome is only supported by few studies. In addition, it is unclear whether negative effects may arise from oral antioxidants intake. Although there are some contrasting reports, oral consumption of compounds with antioxidant activity appears to improve sperm parameters, such as motility and concentration, and decrease DNA damage, but there is not sufficient evidence that fertility rates and live birth really improve after antioxidants intake. Moreover, it depends on the type of antioxidants, treatment duration, and even the diagnostics of the man's fertility, among other factors. Literature also suggests that the main advantage of antioxidant therapy is to extend sperm preservation to be used during ART. Herein, we discuss ROS production and its relevance in male fertility and antioxidant therapy with focus on molecular mechanisms and clinical evidence.

Keywords: assisted reproductive technologies; sperm ROS; pregnancy; infertility; antioxidants therapy; reproductive outcome

1. Introduction

The mammalian spermatozoon is a cell with a high demand for energy to perform its function. Spermatozoa obtain their energy by two main metabolic pathways: glycolysis that occurs in the principal piece of the flagellum and oxidative phosphorylation (OXPHOS) that takes place on mitochondria located at the midpiece of the flagellum [1]. Spermatozoa contain between 50 and 75 mitochondria [2] and as with any other kind of cell that performs aerobic metabolism, is associated with the production of free radicals named reactive oxygen species (ROS) that include the hydroxyl radicals (•OH), superoxide anion (•O_2^-), hydrogen peroxide (H_2O_2), and nitric oxide (NO). These ROS are highly reactive molecules due to the presence of an unpaired electron in their outer shell. In addition, they have a very short half-life in the range of nanoseconds to milliseconds. ROS are produced as a consequence of natural cell machinery and participate in the normal function of a cell. However, when ROS production overcomes cellular antioxidant defenses surpassing a physiological range, they cause deleterious effects due to oxidative stress (OS) that results in oxidation of lipids, proteins, carbohydrates, and nucleotides [3].

Male subfertility and infertility have been associated with OS. Moreover, since infertile men have lower seminal plasma antioxidant capacity in comparison with fertile men, when higher levels of ROS occur, they led to an increase of lipid peroxidation (LPO) [4]. It is well described that when ROS overproduction occurs, it induces sperm DNA damage, although they have the potential to fertilize embryo development and fertility might be disturbed [5,6]. It is unclear how this is related with the fact that nowadays infertility is becoming a worldwide health problem, where one out of six couples are under fertility treatment and thus the use of assisted reproductive technologies (ART) to overcome this problem is growing exponentially. Nevertheless, ART is not harmless and is also associated with an increase of ROS production [7]. Although there is literature focused on the effects of consumption of oral substances with antioxidant properties on sperm parameters, the purpose of this review is to discuss the efficiency of antioxidant intake as a dietary supplement as well as an additive through ART procedures to counteract excessive ROS production that leads to infertility. We will also focus on the molecular mechanisms of action of those compounds with antioxidant activity in the male reproductive system, mainly reviewing literature that relates antioxidant treatment with ART, clinical pregnancy, and live birth as final outcomes.

2. Sources of ROS in Spermatozoa

Several situations result in nonphysiological levels of ROS overwhelming the natural scavenger systems (Figure 1). For example, lifestyle habits, such as alcohol consumption, smoking, exposure to toxicants, or pathologies such as obesity, varicocele, stress, and ageing have been associated with increased production of ROS in seminal plasma [8]. Presence of leucocytes in semen, as well as high percentage of spermatozoa with morphological anomalies [9] or immature spermatozoa with cytoplasmatic droplets containing high amount of enzymes are some examples associated to high ROS levels [9–12].

Currently, human infertility is a global health problem that has led to an exponential grown in the use of ART in the last years to overcome fertility problems. However, ART protocols imply sample centrifugation, light exposure, change of oxygen concentration, pH, or temperature, and the use of culture media with metals content that can produce hydroxyl radicals by Haber–Weiss and Fenton reactions (see explanation below). Hence, optimization of ART protocols has been proposed to minimize artificial ROS production, for instance, by decreasing g-force during sperm selection [13], decreasing spermatozoa incubation time during in vitro fertilization (IVF), which in turn decreases the time where aberrant spermatozoa that produce more ROS are in contact with the oocyte, and by decreasing sperm concentration or atmospheric oxygen concentration during embryo culture under in vitro conditions [7]. In order to reduce human leucocyte contamination on raw semen, paramagnetic bead technology (Dynabead®) can be used. Thus, magnetic beads coated with leukocyte antigen CD45 decrease leukocyte contamination [14,15], doubling the percentage of spermatozoa–oocyte penetration, as shown in a heterologous assay using hamster oocytes [16].

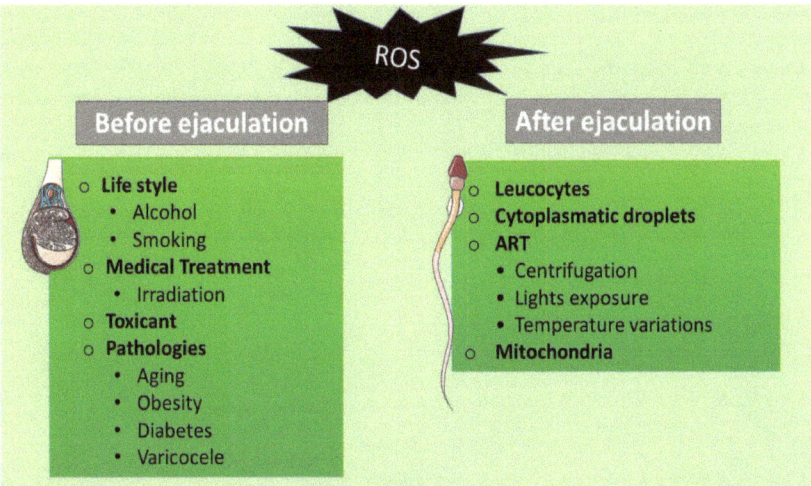

Figure 1. Potential stimuli that cause reactive oxygen species (ROS) production in spermatozoa.

3. Bivalent Role of ROS on Sperm Function

Mammalian spermatozoa are extraordinary cells able to survive in a different body from where they were created. They are very specialized cells having the sole purpose to deliver the paternal genome into the oocyte. However, after ejaculation, spermatozoa must undergo a complex process within the female reproductive tract named capacitation, which allows spermatozoa to fertilize the oocyte [17,18]. Capacitation is a cascade of different cellular events that imply high production and consumption of energy. Although there is controversy on the preponderant metabolic pathways, glycolysis or OXPHOS, used by spermatozoa to generate energy in the form of ATP, it seems that there are sperm species preferences [1]. OXPHOS is the most efficient pathway, obtaining about 30 molecules of ATP by oxidizing one molecule of glucose, while during glycolysis, only two molecules of ATP are obtained per molecule of glucose. It has been described that OXPHOS is the major source of ROS in spermatozoa [19]. Furthermore, ROS might play a bivalent role in sperm function: mild ROS levels boost different intracellular events that culminate on oocyte fertilization, while higher ROS levels induce sperm DNA damage and embryo miscarriage [20,21]. In a comprehensive review, Ford summarized ROS physiological functions on sperm capacitation [22]. It is known that soluble adenylyl cyclase (sAC) is activated by bicarbonate and Ca^{2+}, converting ATP into cAMP, subsequently activating the PKA pathway that mediates the phosphorylation of protein in tyrosine residues, which is used as a hallmark of sperm capacitation [23,24]. It has been proposed that ROS participate in the activation of the cAMP/PKA pathway by increasing cAMP levels, although the mechanism of cAMP production is still not clear in spermatozoa [22]. In adipocytes, it has been proposed that the mechanism of action is through inhibition of phosphodiesterase activity [25]. In human spermatozoa, it was proven that ROS action is mediated by PKA [26]. Thus, the induction of tyrosine phosphorylation was suppressed by a PKA inhibitor (H89) and the responsiveness to progesterone (sperm-oocyte fusion) when spermatozoa were coincubated with NADPH proved it to be a ROS generator [26]. In a different study, capacitated human spermatozoa showed increased levels of cAMP that was mimicked in vitro by exposure of spermatozoa to superoxide anions (O_2^-). Superoxide dismutase (SOD) addition inhibited cAMP levels and the sperm acrosome reaction in a concentration-dependent manner [27]. These results were confirmed by others where superoxide anions increased cAMP concentration and capacitated spermatozoa produced H_2O_2, leading to an increase in protein tyrosine phosphorylation [28]. Nevertheless, when ROS production overcomes antioxidant defenses, detrimental effects on spermatozoa can be summarized as increased LPO and DNA damage and reduction of sperm motility, which are associated with lower sperm fertility (Reviewed by [29]). Thus, ROS homeostasis is pivotal for male reproductive potential as they mediate important functions of sperm, such as capacitation, but when ROS levels surpass these biological levels, they readily oxidize lipids and proteins at membranes and compromise sperm quality and fertilization capacity (Figure 2).

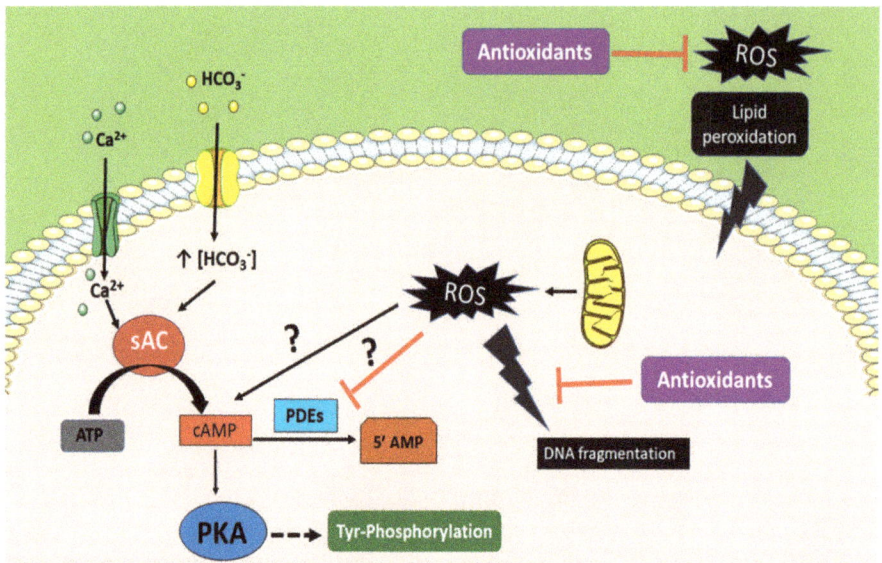

Figure 2. Proposed model of the bivalent role of reactive oxygen species (ROS) in sperm. (i) High levels of ROS concentration induced by different factors such as assisted reproductive technologies (ART), diseases, medical treatment, life style, etc., overwhelming the sperm antioxidant system induce plasma membrane lipid peroxidation and DNA damage. (ii) Physiological ROS level produced mainly by mitochondria induce production of high levels of cAMP by an undetermined mechanism, activating the PKA pathway, and leading to tyrosine phosphorylation, a hallmark of sperm capacitation.

4. Mechanism of ROS Defense in Spermatozoa

Spermatozoa differentiation is achieved during spermiogenesis as they gradually lose their cytoplasm. By the end of the process, the cytoplasm content is very small compared to other cells, where most of the space is occupied by DNA (sperm head). This special feature results in spermatozoa possessing low intracellular antioxidant activity consisting of superoxide dismutase (SOD), nuclear glutathione peroxidase (GPx), peroxiredoxin (PRDX), thioredoxin (TRX), and thioredoxin reductase (TRD) [30]. Therefore, sperm ROS scavenger activity basically depends on the antioxidant content of the seminal plasma, which is formed mainly by a trio of enzymes where SOD converts superoxide anion (O_2^-.) to hydrogen peroxide (H_2O_2), preventing the formation of hydroxyl radical that is an inductor of LPO. However, the H_2O_2 generated is a strong membrane oxidant that is rapidly eliminated either by catalase (CAT) or GPx activities, giving H_2O as a product. Finally, seminal plasma also contains nonenzymatic antioxidant components such as α-tocopherol (vitamin E), ascorbic acid (vitamin C), pyruvate, urate, taurine, and hypotaurine [31].

It should be noted that most ART involves washing steps, meaning that all the natural antioxidant defenses contained in seminal plasma are removed. Likewise, this also happens after natural insemination. During ejaculation, spermatozoa are surrounded by antioxidant molecules coming from seminal plasma but once the ejaculate reaches the vagina, seminal plasma is diluted, leading in both cases to spermatozoa facing ROS. Although spermatozoa possess antioxidant scavenger systems, it seems that they are not strong enough when ROS levels exceed physiological levels, subsequently making spermatozoa highly susceptible to OS.

5. Lipid Peroxidation

The sperm plasma membrane contains a high proportion of polyunsaturated fatty acid (PUFAs) to generate the fluidity needed in order to accomplish the membrane fusion events associated with fertilization. This high PUFAs content makes spermatozoa especially susceptible to suffer LPO [32,33]. The highly reactive hydroxyl radical (OH^-) is an inductor of LPO produced through two consecutive reactions (Figure 3): the first is the Haber–Weiss reaction in which a ferric ion (Fe^{3+}) in the presence of a superoxide radical (O_2^-) is reduced to ferrous ion (Fe^{2+}), followed by Fenton reaction, where Fe^{2+} reacts with hydrogen peroxide (H_2O_2), forming Fe^{3+} and a hydroxyl radical (OH^-).

$$Fe^{3+} + \cdot O_2^- \longrightarrow Fe^{2+} + O_2$$

$$Fe^{2+} + H_2O_2 \longrightarrow Fe^{3+} + OH^- + \cdot OH$$

Total Reaction net:

$$\cdot O_2^- + H_2O_2 \longrightarrow O_2 + OH^- + \cdot OH$$

Figure 3. Haber–Weiss Reaction and Fenton reaction.

Secondary products are formed during LPO: malondialdehyde (MDA), propanol, hexanol, and 4-hydroxynonenal (4-HNE) [34], which are highly reactive and may attack other nearby PUFAs, thus initiating a chain reaction with harmful effects that eventually disrupts membrane fluidity. These secondary products are used as lipid oxidative stress biomarkers.

Nowadays, cryopreservation is becoming an important issue for the success of ART in humans and livestock. Although cryopreservation is routinely used, it is a tough procedure associated with deleterious effects on sperm function due to an increase of ROS production linked to LPO and thus an increase of membrane permeability [35–37]. In this context, the use of antioxidants as additives during cryopreservation/thawing procedure is a common strategy to counteract negative effects of ROS on sperm function.

6. Effects of Oral Antioxidant Intake on Male Reproductive Outcome

Currently, there is a growing trend of oral antioxidant intake to counteract high levels of ROS found in spermatozoa and seminal plasma of subfertile or infertile men. This hypothesis is supported by several works that describe an improvement of sperm parameters after oral antioxidant intake. Among those improvements, sperm concentration, motility, or decrease of DNA damaged are reported (Reviewed by [38]). However, only a few works have shown the effect of antioxidant therapy on fertility outcomes. Here, we discuss the major findings of oral antioxidant intake in reproduction outcome and its endpoints, such as fertility and live birth (summarized in Table 1).

Table 1. Effects of oral antioxidant intake on infertile men's reproductive outcome.

Antioxidant Type and Daily Dose	Period Intervention (months)	ART	Relevant Findings	Participants	Problem	Reference
Astaxantin (16 mg)	3	NI and IUI	↑ Pregnancy rate 54.5% (5/11) vs. 10.5% (2/19) placebo group	30	Infertile	[39]
LC (1 g twice) LAC (0.5 g twice)	3		↓ ROS levels ↑ Pregnancy (11.7%) in patients with abacterial-PVE with normal values of leucocytes It didn't improve pregnancy (0%) in abacterial-PVE patients with high levels of leucocytes	54	PVE	[40]
Nonsteroidal anti-inflammatory + carnitine (Carnitene, 2 g + Nicetile 1 g) Carnitine (Carnitene, 2 g + Nicetile 1 g) Nonsteroidal anti-inflammatory Nonsteroidal anti-inflammatory + carnitine (Carnitene, 2 g + Nicetile 1 g)	2 + 2 4 4 4		23.1% pregnancy 0% pregnancy 6.2% pregnancy 3.8% pregnancy	98	PVE with ↑ levels of leucocytes	[41]
LC (3 g), LAC (3 g), LC (2 g) + LAC (1 g)	6	NI	↑ Total oxyradicals scavenging capacity of seminal fluid ↑ Sperm motility and concentration. Pregnancy rate was not modified	60	Asthenozoospermic	[42]
LC (1 mg), fumarate (725 mg), LAC (500 mg), Fructose (1000 mg), CoQ10 (20 mg), Vitamin C (90 mg), Zinc (10 mg), Folic acid (200 µg), Vitamin B12 (1.5 µg)	6	NI	↑ Achieved pregnancy in treated men 22.2% (10/45) vs. 4.1% (2/49) non treated group	104	Oligo- and/or astheno- and/or teratozoospermia	[43]
LC fumarate (2 g), LAC (1 g) Clomiphene citrate (50 mg) and a complex of vitamins and microelements	3–4	NI	↑ Sperm concentration No modification in pregnancy rates	173	Oligo- and/or asteno- and/or teratozoospermia	[44]

Table 1. Cont.

Antioxidant Type and Daily Dose	Period Intervention (months)	ART	Relevant Findings	Participants	Problem	Reference
LC fumarate (1 g), Acetyl-L-carnitine HCl (0.5 g), Fructose (1 g), Citric acid (50 mg), Vitamin C (90 mg), Zinc (10 mg), Folic acid (200 µg), Selenium (50 µg), Coenzyme Q-10 (20 mg), Vitamin B12 (1.5 µg)	6	NI	↑ Sperm concentration, % of sperm motile or progressive motility as well as sperm with normal morphology. Treated men achieved 29% pregnancy versus 17.9% in the placebo group	90	After performed a varicocelectomy	[45]
Vitamin E (600 mg)	3	IVF	Improvement of zona pellucida binding test. No effect on ROS levels. No alteration on seminal plasma vitamin E levels	30	Infertile	[46]
Vitamin E (300 mg)	3	NI	21% of men had improved sperm motility and achieved pregnancy where 81.8% of pregnancies finished with a live birth	52	Asthenospermic	[47]
Vitamin E (200 mg)	1	IVF	↓ Sperm LPO. ↑ Fertility rate: 19.3 ± 23.3 pre-treatment versus 29.1 ± 22.2 post-treatment	15	Normospermic infertile	[48]
Vitamin E (1 g), Vitamin D (1 g)	2	ICSI	76.3% respond to the treatment with ↓DNA damage. ↑ Pregnancy rate (6.9 vs. 49.3%). ↑ Implantation rate (2.2 vs. 19.2%). Equal embryo quality	38	Infertile men non responding to ICSI	[49]
Vitamin E (400 IU) Selenium (200 µg)	3.5	NI	10.8% pregnancy	690	Infertile	[50]
Vitamin E (400 IU), Vitamin C (100 mg), Lycopene (6 mg), Zinc (25 mg), Selenium (26 µg), Folate (0.5 mg), Garlic (1000 mg)	3	IVF-ICSI	Doubled pregnancy rate (63.9 vs. 37.5%). Doubled implantation rate (46.2 vs. 24%). Doubled viable pregnancy rate (38.5 vs. 16%)	60	Infertile men with ↑ levels of DNA fragmentation and poor motility and membrane integrity	[51]
Zinc sulphate (220 mg)	4	NI	21.4% (3/14) of patients achieved pregnancy. Zinc levels were increased in seminal plasma	14	Human	[52]
Zinc sulphate (500 mg)	3	NI	Improved pregnancy (22.5%) vs. placebo (4.3%). Zinc levels were not modified on seminal plasma	100	Asthenozoospermic	[53]

NI: natural insemination, IVF: in vitro fertilization, ICSI: intracytoplasmic sperm injection, IU: international unit, PVE: prostate-vesiculo-epididymitis, LC: L-carnitine, LAC: L-acetyl-carnitine, LPO: lipid peroxidation, ↑ increase, ↓ decrease.

6.1. Carnitines

Carnitines are synthetized by the organism and found in seminal plasma at higher concentration than in spermatozoa. The L-carnitine (LC) isomer is the bioactive form [54] with a pivotal role in mitochondrial β-oxidation, acting as a shuttle of the activated long-chain fatty acids into the mitochondria [55] where L-acetyl-carnitine (LAC) is an acyl derivative of LC. Long-chain fatty acids provide energy to mature spermatozoa (with positive effects on sperm motility) and during maturation and the spermatogenic process [56]. Oral intake of LC (1 g twice/day) and LAC (0.5 g twice/day) for three months reduced ROS levels in spermatozoa and improved pregnancy (11.7%) in patients with abacterial prostate-vesiculo-epididymitis (PVE) with normal values of leucocytes, but it did not improve pregnancy at all (0%) in those PVE patients with high levels of leucocytes [40]. A year later, the same group tested patients diagnosed with abacterial PVE concomitant with high levels of leucocytes and showed that pretreatment for two months with a nonsteroidal anti-inflammatory followed by two months of carnitine oral intake achieved 23.1% pregnancy in comparison with the four-month carnitine intake group (0%), nonsteroidal anti-inflammatory group (6.2%), and the group receiving four-month nonsteroidal anti-inflammatory compounds and carnitines (3.8%) [41]. In another study, the effect of daily intake of LC (3 g), LAC (3 g), or a combination of LC (2 g) and LAC (1 g) was discriminated over six months and results were followed up 9 months after intervention in idiopathic asthenozoospermic men ($n = 60$) [42]. Treated men improved their total oxyradicals scavenging capacity of seminal fluid [42]. Overall, LAC or the combination of LAC + LC treatment had better improvement of sperm motility and concentration. Nevertheless, those patients with lower basal values of sperm motility had higher probability to respond to the treatment but pregnancy rate was not improved by any treatment in comparison with placebo control group [42]. Recently, coadministration of LC fumarate (2 g), LAC (1 g), and clomiphene citrate (50 mg) concurrently with vitamins and minerals in patients with idiopathic oligo- and/or asteno- and/or teratozoospermia ($n = 173$) enhanced sperm concentration specially in those patients with multiple impairment semen parameters (oligoasthenoteratozoospermic patients), but did not improve the morphology, progressive sperm motility neither pregnancy rates in comparison with control group [44]. A meta-analysis concerning carnitine used as an oral antioxidant therapy concluded that this molecule might be effective for improving pregnancy rates regarding the limits of patient inclusion criteria and the lower number of men evaluated in each study [57].

6.2. Vitamins

The interest of vitamin E and its use as antioxidant is due to its protective activity against ROS which subsequently decreases LPO, and therefore exerts positive effects on sperm functions, such as sperm concentration and motility [58]. However, its effects in fertility are less clear. For example, in a small clinical trial ($n = 30$), oral administration of vitamin E (300 mg twice daily) for three months raised the levels of vitamin E in blood serum, although human seminal plasma levels were not modified, questioning its possible effects on reproductive parameters [50]. Nevertheless, in this clinical trial, vitamin E treatment achieved an improvement of the zona pellucida binding test without any other improvement described, including ROS level [50]. Similarly, 15 normospermic infertile men after one month of daily consumption of 200 mg of vitamin E improved their fertilization rate (19.3 ± 23.3 pretreatment versus 29.1 ± 22.2 post-treatment) after IVF. Those results were associated with lower sperm LPO levels in comparison with preintervention values [52]. In another work, oral administration of vitamin E (100 mg thrice daily) to patients with asthenospermia ($n = 52$) established three different groups of men according to the results: (i) men without improvement of their sperm motility (40%); (ii) men with improved sperm motility but did not achieve pregnancy (39%); (iii) men with improved motility and achieved pregnancy (21%), of which 81.8% of pregnancies finished in live birth. The placebo control group did not achieve any pregnancies [47]. Later, daily intake of a combination of vitamin E and C (1 mg of each component) for two months in patients where intracytoplasmic sperm injection (ICSI) had previously failed was studied ($n = 38$). The results showed two different populations: (i) those where the antioxidant treatment decreased the percentage of sperm

DNA damage (n = 29) and (ii) those where the treatment did not affect this parameter (n = 9) [49]. The most interesting result was observed in the responsive group that after ICSI, the pregnancy rate (6.9 vs. 49.3%) and implantation rate (2.2 vs. 19.2%) were improved compared with the pretreatment group, although no differences were found in embryo quality [49]. In a nonplacebo-controlled and nondouble-blind design trial, daily intake of a combination of selenium (200 µg) and vitamin E (400 UI) followed for 3.5 months by infertile men (n = 690) achieved 10.8% spontaneous pregnancy [50].

Several studies have been performed looking for beneficial effects from a combination compounds with antioxidant activity. For example, a formulation using a mix of several compounds with antioxidant activity (vitamin C, vitamin E, carnitine, folic acid, lycopene, selenium, and zinc) was evaluated using a mouse Gpx5 knock-out (KO) subjected to a second stress: scrotal heat (KO + SH) (42 °C for 30 min) [58]. Although the exact ingestion quantity of this antioxidant combination could not be determined, their effects include the reversion of sperm DNA oxidation induced in KO + SH animals and protection of seminiferous tubules. The results showed that animals supplemented with KO + SH versus the nonsupplemented animals had double the fertilization rate (73.7 vs. 35.2%) and fetus reabsorption was halved (8.9 vs. 17.8%) [58]. In another trial, infertile human patients with oligo- and/ or astheno- and/or teratozoospermia with or without varicocele (n = 104) using a combination of antioxidants (vitamin C 90 mg, vitamin B12 1.5 µg, LC 1mg, fumarate 725 mg, LAC 500 mg, fructose 1000 mg, CoQ_{10} 20 mg, zinc 10 mg, and folic acid 200 µg) were studied for six months. The results showed that the individuals from the treated group, regardless of whether they suffered from varicocele or not, presented improved sperm concentration total sperm motility [43]. Moreover, after treatment, 22.2% (10/45) of supplemented patients achieved pregnancy, while in the control group, only 4.1% (2/49) of the couples were pregnant [43]. A close analysis of the men from the supplemented group revealed that only 4.8% (1/21) of patients suffering varicocele improved after treatment, while the nonvaricocele group achieved 37.5% (9/24) pregnancy [43]. A different group studied the effect of a commercial multiantioxidant supplement (vitamin E 400 IU, vitamin C 100 mg, lycopene 6 mg, zinc 25 mg, selenium 26 µg, folate 0.5 mg, garlic 1000 mg) for three months on 60 men with high levels of DNA fragmentation and poor sperm motility and membrane integrity [51]. The treatment achieved doubled pregnancy rate (63.9 vs. 37.5%), implantation rate (46.2 vs. 24%), and viable pregnancy rate (38.5 vs. 16%) versus the placebo group without any modification of any sperm parameters, fertilization, or embryo quality rates [51]. However, this work was later criticized because of the experimental design, particularly the low number of individuals in the trial, unequal distribution of individuals between the placebo (n = 16) and treatment groups (n = 36) and the suitability of the statistical analysis used [59].

Contradictory results were found when men were supplemented with different oral antioxidants after varicocelectomy. Oral intake of vitamin E (300 mg twice/day) for 12 months (n = 40) improved the sperm parameters of sperm concentration and the percentage of motile spermatozoa, although these data were not significant compared with control [60]. Recently, a multiple antioxidant combo was tested (L-carnitine fumarate 1 g, acetyl-L-carnitine HCl 0.5 g, fructose 1 g, citric acid 50 mg, vitamin C 90 mg, zinc 10 mg, folic acid 200 µg, selenium 50 µg, coenzyme Q-10 20 mg, and vitamin B12 1.5 µg) after varicocelectomy (n = 90) for six months [45]. Surgery improved the following sperm parameters: sperm concentration, percentage of motile spermatozoa or progressive motility, and spermatozoa with normal morphology. Moreover, treated men achieved 29% pregnancy versus 17.9% in the placebo group [45].

6.3. Zinc

Zinc is a metalloprotein cofactor for DNA transcription and protein synthesis. Moreover, zinc is necessary for the maintenance of spermatogenesis and optimal function of the testis, prostate, and epididymis [61], in addition to their antioxidant properties preventing LPO [62]. A trial using zinc sulphate as an antioxidant therapy administrated orally (250 mg twice daily) for three months reported an improvement in the reproductive outcome of asthenozoospermic men (n = 100), particularly in the sperm parameters of concentration, motility, and sperm membrane integrity (hypoosmotic swelling test). It was also noticed a decrease of antisperm antibodies on seminal plasma without modification of zinc levels on seminal plasma [53]. Pregnancies were also improved in couples where men underwent treatment when compared with placebo, 22.5% (11/49) versus 4.3% (2/48), respectively [53]. In another trial with only 14 patients and no control group, sperm parameters were improved after zinc treatment (220 mg daily for four months) and 21.4% (3/14) of patients achieved pregnancy and increase zinc levels on seminal plasma [52]. Although beneficial evidence has been found on reproductive outcome after zinc intake, the lower number of studies and subjects under treatment without a proper control does not allow further discussion of the possible positive effects of zinc intake on reproduction outcome.

6.4. Natural Compounds—Traditional Medicine

Natural compounds have been used traditionally to treat diseases. For instance, beneficial effects on reproductive outcome have been reported using products derived from tea (*Camelia sinensis* (L.)), which is the second most consumed beverage after water [63]. For example, an in vitro experiment using green tea extract or epigallocatechin-3-gallate (EGCG) added to human spermatozoa media improved sperm capacitation hallmarks, such as tyrosine phosphorylation and cholesterol efflux, through the estrogen receptor pathway [64]. EGCG has been shown to have beneficial effects when extreme stresses are applied to male mice [65,66]. Interestingly, adverse effects induced by artificial testicular hyperthermia were ameliorated by oral administration of green tea extract [65]. Positive effects were visible after 28 days of heat stress induction, improving sperm concentration, percentage of motile and progressive spermatozoa, and sperm membrane integrity [65]. Another example of the beneficial effects of EGCG were described when intraperitoneal administration (50 mg/kg) protected against testicular injury induced by ionizing radiation in rats [66]. Thus, treated animals restored testicular function with an improvement in the number of pups by littler reducing LPO (TBARs) and protein carbonyl levels [66]. EGCG's mechanism of action is via the mitogen-activated protein kinase/BCL2 family/caspase 3 pathway [66]. In another work, the combination of two different tea extracts, white and green, where evaluated as additives to improve ART sperm of rats stored at room temperature. The authors found doubled levels of epigallocatechin (EGC) and EGCG in white tea in comparison with green tea [67], highlighting the variability associated with the type of tea extract used. Moreover, although both extracts had positive effects, the white tea extract had better ferric reducing antioxidant power than the green tea extract and the control. The beneficial effects were proportional to the concentration used, with 1 mg/mL of white tea extract being the best concentration tested for improving sperm survival and decreasing LPO over 72 hours of storage at room temperature [67]. Encouraged by the antioxidant effects on sperm parameters of white tea, the same group explored the oral administration potential of the extract to improve prediabetic type II (PreDM) male reproduction features known to be decreased due to oxidative stress [68]. PreDM is characterized by mild hyperglycemia, glucose intolerance, and insulin resistance and has been related with infertility or subfertility problems in males [69]. Consequently, using rat as an animal model, drinking white tea counteracted the negative effects of PreDM on the male reproductive tract. For example, white tea consumption improved testicular antioxidant power and decreased lipid peroxidation and protein oxidation [68]. Ingestion of white tea also restored sperm motility and restored sperm showing morpho-anomalies to normal levels [68].

7. Antioxidants as a Tool to Improve Male ART Outcomes

Human infertility already affects one of six couples worldwide [70] and male factors contribute to 20–50% of infertility [71]. Infertile men tend to have higher ROS levels than fertile men. To counteract fertility problems, different ART have been developed, mainly IVF and ICSI. In both cases, gametes are extracted from the body and incubated in in vitro conditions and, after a while, an embryo is transferred into the uterus. It should be noted that due to legislation and ethical issues, it is easier to perform experiments in animal models than in humans to test antioxidant effects on different ART. The interest in the use of antioxidants to improve sperm parameters is not new. As early as 1943, in a study focused on sperm metabolism and oxygen consumption, MacLeod showed that sperm produce hydrogen peroxide, which has a deleterious effect on sperm motility, and it can be counterbalanced by addition of catalase to the media [72]. Later, some authors followed the same rationale and tried to adapt MacLeod's hypothesis to different ART, such as cryopreservation, IVF, and ICSI.

Sperm conservation for long periods of time in liquid nitrogen (cryopreservation) is designed to keep sperm viable. From a practical point of view, cryopreservation is a tool to enable male fertility before, for example, chemotherapy, radiotherapy, vasectomy, or exposure to toxicants, or just to have time to screen donors for infectious agents, such as the human immunodeficiency or hepatitis B viruses [73]. On the other hand, from the animal industry point of view, the use of cryopreservation aims to maximize the number of services (inseminations) that can be performed from a simple ejaculation, ensuring the quality of genetical material preserved, or allowing the transportation of this genetical material to distant places. Cryopreservation is also of special interest to preserve endangered species. However, cryopreservation is not a harmless technique, inducing DNA and LPO damage and other adverse effects [74]. Moreover, cryopreservation, like ART, involves centrifugation, which is associated with production of ROS [13] and removal of seminal plasma which contains the main sperm antioxidant scavenger systems.

Antioxidant supplementation to cryopreservation media has been proposed as a way to overcome ROS production and OS status in spermatozoa (summarized in Table 2). For example, supplementation with a synthetic phenolic antioxidant, butylated hydroxytoluene (BHT), during boar sperm cryopreservation improved post-thawing sperm survival, decreased MDA levels at the concentration of 0.4 mM BHT, and embryo development was improved (28.8% vs. 15.8%) without modification of embryo cleave percentage in comparison to the control [75]. Later, it was described that 1 mM BHT improved antioxidant sperm activity, pregnancy rate (86.7 vs. 63.6%), the number of gilts farrowing (86.7 vs. 45.4%), and the number of piglets born (10.8 ± 1.6 vs. 8.2 ± 2.2) after performing intrauterine artificial insemination (IUI) using cryopreserved sperm versus control [76]. Subsequently, in a multitest in which four different compounds with antioxidant activity (BHT 2 mM, ascorbic acid 8.5 mg/mL, hypotaurine 10 mM, and cysteine 5 mM) were added during goat sperm cryopreservation, LPO was decreased but only ascorbic acid and BHT significantly improved fertility in comparison with control after performing artificial insemination (AI) [77].

Table 2. Antioxidants used as additives in different ART and their reproduction outcomes.

Antioxidant Type and Dose	Administration	Procedure	Principal Results Found	Stress	Specie	Reference
BHT 0.4 mM	In vitro	IVF	↑Sperm survival ↓ Sperm MDA levels at the concentration ↑ Embryo develop 28.8% treated vs. 15.8% control	Cryopreservation	Boar	[75]
BHT 1 mM BHT	In vitro	IUI	↑ Pregnancy rate (86.7 vs. 63.6%), ↑ n° of gilts farrowing (86.7 vs. 45.4%) ↑ n° of piglets born (10.8 ± 1.6 vs. 8.2 ± 2.2)	Cryopreservation	Boar	[76]
BHT (2 mM), Ascorbic acid (8.5 mg/mL), Cysteine (5 mM), Hypotaurine (10 mM)	In vitro	AI	↓Sperm LPO ↑Fertility: ascorbic acid (42.85%), BHT (35.71%), control (26.38%)	Cryopreservation	Goat	[77]
Caffeine (1.15 mM), β-mercaptoethanol (50 μM)	In vitro	AI	No effect on pregnancy rate ↑ Litter size in treated samples (10.0 ±1.0) vs. control (5.7 ± 1.5)	Cryopreservation	Boar	[78]
CAT (200 IU/mL)	In vitro		No differences on sperm parameters ↓ 2 pronucleus zygote (25.5% control vs. 13.2% treated)↓ Cleaved embryos: 7.6% treated vs. 16.7% control	Cryopreservation	Ram	[79]
Carnitine, Folic acid, Lycopene, Selenium, Vitamin C, Vitamin E, Zinc	Oral	NI	Duplicate fertilization rate (73.7 vs. 35.2%) Halved fetus reabsorption (9 vs. 18%)	Gpx5 knockout (KO) + Scrotal heat stress (KO + HS)	Mouse	[58]
Cysteine (2 mM)	In vitro	IUI	↑ SOD and CAT levels and = MDA levels ↑Sperm total motility ↓ acrosome abnormalities Slight tendency to improve (p >0.05) non-return rate 74.54 (41/55) in comparison to control 57.14 (28/49)	Cryopreservation	Bull	[80]
Cysteine (10 mM), Rosemary extract (Rosmarinus officinalis). or a combination of both	In vitro	IVF	↑% sperm motility and progressive motility ↓ Acrosome membrane damaged Rosemary yielded better cleave% without affects blastocysts	Cryopreservation	Boar	[81]

Table 2. Cont.

Antioxidant Type and Dose	Administration	Procedure	Principal Results Found	Stress	Specie	Reference
Cysteine (5 mM) Trehalose (25 mM)	In vitro	IUI	No improvement of antioxidants features No differences on non-return rate was found after IUI	Cryopreservation	Bull	[82]
Cysteamine (5 µM), Lycopene (500 µg/mL)	In vitro	IUI	No differences on non-returned rate	Cryopreservation	Bull	[83]
EGCG (50 mg/kg)	Intraperitoneal		Restore testicular function ↓ LPO and protein carbonyl levels ↑ Number of pups by littler	Ionizing radiation	Rat	[66]
GSH (0.5 and 1.0 mM) GSH 0.5 mM + SOD 100 U/mL	In vitro	IUI	Equal nonreturn rates	Cryopreservation	Bull	[84]
Melatonin (1 mM)	In vitro	IVF	↑ Sperm viability rates ↑% of total motile and progressive motile spermatozoa ↑ DNA integrity Faster first embryonic division	Cryopreservation	Ram	[85]
Metformin (50 to 5000 µM)	In vitro	IVF	Duplicate fertilization rate and embryo development	Cryopreservation	Mouse	[86]
NAC (1–10 mM)	In vitro	ICSI	Decrease ROS ICSI outcome wasn't modified	Thawing + H_2O_2	Bull	[87]
NAC (10 µM), LAC (10 µM), α-Lipoic Acid (5 µM)	In vitro	IVF	↓ Embryo intracellular levels of H_2O_2 Accelerated embryo development and blastocysts ↑ TE and ICM cell numbers	Incubation under 20% O_2	Mouse	[88]
Taurine (2 mM)	In vitro	IUI	↓ GSH and SOD levels but ↑ five-fold CAT levels ↑ MDA levels = nonreturn rates	Cryopreservation	Bull	[80]
Zinc chloride (10 µg/mL), D-aspartic acid (500 µg/mL) Coenzyme Q10 (40 µg/mL)	In vitro	IVF	↑% of total spermatozoa motile and progressive motility ↓ Sperm and blastomeres DNA fragmentation ↑ 8-cells blastocyst: 51.4% treatment vs. 37.1% control	Cryopreservation	Bull	[89]

IVF: in vitro fertilization; AI: Artificial Insemination; IU: international unit; NI: natural insemination; IUI: intrauterine insemination; TE: trophectoderm; ICM: inner cell mass; BHT: butylated hydroxytoluene; CAT: catalase; GSH: reduced glutathione; NAC: N-acetyl-L-cysteine; LAC: L-acetyl-carnitine; EGCG: epigallocatechin-3-gallate, ↑ increase, ↓ decrease.

The importance and the use of the amino acid cysteine in the fight against ROS impacts on the cell is due to the fact it is a limiting substrate for glutathione synthesis [90]. Cysteine (2 mM) and taurine (2 mM) (a cysteine derived) antioxidant properties were controversial when they were used during the cryopreservation procedure of bull spermatozoa [80]. Taurine decreased GSH and SOD levels, while CAT levels were five times higher than control, but MDA levels were also higher. However, cysteine increased SOD and CAT levels without an effect on MDA levels [80]. The nonreturn rate was not modified when IUI were performed by neither of the compounds; however, a nonsignificant ($p > 0.05$) tendency of improvement was observed in cysteine-treated straws 74.54% (41/55) in comparison to control 57.14% (28/49) [80]. Similar results were obtained when a higher concentration of cysteine (5 mM) and trehalose (25 mM) were added again to bull cryopreservation media. Thus, the antioxidant features of these compounds were not proved; neither MDA nor GPx levels were enhanced [82]. Furthermore, no improvement on the nonreturn rate was found after IUI [82]. Similarly, using cysteamine (5 µM), a decarboxylated derivative of cysteine and lycopene (500 µg/mL) during bull sperm cryopreservation, no differences were found in the nonreturn rate [83]. In other study, the authors used N-acetyl-L-cysteine (NAC), an acetylated cysteine residue which has been shown to effectively reduce ROS formation when H_2O_2 stress were used in thawed bull spermatozoa [87]. However, neither sperm DNA, nor the number of blastocysts were not improved after performing ICSI using spermatozoa cryopreserved in the presence of NAC [87]. Nevertheless, in an IVF study on mice using fresh spermatozoa, where gametes and embryos were stressed by incubation under 20% oxygen atmosphere (over physiological levels on oviduct and uterine from 2–8% [91]), a combination of substances with antioxidant activity were tested (LAC 10 µM, NAC 10 µM, α-Lipoic Acid 5 µM) in either IVF media, embryo culture media, or both. Treated samples had lower intracellular levels of H_2O_2, accelerated embryo development, and significantly increased trophectoderm (TE) cell numbers, inner cell mass (ICM), and total cell numbers [88]. All these effects were exacerbated when the antioxidant combo were added during the whole process [88].

Positive effects were also described when thawed bull spermatozoa were supplemented with an antioxidant combination (zinc chloride 10 µg/mL, D-aspartic acid 500 µg/mL, and coenzyme-Q10 40 µg/mL), obtaining a better percentage of total sperm motile and progressive motility and a decrease of DNA fragmentation through sperm incubation [89]. Moreover, antioxidant supplementation improved embryo development. Although no differences were found in the cleave percentage, the number of blastocysts that reached the eight-cell stage was 37.1% in the control versus 51.7% in the treated group [89].

Following the rationale of MacLeod [72], adding antioxidant enzymes to counteract the adverse effects of ROS on spermatozoa was used to improve sperm cryopreservation. Enzymes with antioxidant properties were added to bull cryopreservation media—0.5 and 1.0 mM of reduced glutathione (GSH) or a combination of 0.5 mM of GSH and 100 U/mL of SOD— but did not modify the nonreturn rates [84]. In another study, the use of CAT (200 IU/mL) was used to cryopreserve ram (*Capra pyrenaica*) epidydimal spermatozoa obtained postmortem [79]. At this concentration, no differences were found in sperm parameters but negative effects were described on fertility: fewer pronucleus zygotes (25.5% control vs. 13.2% treated) and cleaved embryos were obtained from treated samples after IVF (16.7% control vs. 7.6% treated) [79].

Natural compounds with antioxidant activity have also been tested in ART. Metformin, a biguanide isolated from *Galea officialis* used worldwide as a treatment for diabetes type II [92], was recently added to the cryopreservation sperm media of chicken due to its antioxidant properties, among other properties [93]. Cryopreserved mouse spermatozoa treated with metformin displayed better motility, sperm viability, doubled fertilization rate and embryo development, and halved DNA fragmentation rate [86]. These promising results of supplementation of cryopreservation media with metformin appeared to be related to the activation of 5'AMP-activated protein kinase (AMPK). However, recently, negative results have been described when metformin (1 and 10 mM) was used to improve boar sperm preservation at 17 °C, decreasing sperm motility and mitochondria potential [94]. In an in vitro study performed in human spermatozoa kept at physiological temperature, metformin (10 mM) induced a reduction of sperm motility, where the mechanism of action was associated with PKA pathway inhibition [95]. Boar spermatozoa were coincubated during cryopreservation with rosemary extract (*Rosmarinus officinalis*) or cysteine (10 mM) or a combination of both [81]. Although both compounds enhance some sperm properties, the most noticeable effects were found by rosemary compound, enhancing total sperm motility, progressive motility, and preventing acrosome membrane damage three hours post-thawing in comparison to control [81]. Rosemary-treated spermatozoa yielded better cleave percentages without affecting blastocyst formation rate after performing IVF [81].

Melatonin (MLT) is a hormone endogenously synthesized mainly by the pineal gland. It has been detected in human seminal fluid [96] and melatonin receptors have been described in sperm of several species [97]. MLT's antioxidant property was tested in cryopreserved human spermatozoa [98]. MLT increased the expression of the antioxidant-related gene Nrf2 as well as its downstream genes SOD2, CAT, HO-1, and GSTM1, leading to lower ROS levels and LPO [98]. On the other hand, MLT (1 µM) used during boar semen preservation at 17 °C only showed a modest membrane protective effect [99]. By contrast, cryopreserved ram sperm supplemented with MLT achieved higher viability rates, higher percentages of total motile and progressive motile spermatozoa, and higher DNA integrity [85]. However, after IVF, only faster first embryonic division without any other embryo output difference was observed in those samples supplemented with MLT [85].

Yamaguchi, et al. [100] showed that thawed boar spermatozoa supplemented with caffeine improved fertility [100]. Later, the same authors tested a combination of caffeine (1.15 mM) with the antioxidant compound β-mercaptoethanol (50 µM) but pregnancy rate was not modified (20 vs. 21% control and treatment respectively) after AI. However, litter size (10.0 ±1.0) almost doubled the data from control samples (5.7 ± 1.5) ($p < 0.07$) [78].

8. Antioxidants as a Therapy to Improve Reproduction Outcome

Sperm produce ROS as consequence of high aerobic metabolism. ROS production at nonphysiological levels overwhelm cellular scavenger systems and result in deleterious effects, such as lipid and protein peroxidation and DNA damage. Infertile men are known to possess pathological ROS levels, leading to sperm DNA fragmentation and lower ART outcome [29]. Thus, to deal with ROS overproduction and their deleterious effects at cellular levels in the male reproductive system, different strategies have been tested: (i) antioxidant oral consumption and (ii) antioxidants used as additives to media during ART.

Literature concerning the use of compounds with antioxidant activity and the improvement of sperm function is extensive. Nevertheless, others have found negative results [101,102], questioning the beneficial impact of antioxidant prescription and arguing that there is not clear evidence supporting prescription of antioxidants [103] or even that the over exposure to antioxidants can lead to other pathologies [104]. Others have found that administration of high doses of antioxidants have harmful effects on health [105,106]. Most trials have the handicap of using a lower number of men or are not double-blind or placebo-controlled. Moreover, the heterogeneity of the treatments and concentrations used as well as the experimental design make it hard to establish solid conclusions. Studies with greater numbers of patients should be performed, including large control groups to address the effects of oral antioxidant consumption on reproductive outcome. Moreover, arbitrary formulations of antioxidants should be avoided and classical pharmacological concentration-dependent experiments should be performed in order to find effective concentrations of antioxidants. Rather than by oral consumption, better reproductive outcome results are described when antioxidants were implemented in ART, especially during cryopreservation-thawing procedures. Antioxidant supplementation decreased LPO and improved reproductive outcome. Antioxidant concentration should be adapted to each form of ART. The future of antioxidant therapy to improve ART involves the development of nonintrusive technologies that can discern between sperm with or without lipid peroxidation or DNA damage, allowing physicians to inject healthy sperm into the oocyte by ICSI.

Author Contributions: Conceptualization, D.M.-H.; M.G.A. and P.F.O.; Writing-Original Draft Preparation, D.M.-H and M.G.A. Writing-Review & Editing, D.M.-H.; M.J.B.; A.R.B.; M.G.A. and P.F.O. Supervision, M.G.A. Funding Acquisition, M.G.A and P.F.O.

Funding: This work was supported by the Portuguese "Fundação para a Ciência e a Tecnologia"—FCT: M.G. Alves (IFCT2015, PTDC/BIM-MET/4712/2014 and PTDC/MEC-AND/28691/2017); P.F. Oliveira (IFCT2015 and PTDC/BBB-BQB/1368/2014); UMIB (Pest-OE/SAU/UI0215/2014) and co-funded by FEDER via Programa Operacional Fatores de Competitividade-COMPETE/QREN & FSE and POPH funds.

Acknowledgments: David Martin Hidalgo is recipient of a post-doctoral fellowship from Junta de Extremadura and Fondo Social Europeo (PO17020).

Conflicts of Interest: The authors declare no conflict of interest.

References

1. Du Plessis, S.S.; Agarwal, A.; Mohanty, G.; van der Linde, M. Oxidative phosphorylation versus glycolysis: What fuel do spermatozoa use? *Asian J. Androl.* **2015**, *17*, 230–235. [CrossRef] [PubMed]
2. Ankel-Simons, F.; Cummins, J.M. Misconceptions about mitochondria and mammalian fertilization: Implications for theories on human evolution. *Proc. Natl. Acad. Sci. USA* **1996**, *93*, 13859–13863. [CrossRef] [PubMed]
3. Birben, E.; Sahiner, U.M.; Sackesen, C.; Erzurum, S.; Kalayci, O. Oxidative stress and antioxidant defense. *World Allergy Organ. J.* **2012**, *5*, 9–19. [CrossRef] [PubMed]
4. Subramanian, V.; Ravichandran, A.; Thiagarajan, N.; Govindarajan, M.; Dhandayuthapani, S.; Suresh, S. Seminal reactive oxygen species and total antioxidant capacity: Correlations with sperm parameters and impact on male infertility. *Clin. Exp. Reprod. Med.* **2018**, *45*, 88–93. [CrossRef] [PubMed]
5. Hammadeh, M.E.; Al Hasani, S.; Rosenbaum, P.; Schmidt, W.; Hammadeh, C.F. Reactive oxygen species, total antioxidant concentration of seminal plasma and their effect on sperm parameters and outcome of IVF/ICSI patients. *Arch. Gynecol. Obstet.* **2008**, *277*, 515–526. [CrossRef]
6. Jurisicova, A.; Varmuza, S.; Casper, R.F. Programmed cell death and human embryo fragmentation. *Mol. Hum. Reprod.* **1996**, *2*, 93–98. [CrossRef]
7. Agarwal, A.; Said, T.M.; Bedaiwy, M.A.; Banerjee, J.; Alvarez, J.G. Oxidative stress in an assisted reproductive techniques setting. *Fertil. Steril.* **2006**, *86*, 503–512. [CrossRef]
8. Kumar, N.; Singh, A.K. Reactive oxygen species in seminal plasma as a cause of male infertility. *J. Gynecol. Obstet. Hum. Reprod.* **2018**, *47*, 565–572. [CrossRef]
9. Aziz, N.; Saleh, R.A.; Sharma, R.K.; Lewis-Jones, I.; Esfandiari, N.; Thomas, A.J., Jr.; Agarwal, A. Novel association between sperm reactive oxygen species production, sperm morphological defects, and the sperm deformity index. *Fertil. Steril.* **2004**, *81*, 349–354. [CrossRef]
10. Cooper, T.G. The epididymis, cytoplasmic droplets and male fertility. *Asian J. Androl.* **2011**, *13*, 130–138. [CrossRef]
11. Gomez, E.; Buckingham, D.W.; Brindle, J.; Lanzafame, F.; Irvine, D.S.; Aitken, R.J. Development of an image analysis system to monitor the retention of residual cytoplasm by human spermatozoa: Correlation with biochemical markers of the cytoplasmic space, oxidative stress, and sperm function. *J. Androl.* **1996**, *17*, 276–287.
12. Huszar, G.; Vigue, L. Correlation between the rate of lipid peroxidation and cellular maturity as measured by creatine kinase activity in human spermatozoa. *J. Androl.* **1994**, *15*, 71–77.
13. Shekarriz, M.; DeWire, D.M.; Thomas, A.J., Jr.; Agarwal, A. A method of human semen centrifugation to minimize the iatrogenic sperm injuries caused by reactive oxygen species. *Eur. Urol.* **1995**, *28*, 31–35. [CrossRef]
14. Whittington, K. Relative contribution of leukocytes and of spermatozoa to reactive oxygen species production in human sperm suspensions. *Int. J. Androl.* **1999**, *22*, 229–235. [CrossRef]
15. Ford, W.C.; Whittington, K.; Williams, A.C. Reactive oxygen species in human sperm suspensions: Production by leukocytes and the generation of NADPH to protect sperm against their effects. *Int. J. Androl.* **1997**, *20* (Suppl. 3), 44–49.
16. Aitken, R.J.; Buckingham, D.W.; West, K.; Brindle, J. On the use of paramagnetic beads and ferrofluids to assess and eliminate the leukocytic contribution to oxygen radical generation by human sperm suspensions. *Am. J. Reprod. Immunol.* **1996**, *35*, 541–551. [CrossRef]

17. Austin, C.R. Observations on the penetration of the sperm in the mammalian egg. *Aust. J. Sci. Res. B* **1951**, *4*, 581–596. [CrossRef]
18. Chang, M.C. Fertilizing capacity of spermatozoa deposited into the fallopian tubes. *Nature* **1951**, *168*, 697–698. [CrossRef]
19. Koppers, A.J.; De Iuliis, G.N.; Finnie, J.M.; McLaughlin, E.A.; Aitken, R.J. Significance of mitochondrial reactive oxygen species in the generation of oxidative stress in spermatozoa. *J. Clin. Endocrinol. Metab.* **2008**, *93*, 3199–3207. [CrossRef]
20. Carrell, D.T.; Liu, L.; Peterson, C.M.; Jones, K.P.; Hatasaka, H.H.; Erickson, L.; Campbell, B. Sperm DNA fragmentation is increased in couples with unexplained recurrent pregnancy loss. *Arch. Androl.* **2003**, *49*, 49–55. [CrossRef]
21. Lewis, S.E.; Aitken, R.J. DNA damage to spermatozoa has impacts on fertilization and pregnancy. *Cell Tissue Res.* **2005**, *322*, 33–41. [CrossRef]
22. Ford, W.C. Regulation of sperm function by reactive oxygen species. *Hum. Reprod. Update* **2004**, *10*, 387–399. [CrossRef]
23. Visconti, P.E.; Bailey, J.L.; Moore, G.D.; Pan, D.; Olds-Clarke, P.; Kopf, G.S. Capacitation of mouse spermatozoa. I. Correlation between the capacitation state and protein tyrosine phosphorylation. *Development (Camb. Engl.)* **1995**, *121*, 1129–1137.
24. Visconti, P.E.; Moore, G.D.; Bailey, J.L.; Leclerc, P.; Connors, S.A.; Pan, D.; Olds-Clarke, P.; Kopf, G.S. Capacitation of mouse spermatozoa. II. Protein tyrosine phosphorylation and capacitation are regulated by a cAMP-dependent pathway. *Development (Camb. Engl.)* **1995**, *121*, 1139–1150.
25. Raimondi, L.; Banchelli, G.; Sgromo, L.; Pirisino, R.; Ner, M.; Parini, A.; Cambon, C. Hydrogen peroxide generation by monoamine oxidases in rat white adipocytes: Role on cAMP production. *Eur. J. Pharmacol.* **2000**, *395*, 177–182. [CrossRef]
26. Aitken, R.J.; Harkiss, D.; Knox, W.; Paterson, M.; Irvine, D.S. A novel signal transduction cascade in capacitating human spermatozoa characterised by a redox-regulated, cAMP-mediated induction of tyrosine phosphorylation. *J. Cell Sci.* **1998**, *111 Pt 5*, 645–656.
27. Zhang, H.; Zheng, R.L. Promotion of human sperm capacitation by superoxide anion. *Free Radic. Res.* **1996**, *24*, 261–268. [CrossRef]
28. Lewis, B.; Aitken, R.J. A redox-regulated tyrosine phosphorylation cascade in rat spermatozoa. *J. Androl.* **2001**, *22*, 611–622.
29. Agarwal, A.; Virk, G.; Ong, C.; du Plessis, S.S. Effect of Oxidative Stress on Male Reproduction. *World J. Men's Health* **2014**, *32*, 1–17. [CrossRef]
30. O'Flaherty, C. Peroxiredoxin 6: The Protector of Male Fertility. *Antioxid* **2018**, *7*, 173. [CrossRef]
31. Saleh, R.A.; Agarwal, A. Oxidative stress and male infertility: From research bench to clinical practice. *J. Androl.* **2002**, *23*, 737–752.
32. Alvarez, J.G.; Storey, B.T. Differential incorporation of fatty acids into and peroxidative loss of fatty acids from phospholipids of human spermatozoa. *Mol. Reprod. Dev.* **1995**, *42*, 334–346. [CrossRef]
33. Aitken, R.J.; Harkiss, D.; Buckingham, D.W. Analysis of lipid peroxidation mechanisms in human spermatozoa. *Mol. Reprod. Dev.* **1993**, *35*, 302–315. [CrossRef]
34. Ayala, A.; Munoz, M.F.; Arguelles, S. Lipid peroxidation: Production, metabolism, and signaling mechanisms of malondialdehyde and 4-hydroxy-2-nonenal. *Oxidative Med. Cell. Longev.* **2014**, *2014*, 360438. [CrossRef]
35. Mazzilli, F.; Rossi, T.; Sabatini, L.; Pulcinelli, F.M.; Rapone, S.; Dondero, F.; Gazzaniga, P.P. Human sperm cryopreservation and reactive oxygen species (ROS) production. *Acta Eur. Fertil.* **1995**, *26*, 145–148.
36. Chatterjee, S.; Gagnon, C. Production of reactive oxygen species by spermatozoa undergoing cooling, freezing, and thawing. *Mol. Reprod. Dev.* **2001**, *59*, 451–458. [CrossRef]
37. Kadirvel, G.; Kumar, S.; Kumaresan, A. Lipid peroxidation, mitochondrial membrane potential and DNA integrity of spermatozoa in relation to intracellular reactive oxygen species in liquid and frozen-thawed buffalo semen. *Anim. Reprod. Sci.* **2009**, *114*, 125–134. [CrossRef]
38. Showell, M.G.; Mackenzie-Proctor, R.; Brown, J.; Yazdani, A.; Stankiewicz, M.T.; Hart, R.J. Antioxidants for male subfertility. *Cochrane Database Syst. Rev.* **2014**, *12*. [CrossRef]
39. Comhaire, F.H.; Garem, Y.E.; Mahmoud, A.; Eertmans, F.; Schoonjans, F. Combined conventional/antioxidant "Astaxanthin" treatment for male infertility: A double blind, randomized trial. *Asian J. Androl.* **2005**, *7*, 257–262. [CrossRef]

40. Vicari, E.; Calogero, A.E. Effects of treatment with carnitines in infertile patients with prostato-vesiculo-epididymitis. *Hum. Reprod. (Oxf. Engl.)* **2001**, *16*, 2338–2342. [CrossRef]
41. Vicari, E.; La Vignera, S.; Calogero, A.E. Antioxidant treatment with carnitines is effective in infertile patients with prostatovesiculoepididymitis and elevated seminal leukocyte concentrations after treatment with nonsteroidal anti-inflammatory compounds. *Fertil. Steril.* **2002**, *78*, 1203–1208. [CrossRef]
42. Balercia, G.; Regoli, F.; Armeni, T.; Koverech, A.; Mantero, F.; Boscaro, M. Placebo-controlled double-blind randomized trial on the use of L-carnitine, L-acetylcarnitine, or combined L-carnitine and L-acetylcarnitine in men with idiopathic asthenozoospermia. *Fertil. Steril.* **2005**, *84*, 662–671. [CrossRef] [PubMed]
43. Busetto, G.M.; Agarwal, A.; Virmani, A.; Antonini, G.; Ragonesi, G.; Del Giudice, F.; Micic, S.; Gentile, V.; De Berardinis, E. Effect of metabolic and antioxidant supplementation on sperm parameters in oligo-astheno-teratozoospermia, with and without varicocele: A double-blind placebo-controlled study. *Andrologia* **2018**, *50*, e12927. [CrossRef] [PubMed]
44. Bozhedomov, V.A.; Lipatova, N.A.; Bozhedomova, G.E.; Rokhlikov, I.M.; Shcherbakova, E.V.; Komarina, R.A. Using L- and acetyl-L-carnitines in combination with clomiphene citrate and antioxidant complex for treating idiopathic male infertility: A prospective randomized trial. *Urologiia (Mosc. Russ. 1999)* **2017**, *3*, 22–32. [CrossRef]
45. Kizilay, F.; Altay, B. Evaluation of the effects of antioxidant treatment on sperm parameters and pregnancy rates in infertile patients after varicocelectomy: A randomized controlled trial. *Int. J. Impot. Res.* **2019**, *1*. [CrossRef]
46. Kessopoulou, E.; Powers, H.J.; Sharma, K.K.; Pearson, M.J.; Russell, J.M.; Cooke, I.D.; Barratt, C.L. A double-blind randomized placebo cross-over controlled trial using the antioxidant vitamin E to treat reactive oxygen species associated male infertility. *Fertil. Steril.* **1995**, *64*, 825–831. [CrossRef]
47. Suleiman, S.A.; Ali, M.E.; Zaki, Z.M.; el-Malik, E.M.; Nasr, M.A. Lipid peroxidation and human sperm motility: Protective role of vitamin E. *J. Androl.* **1996**, *17*, 530–537.
48. Geva, E.; Bartoov, B.; Zabludovsky, N.; Lessing, J.B.; Lerner-Geva, L.; Amit, A. The effect of antioxidant treatment on human spermatozoa and fertilization rate in an in vitro fertilization program. *Fertil. Steril.* **1996**, *66*, 430–434. [CrossRef]
49. Greco, E.; Romano, S.; Iacobelli, M.; Ferrero, S.; Baroni, E.; Minasi, M.G.; Ubaldi, F.; Rienzi, L.; Tesarik, J. ICSI in cases of sperm DNA damage: Beneficial effect of oral antioxidant treatment. *Hum. Reprod. (Oxf. Engl.)* **2005**, *20*, 2590–2594. [CrossRef]
50. Moslemi, M.K.; Tavanbakhsh, S. Selenium-vitamin E supplementation in infertile men: Effects on semen parameters and pregnancy rate. *Int. J. Gen. Med.* **2011**, *4*, 99–104. [CrossRef]
51. Tremellen, K.; Miari, G.; Froiland, D.; Thompson, J. A randomised control trial examining the effect of an antioxidant (Menevit) on pregnancy outcome during IVF-ICSI treatment. *Aust. N. Z. J. Obstet. Gynaecol.* **2007**, *47*, 216–221. [CrossRef] [PubMed]
52. Tikkiwal, M.; Ajmera, R.L.; Mathur, N.K. Effect of zinc administration on seminal zinc and fertility of oligospermic males. *Indian J. Physiol. Pharmacol.* **1987**, *31*, 30–34. [PubMed]
53. Omu, A.E.; Dashti, H.; Al-Othman, S. Treatment of asthenozoospermia with zinc sulphate: Andrological, immunological and obstetric outcome. *Eur. J. Obstet. Gynecol. Reprod. Biol.* **1998**, *79*, 179–184. [CrossRef]
54. Kerner, J.; Hoppel, C. Genetic disorders of carnitine metabolism and their nutritional management. *Annu. Rev. Nutr.* **1998**, *18*, 179–206. [CrossRef] [PubMed]
55. Jeulin, C.; Lewin, L.M. Role of free L-carnitine and acetyl-L-carnitine in post-gonadal maturation of mammalian spermatozoa. *Hum. Reprod. Update* **1996**, *2*, 87–102. [CrossRef] [PubMed]
56. Agarwal, A.; Said, T.M. Carnitines and male infertility. *Reprod. Biomed. Online* **2004**, *8*, 376–384. [CrossRef]
57. Zhou, X.; Liu, F.; Zhai, S. Effect of L-carnitine and/or L-acetyl-carnitine in nutrition treatment for male infertility: A systematic review. *Asia Pac. J. Clin. Nutr.* **2007**, *16* (Suppl. 1), 383–390.
58. Gharagozloo, P.; Gutierrez-Adan, A.; Champroux, A.; Noblanc, A.; Kocer, A.; Calle, A.; Perez-Cerezales, S.; Pericuesta, E.; Polhemus, A.; Moazamian, A.; et al. A novel antioxidant formulation designed to treat male infertility associated with oxidative stress: Promising preclinical evidence from animal models. *Hum. Reprod. (Oxf. Engl.)* **2016**, *31*, 252–262. [CrossRef]
59. Baker, H.W.; Edgar, D. Trials of antioxidants for male infertility. *Aust. N. Z. J. Obstet. Gynaecol.* **2008**, *48*, 125–126. [CrossRef]

60. Ener, K.; Aldemir, M.; Isik, E.; Okulu, E.; Ozcan, M.F.; Ugurlu, M.; Tangal, S.; Ozayar, A. The impact of vitamin E supplementation on semen parameters and pregnancy rates after varicocelectomy: A randomised controlled study. *Andrologia* **2016**, *48*, 829–834. [CrossRef]
61. Fallah, A.; Mohammad-Hasani, A.; Colagar, A.H. Zinc is an Essential Element for Male Fertility: A Review of Zn Roles in Men's Health, Germination, Sperm Quality, and Fertilization. *J. Reprod. Infertil.* **2018**, *19*, 69–81.
62. Zago, M.P.; Oteiza, P.I. The antioxidant properties of zinc: Interactions with iron and antioxidants. *Free Radic. Biol. Med.* **2001**, *31*, 266–274. [CrossRef]
63. Martins, A.D.; Alves, M.G.; Bernardino, R.L.; Dias, T.R.; Silva, B.M.; Oliveira, P.F. Effect of white tea (*Camellia sinensis* (L.)) extract in the glycolytic profile of Sertoli cell. *Eur. J. Nutr.* **2014**, *53*, 1383–1391. [CrossRef]
64. De Amicis, F.; Santoro, M.; Guido, C.; Russo, A.; Aquila, S. Epigallocatechin gallate affects survival and metabolism of human sperm. *Mol. Nutr. Food Res.* **2012**, *56*, 1655–1664. [CrossRef]
65. Abshenas, J.; Babaei, H.; Zare, M.-H.; Allahbakhshi, A.; Sharififar, F. The effects of green tea (*Camellia sinensis*) extract on mouse semen quality after scrotal heat stress. *Vet. Res. Forum* **2011**, *2*, 242–247.
66. Ding, J.; Wang, H.; Wu, Z.B.; Zhao, J.; Zhang, S.; Li, W. Protection of murine spermatogenesis against ionizing radiation-induced testicular injury by a green tea polyphenol. *Biol. Reprod.* **2015**, *92*, 1–13. [CrossRef]
67. Dias, T.R.; Alves, M.G.; Tomas, G.D.; Socorro, S.; Silva, B.M.; Oliveira, P.F. White tea as a promising antioxidant medium additive for sperm storage at room temperature: A comparative study with green tea. *J. Agric. Food Chem.* **2014**, *62*, 608–617. [CrossRef]
68. Oliveira, P.F.; Tomas, G.D.; Dias, T.R.; Martins, A.D.; Rato, L.; Alves, M.G.; Silva, B.M. White tea consumption restores sperm quality in prediabetic rats preventing testicular oxidative damage. *Reprod. Biomed. Online* **2015**, *31*, 544–556. [CrossRef]
69. Rato, L.; Alves, M.G.; Dias, T.R.; Lopes, G.; Cavaco, J.E.; Socorro, S.; Oliveira, P.F. High-energy diets may induce a pre-diabetic state altering testicular glycolytic metabolic profile and male reproductive parameters. *Andrology* **2013**, *1*, 495–504. [CrossRef]
70. Sharlip, I.D.; Jarow, J.P.; Belker, A.M.; Lipshultz, L.I.; Sigman, M.; Thomas, A.J.; Schlegel, P.N.; Howards, S.S.; Nehra, A.; Damewood, M.D.; et al. Best practice policies for male infertility. *Fertil. Steril.* **2002**, *77*, 873–882. [CrossRef]
71. Jarow, J.P. Diagnostic approach to the infertile male patient. *Endocrinol. Metab. Clin. N. Am.* **2007**, *36*, 297–311. [CrossRef]
72. MacLeod, J. The Role of Oxygen in the Metabolism and Motility of Human Spermatozoa. *Am. J. Physiol.-Leg. Content* **1943**, *138*, 512–518. [CrossRef]
73. Di Santo, M.; Tarozzi, N.; Nadalini, M.; Borini, A. Human Sperm Cryopreservation: Update on Techniques, Effect on DNA Integrity, and Implications for ART. *Adv. Urol.* **2012**, *2012*, 854837. [CrossRef]
74. Yeste, M. Sperm cryopreservation update: Cryodamage, markers, and factors affecting the sperm freezability in pigs. *Theriogenology* **2016**, *85*, 47–64. [CrossRef]
75. Roca, J.; Gil, M.A.; Hernandez, M.; Parrilla, I.; Vazquez, J.M.; Martinez, E.A. Survival and fertility of boar spermatozoa after freeze-thawing in extender supplemented with butylated hydroxytoluene. *J. Androl.* **2004**, *25*, 397–405. [CrossRef]
76. Trzcinska, M.; Bryla, M.; Gajda, B.; Gogol, P. Fertility of boar semen cryopreserved in extender supplemented with butylated hydroxytoluene. *Theriogenology* **2015**, *83*, 307–313. [CrossRef]
77. Memon, A.A.; Wahid, H.; Rosnina, Y.; Goh, Y.M.; Ebrahimi, M.; Nadia, F.M. Effect of antioxidants on post thaw microscopic, oxidative stress parameter and fertility of Boer goat spermatozoa in Tris egg yolk glycerol extender. *Anim. Reprod. Sci.* **2012**, *136*, 55–60. [CrossRef]
78. Yamaguchi, S.; Funahashi, H. Effect of the addition of beta-mercaptoethanol to a thawing solution supplemented with caffeine on the function of frozen-thawed boar sperm and on the fertility of sows after artificial insemination. *Theriogenology* **2012**, *77*, 926–932. [CrossRef]
79. Lopez-Saucedo, J.; Paramio, M.T.; Fierro, R.; Izquierdo, D.; Catala, M.G.; Coloma, M.A.; Toledano-Diaz, A.; Lopez-Sebastian, A.; Santiago-Moreno, J. Sperm characteristics and heterologous in vitro fertilisation capacity of Iberian ibex (*Capra pyrenaica*) epididymal sperm, frozen in the presence of the enzymatic antioxidant catalase. *Cryobiology* **2014**, *68*, 389–394. [CrossRef]
80. Sariozkan, S.; Bucak, M.N.; Tuncer, P.B.; Ulutas, P.A.; Bilgen, A. The influence of cysteine and taurine on microscopic-oxidative stress parameters and fertilizing ability of bull semen following cryopreservation. *Cryobiology* **2009**, *58*, 134–138. [CrossRef]

81. Malo, C.; Gil, L.; Gonzalez, N.; Martinez, F.; Cano, R.; de Blas, I.; Espinosa, E. Anti-oxidant supplementation improves boar sperm characteristics and fertility after cryopreservation: Comparison between cysteine and rosemary (*Rosmarinus officinalis*). *Cryobiology* **2010**, *61*, 142–147. [CrossRef] [PubMed]
82. Buyukleblebici, S.; Tuncer, P.B.; Bucak, M.N.; Eken, A.; Sariozkan, S.; Tasdemir, U.; Endirlik, B.U. Cryopreservation of bull sperm: Effects of extender supplemented with different cryoprotectants and antioxidants on sperm motility, antioxidant capacity and fertility results. *Anim. Reprod. Sci.* **2014**, *150*, 77–83. [CrossRef] [PubMed]
83. Tuncer, P.B.; Buyukleblebici, S.; Eken, A.; Tasdemir, U.; Durmaz, E.; Buyukleblebici, O.; Coskun, E. Comparison of cryoprotective effects of lycopene and cysteamine in different cryoprotectants on bull semen and fertility results. *Reprod. Domest. Anim. Zuchthyg.* **2014**, *49*, 746–752. [CrossRef] [PubMed]
84. Foote, R.H.; Brockett, C.C.; Kaproth, M.T. Motility and fertility of bull sperm in whole milk extender containing antioxidants. *Anim. Reprod. Sci.* **2002**, *71*, 13–23. [CrossRef]
85. Succu, S.; Berlinguer, F.; Pasciu, V.; Satta, V.; Leoni, G.G.; Naitana, S. Melatonin protects ram spermatozoa from cryopreservation injuries in a dose-dependent manner. *J. Pineal Res.* **2011**, *50*, 310–318. [CrossRef] [PubMed]
86. Bertoldo, M.J.; Guibert, E.; Tartarin, P.; Guillory, V.; Froment, P. Effect of metformin on the fertilizing ability of mouse spermatozoa. *Cryobiology* **2014**, *68*, 262–268. [CrossRef] [PubMed]
87. Perez, L.; Arias, M.E.; Sanchez, R.; Felmer, R. N-acetyl-L-cysteine pre-treatment protects cryopreserved bovine spermatozoa from reactive oxygen species without compromising the in vitro developmental potential of intracytoplasmic sperm injection embryos. *Andrologia* **2015**, *47*, 1196–1201. [CrossRef]
88. Truong, T.; Gardner, D.K. Antioxidants improve IVF outcome and subsequent embryo development in the mouse. *Hum. Reprod. (Oxf. Engl.)* **2017**, *32*, 2404–2413. [CrossRef]
89. Gualtieri, R.; Barbato, V.; Fiorentino, I.; Braun, S.; Rizos, D.; Longobardi, S.; Talevi, R. Treatment with zinc, d-aspartate, and coenzyme Q10 protects bull sperm against damage and improves their ability to support embryo development. *Theriogenology* **2014**, *82*, 592–598. [CrossRef]
90. Stipanuk, M.H.; Dominy, J.E., Jr.; Lee, J.I.; Coloso, R.M. Mammalian cysteine metabolism: New insights into regulation of cysteine metabolism. *J. Nutr.* **2006**, *136*, 1652s–1659s. [CrossRef]
91. Fischer, B.; Bavister, B.D. Oxygen tension in the oviduct and uterus of rhesus monkeys, hamsters and rabbits. *J. Reprod. Fertil.* **1993**, *99*, 673–679. [CrossRef]
92. Hardie, D.G.; Alessi, D.R. LKB1 and AMPK and the cancer-metabolism link-ten years after. *BMC Biol.* **2013**, *11*, 36. [CrossRef]
93. Nguyen, T.M.; Seigneurin, F.; Froment, P.; Combarnous, Y.; Blesbois, E. The 5′-AMP-Activated Protein Kinase (AMPK) Is Involved in the Augmentation of Antioxidant Defenses in Cryopreserved Chicken Sperm. *PLoS ONE* **2015**, *10*, e0134420. [CrossRef]
94. Hurtado de Llera, A.; Martin-Hidalgo, D.; Garcia-Marin, L.J.; Bragado, M.J. Metformin blocks mitochondrial membrane potential and inhibits sperm motility in fresh and refrigerated boar spermatozoa. *Reprod. Domest. Anim.* **2018**, *53*, 733–741. [CrossRef]
95. Calle-Guisado, V.; Gonzalez-Fernandez, L.; Martin-Hidalgo, D.; Garcia-Marin, L.J.; Bragado, M.J. Metformin inhibits human spermatozoa motility and signalling pathways mediated by protein kinase A and tyrosine phosphorylation without affecting mitochondrial function. *Reprod. Fertil. Dev.* **2018**, *31*, 787–795. [CrossRef]
96. Luboshitzky, R.; Shen-Orr, Z.; Herer, P. Seminal plasma melatonin and gonadal steroids concentrations in normal men. *Arch. Androl.* **2002**, *48*, 225–232. [CrossRef]
97. Gonzalez-Arto, M.; Vicente-Carrillo, A.; Martinez-Pastor, F.; Fernandez-Alegre, E.; Roca, J.; Miro, J.; Rigau, T.; Rodriguez-Gil, J.E.; Perez-Pe, R.; Muino-Blanco, T.; et al. Melatonin receptors MT1 and MT2 are expressed in spermatozoa from several seasonal and nonseasonal breeder species. *Theriogenology* **2016**, *86*, 1958–1968. [CrossRef]
98. Deng, S.L.; Sun, T.C.; Yu, K.; Wang, Z.P.; Zhang, B.L.; Zhang, Y.; Wang, X.X.; Lian, Z.X.; Liu, Y.X. Melatonin reduces oxidative damage and upregulates heat shock protein 90 expression in cryopreserved human semen. *Free Radic. Biol. Med.* **2017**, *113*, 347–354. [CrossRef]
99. Martin-Hidalgo, D.; Baron, F.J.; Bragado, M.J.; Carmona, P.; Robina, A.; Garcia-Marin, L.J.; Gil, M.C. The effect of melatonin on the quality of extended boar semen after long-term storage at 17 degrees C. *Theriogenology* **2011**, *75*, 1550–1560. [CrossRef]

100. Yamaguchi, S.; Funahashi, H.; Murakami, T. Improved fertility in gilts and sows after artificial insemination of frozen-thawed boar semen by supplementation of semen extender with caffeine and CaCl2. *J. Reprod. Dev.* **2009**, *55*, 645–649. [CrossRef]
101. Donnelly, E.T.; McClure, N.; Lewis, S.E. Antioxidant supplementation in vitro does not improve human sperm motility. *Fertil. Steril.* **1999**, *72*, 484–495. [CrossRef]
102. Menezo, Y.J.; Hazout, A.; Panteix, G.; Robert, F.; Rollet, J.; Cohen-Bacrie, P.; Chapuis, F.; Clement, P.; Benkhalifa, M. Antioxidants to reduce sperm DNA fragmentation: An unexpected adverse effect. *Reprod. Biomed. Online* **2007**, *14*, 418–421. [CrossRef]
103. Menezo, Y.; Entezami, F.; Lichtblau, I.; Belloc, S.; Cohen, M.; Dale, B. Oxidative stress and fertility: Incorrect assumptions and ineffective solutions? *Zygote (Camb. Engl.)* **2014**, *22*, 80–90. [CrossRef]
104. Klein, E.A.; Thompson, I.M., Jr.; Tangen, C.M.; Crowley, J.J.; Lucia, M.S.; Goodman, P.J.; Minasian, L.M.; Ford, L.G.; Parnes, H.L.; Gaziano, J.M.; et al. Vitamin E and the risk of prostate cancer: The Selenium and Vitamin E Cancer Prevention Trial (SELECT). *JAMA* **2011**, *306*, 1549–1556. [CrossRef]
105. Miller, E.R., 3rd; Pastor-Barriuso, R.; Dalal, D.; Riemersma, R.A.; Appel, L.J.; Guallar, E. Meta-analysis: High-dosage vitamin E supplementation may increase all-cause mortality. *Ann. Intern. Med.* **2005**, *142*, 37–46. [CrossRef]
106. Aruoma, O.I.; Halliwell, B.; Gajewski, E.; Dizdaroglu, M. Copper-ion-dependent damage to the bases in DNA in the presence of hydrogen peroxide. *Biochem. J.* **1991**, *273 Pt 3*, 601–604. [CrossRef]

© 2019 by the authors. Licensee MDPI, Basel, Switzerland. This article is an open access article distributed under the terms and conditions of the Creative Commons Attribution (CC BY) license (http://creativecommons.org/licenses/by/4.0/).

MDPI
St. Alban-Anlage 66
4052 Basel
Switzerland
Tel. +41 61 683 77 34
Fax +41 61 302 89 18
www.mdpi.com

Antioxidants Editorial Office
E-mail: antioxidants@mdpi.com
www.mdpi.com/journal/antioxidants